Electric Charge Accumulation in Dielectrics: Measurement and Analysis

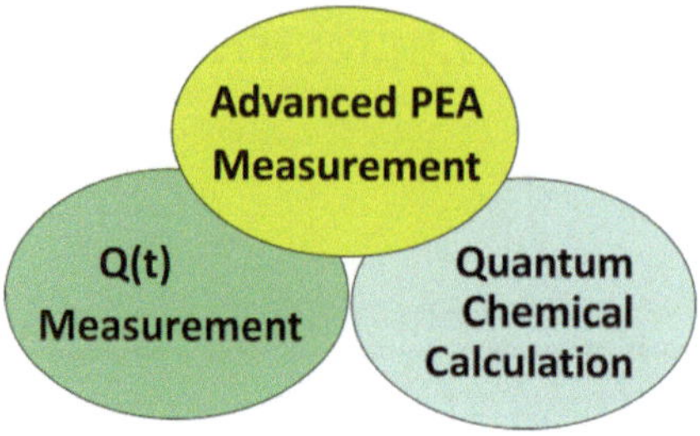
Advanced PEA Measurement
Q(t) Measurement
Quantum Chemical Calculation

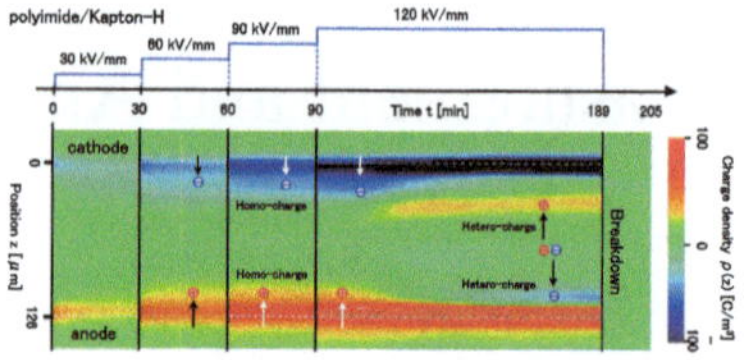
polyimide/Kapton-H
30 kV/mm
60 kV/mm
90 kV/mm
120 kV/mm
Time t [min]
cathode
anode
Homo-charge
Hetero-charge
Breakdown
Position z [μm]
Charge density ρ (z) [C/m³]

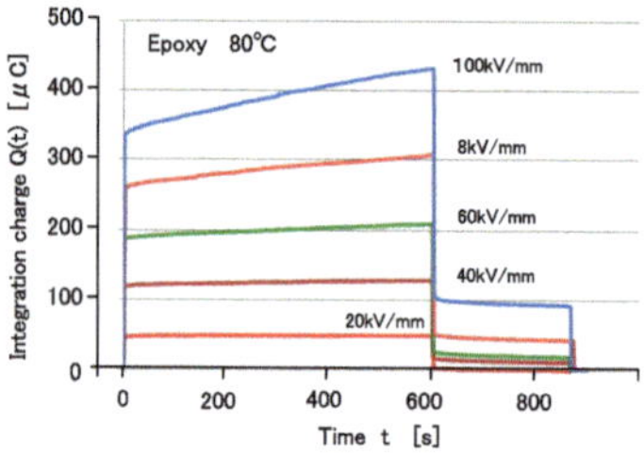
Epoxy 80℃
100kV/mm
8kV/mm
60kV/mm
40kV/mm
20kV/mm
Integration charge Q(t) [μC]
Time t [s]

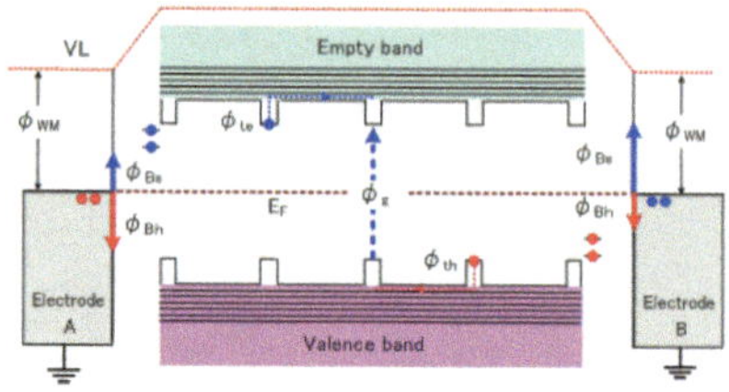
VL
Empty band
Valence band
Electrode A
Electrode B
E_F

Tatsuo Takada · Hanwen Ren · Jin Li ·
Weiwang Wang · Xiangrong Chen · Qingmin Li

Electric Charge Accumulation in Dielectrics: Measurement and Analysis

Science Press
Beijing

Springer

Tatsuo Takada
Tokyo City University
Tokyo, Japan

Hanwen Ren
North China Electric Power University
Beijing, China

Jin Li
Tianjin University
Tianjin, China

Weiwang Wang
Xi'an Jiaotong University
Xi'an, Shaanxi, China

Xiangrong Chen
Zhejiang University
Hangzhou, Zhejiang, China

Qingmin Li
North China Electric Power University
Beijing, China

ISBN 978-981-19-6155-7 ISBN 978-981-19-6156-4 (eBook)
https://doi.org/10.1007/978-981-19-6156-4

Jointly published with Science Press
The print edition is not for sale in China (Mainland). Customers from China (Mainland) please order the print book from: Science Press.

This Springer imprint is published by the registered company Springer Nature Singapore Pte Ltd.
The registered company address is: 152 Beach Road, #21-01/04 Gateway East, Singapore 189721, Singapore

Preface

This text was accomplished on my 81st birthday. When I retired from the university, my research question "Where is the space charge trapped in the chemical structure of the dielectric?" was still not solved. Since charge accumulation is a research world in an atomic scale, it is difficult to answer this question directly. However, I still took on this new challenge.

During my active career, I had been working on the measurement of the charge accumulated in dielectric materials. At present, the measurement targets have been expanded widely, such as the accumulated charge inside solids, the electric field inside liquids, and the charge density on the surface of solid dielectrics. Therefore, measurement technologies also contain a wide range now.

Solid: Elastic pressure wave measurement technology under nanosecond pulse excitation.
Liquid: Polarization phase difference measurement technology based on electro-optical Kerr effect.
Solid surface: Polarization phase difference measurement technology based on electro-optical Pockels effect.

These research results are summarized in the following textbooks in Japanese.

Space Charge Accumulation Measurements and Its Analysis, Vol. I (in Japanese, 340 pages)

Part 1 Physics of pulsed elastic wave
Part 2 Principle of space charge distribution measurement
Part 3 Design and assembling of space charge measurement equipment

Space Charge Accumulation Measurements and Its Analysis, Vol. II (in Japanese, 351 pages)

Part 4 Analysis of charge accumulation in dielectrics using Quantum Chemical Calculation

Space Charge Accumulation Measurements and Its Analysis, Vol. III (in Japanese, 299 pages)

Part 4 Analysis of charge accumulation in dielectrics using Quantum Chemical Calculation
Part 5 Quantum Chemical Calculation

Space Charge Accumulation Measurements and Its Analysis, Vol. IV (in Japanese, 321 pages)

Part 6 Fundamentals of polarized electro-magnetic wave light and its application to electric charge measurement
Part 7 Electric field distribution measurement in liquid insulation, shown in Figure
Part 8 Surface charge distribution measurement on solid dielectric, shown in Figure

The sentences "see Takada text ..." in the following contents correspond to the above textbooks.

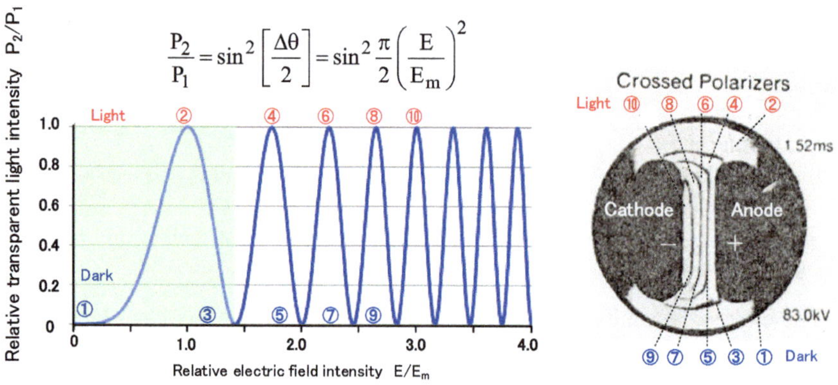

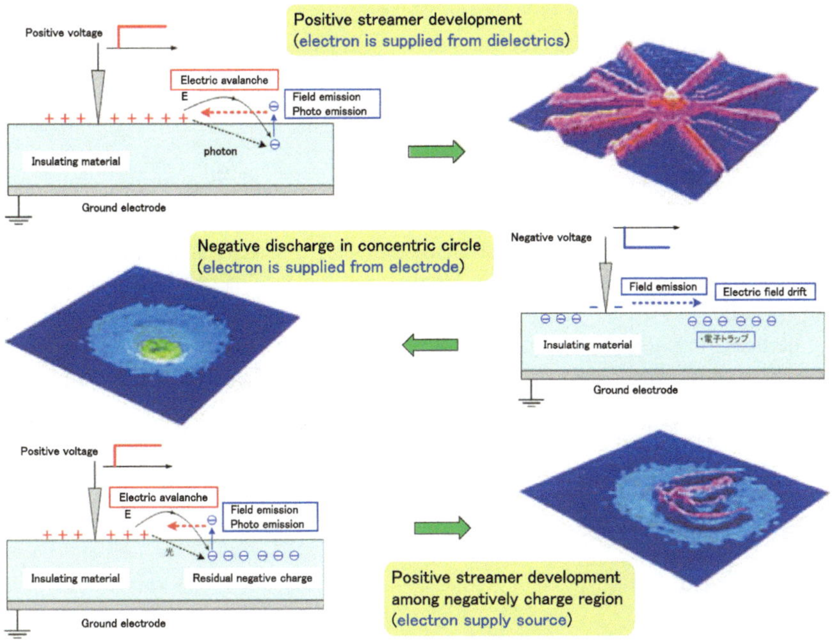

Professor Yasuhiro Tanaka's laboratory, who took over the laboratory after me, has further developed these measurement technologies and shown many interesting charge accumulation phenomena that attract the attention of researchers all around the world. I tried to explain the reason for charge accumulation, but I couldn't give a satisfactory explanation just by qualitative imagination.

The advancement of research needs the connection of new technologies in different fields. Fortunately, Prof. Masafumi Yoshida of Tokyo City University specializes in Quantum Chemical Calculation, and I was immediately instructed by him. I learned with a graduate student (Yuji Hayase) how to calculate the energy level of the chemical structure of a dielectric material and the distribution of electrostatic potential in a molecule by Gaussian software. 12 years have passed since then, and I have been able to understand the "space charge trap in a dielectric" by drawing a diagram based on the calculation.

(1) A molecule is a bond of multiple atoms. Each atom has electronegativity, and the center of the bonded electrons moves from the atom with a low electronegativity to the neighbor atom with a high value. Therefore, some atoms are positively charged and others are negatively charged, and the atoms are bonded in the potential field.
(2) The potentials of many charged atoms form a complex $V(x, y, z)$ distribution in the molecule.
(3) As the Schrödinger equation shows, the electron wave can be a unique standing wave in the potential field $V(x, y, z)$. This intrinsic energy level can be easily

obtained by Gaussian calculation. At the same time, the information on the molecular orbital of the electron wave of each intrinsic energy level can also be obtained.

(4) Electron traps and hole traps are formed in this localized $V(x, y, z)$ distribution space.

Part I in this textbook describes Q(t) measurements. Part II describes PEA measurements, and Part III describes Quantum Chemical Calculations. These achievements also cover the research results and discussions by the co-authors, and they have helped complete the English textbook. I'm very grateful for their cooperation.

I dedicate this textbook to the late Prof. Ziyu Liu of Xi'an Jiaotong University, who initiated our international joint research program of the US, China, and Japan laboratories.

Tokyo, Japan Tatsuo Takada

Acknowledgments

This book *Electric Charge Accumulation in Dielectrics: Measurement and Analysis* is a collaboration of the following six authors. The word count of the whole book is about 300 thousand.

Professor Tatsuo Takada is responsible for completing Chaps. 1–4 in Part I and Chaps. 1–3 in Part III, the word count of which is about 160 thousand. Dr. Hanwen Ren is responsible for completing Chaps. 5 and 6 in Part I and Chaps. 1–5 in Part II, the word count of which is about 120 thousand. Dr. Jin Li has contributed to the contents of Chaps. 4 and 6 in Part I, Chap. 5 in Part II, and Chap. 3 in Part III. Dr. Weiwang Wang has contributed to the contents of Chap. 6 in Part I, and Chaps. 1–3 in Part III. Professor Xiangrong Chen has contributed to the contents of Chap. 5 in Part II. Professor Qingmin Li has contributed to the space charge research results in Part I and Part II, who is also responsible for the content revision, polishing, and editing of the whole book.

Meanwhile, the book publishment is supported by grants from the National Natural Science Foundation of China (No. 51737005, 52127812, 51929701, 52177147, 52207153) and National Key Research and Development Program of China (No. 2021YFB2601404). The authors express deep gratitude for their support.

Contents

About the Authors

Tatsuo Takada Emeritus Professor, Tokyo City University, Tokyo, Japan.

He was born in Yamanashi, Japan, on August 8, 1939. He received his B.E. degree in electrical engineering from Musashi Institute of Technology, Japan, in 1963 and M.E. and Ph.D. degrees from Tohoku University, Japan, in 1966 and 1975, respectively. He became a Lecturer, an Associate Professor, and a Professor at Musashi Institute of Technology, in 1967, 1974, and 1987, respectively. He was a Visiting Scientist at MIT (USA) from 1981 to 1983, and he is a consulting professor at Xi'an Jiaotong University, China, 1995.

He has now undertaken several research projects on the development of acoustic and optical advanced methods for measuring electric charge distribution in dielectrics. For example, in order to investigate the space charge effect in solid dielectric material, the dynamic surface charge distribution on solid dielectric materials, and the electric field vector distribution in liquid insulating materials. He received the progress award from the IEE of Japan in 1996. He has also received Whitehead Memorial Award in 1999 for his pioneering work on PEA, and he gave an Award Lecture in IEEE CEIDP (Conference on Electrical Insulation and Dielectric Phenomena). He further received Chen Jidan Award for his contributed works in charge measurement and quantum chemical calculation and he gave a plenary speech in IEEE ICEMPE (International Conference on Electrical Materials and Power Equipment). He is a Life Fellow member of IEEE. He can be reached by email at takada@a03.itscom.net.

Hanwen Ren Lecturer, North China Electric Power University, Beijing, China.

He was born in Shanxi, China, in 1994. He received his B.Sc. and Ph.D. degrees in electrical engineering from North China Electric Power University, China, in 2016 and 2021. Currently, he is a lecturer and post-doctor in North China Electric Power University. He has published over 30 academic papers in journals and international conferences. Now he is mainly researching in the field of the space charge inside solid insulation and its measurement technology. He can be reached by email at rhwncepu@ncepu.edu.cn.

Jin Li Associate Professor, Tianjin University, Tianjin, China.

He was born in Tangshan, China, on August 8, 1988. He received his B.S. and Ph.D. degrees in Electrical Engineering from Tianjin University in 2012 and 2017. From 2017 to 2019, he worked as a post-doctor in Tianjin University with the support of National Postdoctoral Program for Innovative Talents. From October 2018 to January 2019, he was a visiting researcher in Measurement and Electric Machine Control Laboratory of Tokyo City University. Now he is an Associate Professor at the School of Electrical and Information Engineering, Tianjin University. His research interests are focused on the designing and improvement of solid dielectrics for HVDC power cable and gas-insulated transmission line.

He has authored or co-authored nearly 40 papers published on IEEE Transactions. He received IEEE ICEMPE Excellent Conference Paper, IEEE ASEMD Best Conference Paper, and Outstanding Contributions to the Development of IEEE Standard 2780. He has also chaired several sessions in IEEE ICHVE and IEEE ICEMPE. He is the associate editor of IET High Voltage and member of CIGRE NGN Group and CIGRE D1.73 Working Group. He can be reached by email at lijin@tju.edu.cn.

Weiwang Wang Associate Professor, Xi'an Jiaotong University, Xi'an, China.

He obtained his Ph.D. degrees in School of Electrical Engineering from Xi'an Jiaotong University (XJTU), China, in 2015. From 2016 to 2018, he was a Lecturer of XJTU. Since 2019, he works as an associate professor in XJTU. From 2016 to 2017, he was a post-doc fellow in the Department of Mechanical System Engineering, Measurement and Electric Machine Control Laboratory in Tokyo City University, Japan. His research interests are dielectric materials, functionalized nanocomposites, space charge measurement and theory, power electronic transformer, and multi-physics coupling simulation in power equipment. He has published over 35 academic papers in journals and international conferences, including the 15 SCI papers and 17 EI papers. He has also co-authored one book chapter in English. He has been awarded the 2018 Excellent Doctoral Dissertation in XJTU, and was selected as the Young Talent supported by Chinese Society of Electrical Engineering. As a program leader, he has engaged in over 10 projects, such as the NSFC project in China. He can be reached by email at weiwwang@xjtu.edu.cn.

Xiangrong Chen Professor, Zhejiang University, Hangzhou, China.

He (Senior Member, IEEE) was born in Hunan, China, in 1982. He received his M.S. and Ph.D. degrees in electrical engineering from Xi'an Jiaotong University, Xi'an, China, in 2008 and 2011, respectively. He was a post-doctor in 2011, Assistant Professor from 2012 to 2016, and Associate Professor in 2016 in high voltage engineering with the Department of Manufacturing and Materials Technology, Chalmers University of Technology, Gothenburg, Sweden. Since January 2017, he has been a One-Hundred Talents Researcher/Professor with the College of Electrical Engineering, Zhejiang University (ZJU), Hangzhou, China. His research interests include advanced electrical materials and high-voltage insulation testing technology, advanced power equipment and new power system, and advanced high-voltage technology. Dr. Chen has been an Associate Editor for the *IEEE Transactions on Dielectrics and Electrical Insulation* since May 2021, was the Vice-Dean of ZJU-UIUC Institute from 2020 to 2021, and has been the Head of the Institute of Power System Automation of ZJU since March 2020. He has published more than 110 papers with over 50 papers in IEEE Transactions, Composites Science and Technology, and other well-recognized Journals. He can be reached by email at chenxiangrongxh@zju.edu.cn.

Qingmin Li Professor, North China Electric Power University, Beijing, China.

He received his B.Sc., M.Sc., and Ph.D. degrees in electrical engineering from Tsinghua University, China, in 1991, 1994, and 1999, respectively. He joined Tsinghua University as a lecturer in 1996, and came to the UK in 2000 as a post-doctoral research fellow at Liverpool University and later at Strathclyde University. From 2003 to 2011, he was with Shandong University, China, as a professor in electrical engineering. In 2010, he also joined Arizona State University, USA, as a visiting professor. He is currently a professor in electrical engineering at North China Electric Power University, China. His special fields of interest include solid dielectrics, GIS/GIL insulation, lightning protection of wind turbines, condition monitoring, and fault diagnostics. He can be reached by email at lqmeee@ncepu.edu.cn.

Part I
Fundamentals and Applications of Q(t) Method

Chapter 1
Classification of Charge Accumulation Measurement

1.1 The Progress of Space Charge Measurement Technology

At present (2020), the PEA method has been widely used for measuring the space charge distribution accumulated in dielectric and insulating materials. Until now, various measurement methods have been proposed, and the PEA method has been developed due to the progress of competition and cooperation while comparing and examining the measurement results with researchers in the world. Here, the basic flow of this measurement technology is introduced.

There was a strong demand in the 1970s to measure the accumulated charge distribution. It was attempted to apply CV cable developed for AC conditions to high voltage DC environments. As a result, when the polarity was reversed after applying a DC voltage, the breakdown voltage of the CV cable was significantly reduced. The presumable cause was "accumulation of space charge". By then, the researchers did not have the technology to measure the accumulated charge distribution. Therefore, a method for measuring the space charge distribution was needed. Also, in the field of electret transducer, the development of polymer functional materials in which charge accumulation is stable for a long time was proposed. Therefore, the development of the measurement of the charge distribution was expected.

For more information on this, see Takada Text Vol. 2, Part 4, Chapt.1 (in Japanese).

(1) **Classification of measurement methods**

Figure 1.1 depicts the development flow of measurement technology up to the PEA method. A description is given along the flow.

(a) ***I*(*t*) measurement: pA meter** (pico-ammeter; Fig. 1.1a).
A DC high voltage V_{dc} is applied to the dielectric, and the current $I(t)$ flowing through the sample is measured. Its value is as small as 10^{-9} to 10^{-13} A. The precautions on how to use this pico-ammeter for measurement are described in Appendix 1.1. From the measured values of the current density J [A/m^2] and

T. Takada et al., *Electric Charge Accumulation in Dielectrics: Measurement and Analysis*,
https://doi.org/10.1007/978-981-19-6156-4_1

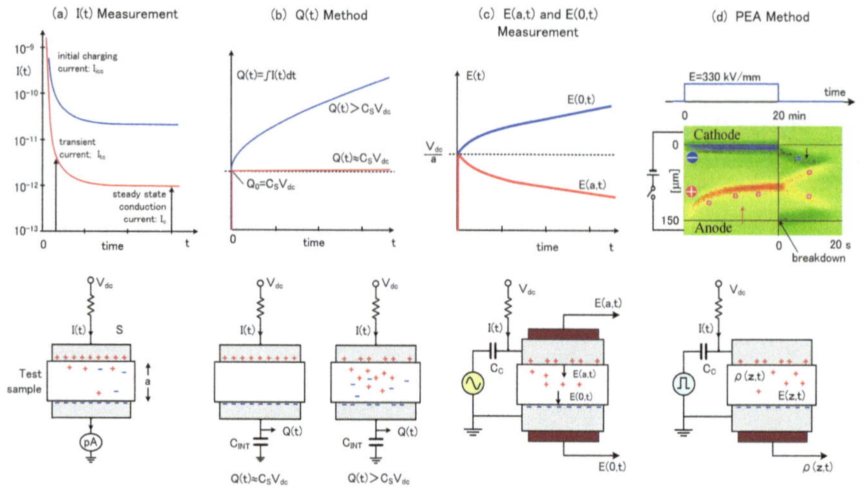

Fig. 1.1 The development process of the measurement methods

the applied electric field E [V/m], the conductivity κ [S/m] of the dielectric can be calculated as J/E. The insulation property was evaluated by the value of the conductivity κ [S/m]. However, from the value of the conductivity, the amount of charge accumulated in the dielectric and the distribution cannot be evaluated.

(b) **Q(t) method** (current integration charge method; Fig. 1.1b) [1].
Replace the pA meter in Fig. 1.1a with an integration capacitor C_{INT}.

$Q(t) = \int I(t)\,\mathrm{d}t$ is obtained by integrating the total current $I(t)$ flowing through the sample with time.

When $Q(t) = Q_0$ $(= C_s V_{dc})$: There is no charge accumulation in the sample. It can be determined that only the charge induced on the electrodes is present. $Q_0 = C_s V_{dc}$ is the initially induced charge immediately after the voltage application. C_s is the capacitance of the sample, V_{dc} is the applied DC voltage.

When $Q(\mathrm{t}) > Q_0$ $(= C_s V_{dc})$, it is determined that there is charge accumulation in the sample.

The Q(t) method can analyse the presence or absence of charge accumulation. However, information on the charge accumulation distribution ($\rho(x, t)$) in the sample cannot be obtained.

(c) ***E*(0,*t*) & *E*(*a*,*t*) measurement** (measurement of electrode surface electric field; Fig. 1.1c) [2, 3].
High-frequency sine wave voltage $v(t) = V_0 \sin\omega t$ is superimposed on DC high voltage V_{dc}. Then, an oscillating wave proportional to the electrode induced charge density σ [C/m^2] is generated. The magnitudes of this vibration wave are measured by a piezoelectric device attached to the back of the anode and the cathode, respectively.

The oscillating wave voltage $v_{anod}(t)$ from the piezoelectric device on the back surface of the anode is proportional to the anode surface charge $\sigma(a, t) = \varepsilon E(a, t)$. On the other side, the oscillating wave voltage $v_{cath}(t)$ from the piezoelectric element on the cathode back is proportional to the cathode surface charge $\sigma(0, t) = \varepsilon E(0, t)$. The following descriptions can be understood from the electric field characteristics of the electrode surface.

- $E(a, t)$ decreases over time $\rightarrow$ $E(a, t)$ decreases because positive charges are injected from the anode.
- $E(0, t)$ increases with time $\rightarrow$ $E(0, t)$ increases as positive charges approach the cathode.

This method can evaluate charge injection from the electrodes. However, information on the charge accumulation distribution ($\rho(x, t)$) in the sample cannot be obtained.

How did we start to measure the electric field on the electrode surfaces?

(1) **Principle of the speaker:** When an audio frequency current ($I_0 \sin\omega t$) is applied to the coil in the magnetic field (B) of the permanent magnet, the coil generates a vertical oscilating force (F) at the audio frequency. $F = I_0 \sin\omega t \times B$.

(2) **Principle of the electret earphone:** When a sound frequency voltage ($V_0 \sin\omega t$) is applied to a piezoelectric device that has been oriented and polarized (P), the device generates a stretching oscilating force (F) at the sound frequency. $F = P \times V_0 \sin\omega t$.

Also, if a voltage ($V_0 \sin\omega t$) is applied to the polymer film in which the electric charge is accumulated, a vibration sound wave should be generated. Since the level of the oscillating sound wave is proportional to the amount of accumulated charge, it is considered that the amount of accumulated charge can be evaluated.

(d) **PEA method** (Pulsed Electro-Acoustic method, $\rho(x, t)$; Fig. 1.1d).
The high-frequency sine wave voltage $v(t) = V_0 \sin\omega t$ is replaced with a nanosecond pulse voltage $v_p(t)$. The pulse pressure wave $p(x, t)$ generated from the accumulated charge is measured by a piezoelectric device attached to the back surface of the ground electrode. Of course, measurement can also be performed with a piezoelectric device attached to the back surface of the high-voltage side. This has not been experimented. The shape and size of the time sweep of the pulse pressure wave $p(x, t)$ are proportional to the shape and size of the accumulated charge distribution $\rho(x, t)$. In this way, the PEA method is finally realized that can measure the charge distribution $\rho(x, t)$ [C/m^3].

The internal electric charge distribution $\rho(x, t)$ shown in Fig. 1.1 can be seen from the measurement results of a low-density polyethylene (LDPE) under a DC high electric field of 330 kV/mm. Positive charges (red) are injected from the anode and move toward the cathode. After 20 mins, the electric field between the positive charge and the cathode rises to nearly 500 kV/mm, and the dielectric breakdown of LDPE has reached. Thus, the PEA method can dynamically and quantitatively measure the charge distribution $\rho(x, t)$.

1.2 Q(t) Method Can Measure Charges in All Trap Levels

(1) Evaluation items for DC insulating materials

Evaluation of DC insulating materials at high electric field and high temperature has become an important issue. The background is the development and evaluation of DC insulation of power electronics used for the control of electric vehicles, home appliances and electric devices. Futhermore, the development and evaluation of insulation and materials for DC power cables also needs an accurate evaluation.

The basic evaluation items of the DC insulating materials are as follows.

Leakage current: How much leakage current can be admitted to maintain the electrical insulation property?

Internal electric field distortion due to charge accumulation: How much internal electric field is distorted from the designed electric field due to charge accumulation?

Accumulated charge is a cause of insulation deterioration: Does the local potential energy of accumulated charge cause insulation deterioration?

(2) Evaluation items for DC insulation of various measurement methods

Table 1.1 summarizes the seven items that can be evaluated by the various measurement methods introduced in Sect. 1.1 above.

Leakage current: Evaluation of ① conductivity and ② leakage current. The measurement methods are I(t) method and Q(t) method.

Charge accumulation: Evaluation of ③ accumulated charge distribution, ④ internal electric field and ⑤ charge accumulation. It can be said from this table is that neither measurement method can evaluate both leakage current and charge accumulation at the same time. In order to correctly evaluate DC insulation materials, it is necessary to use both the Q(t) method and the PEA method.

Table 1.1 Evaluating items of DC insulation for various measurement methods

Evaluation items for DC insulation materials	I(t)	Q(t) Method	PEA Method
① Conductivity κ [S/m]	◎	◎	–
② Leakage current I_{leak} [A]	◎	◎	–
③ Accumulated charge distribution $\rho(x, t)$ [C/m^3]	–	–	◎
④ Internal electric field $E(x, t)$ [V/m]	–	–	◎
⑤ Charge accumulation $Q(t)/Q_0$	–	◎	◎
⑥ Shallow trapped charge	◎	◎	–
⑦ Deep trapped charge	◎	◎	◎

(3) Q(t) method measures all trapped charges

The I(t) and the Q(t) methods measure all the charges including the shallow trapped charges with a fast movement and the accumulated charges inside the deep traps. In particular, the Q(t) method has a feature that the accumulated charge and the leakage charge can be evaluated by the $Q(t)/Q_0$ ratio.

(4) Shallow trapped charge measurement is difficult in the PEA method (see Sect. 1.3)

The PEA method is an excellent measurement method which can measure the distribution of accumulated charges ($\rho(x, t)$) and the internal electric field distribution ($E(x, t)$). It can be used to develop DC insulating materials and evaluate internal electric field distortion. However, since the shallow trapped charge migrates within the measurement time of the pulse voltage application, it is difficult for the PEA method to measure the shallow trapped charge.

(5) Summary

As summarized in Table 1.1, it is necessary to use the Q(t) and the PEA methods together for the evaluation of DC insulation. At present time (2020), there is no data to establish criteria for the following items.

② Leakage current: How much leakage current can admitted to maintain the electrical insulation property.

④ Internal electric field distortion due to charge accumulation: How much distortion of internal electric field due to charge accumulation is acceptable?

It is still unclear how shallow the trapped charge should be considered for the evaluation of internal electric field.

Therefore, it is necessary to accumulate data for clarifying the above criteria.

1.3 PEA Method is Difficult to Measure Shallow Trapped Charge

The PEA method can measure the charge accumulation distribution and its characteristics over time. That is, it is an excellent measuring instrument that can measure the dynamic charge accumulation distribution $\rho(x, t)$. However, there is a point to note here.

The PEA measurement method cannot measure shallow trapped charges.

This is because the shallow trapped charge continuously migrates within measurement time and can't stay in the same place. Furthermore, it is still unclear how to evaluate the electric field distortion casued by shallow trapped charges. As described in Sect. 1.2, "combined use of the PEA method and the Q (t) method" is required to correctly evaluate the accumulation of space charge.

The measurement examples are introduced below.

(1) **Comparison of PEA data and Q (t) data**

(a) **Specimen A**: PC (polycarbonate).
PEA data: Anode charge (red) and cathode charge (blue) are observed, and $\rho(x, t) = 0$ (green) in the sample with no charge accumulation.
Q(t) data: Q(t) data is equal to the initial charge amount Q_0. Since $Q(t) = Q_0$, there is no charge accumulation (Fig. 1.2).

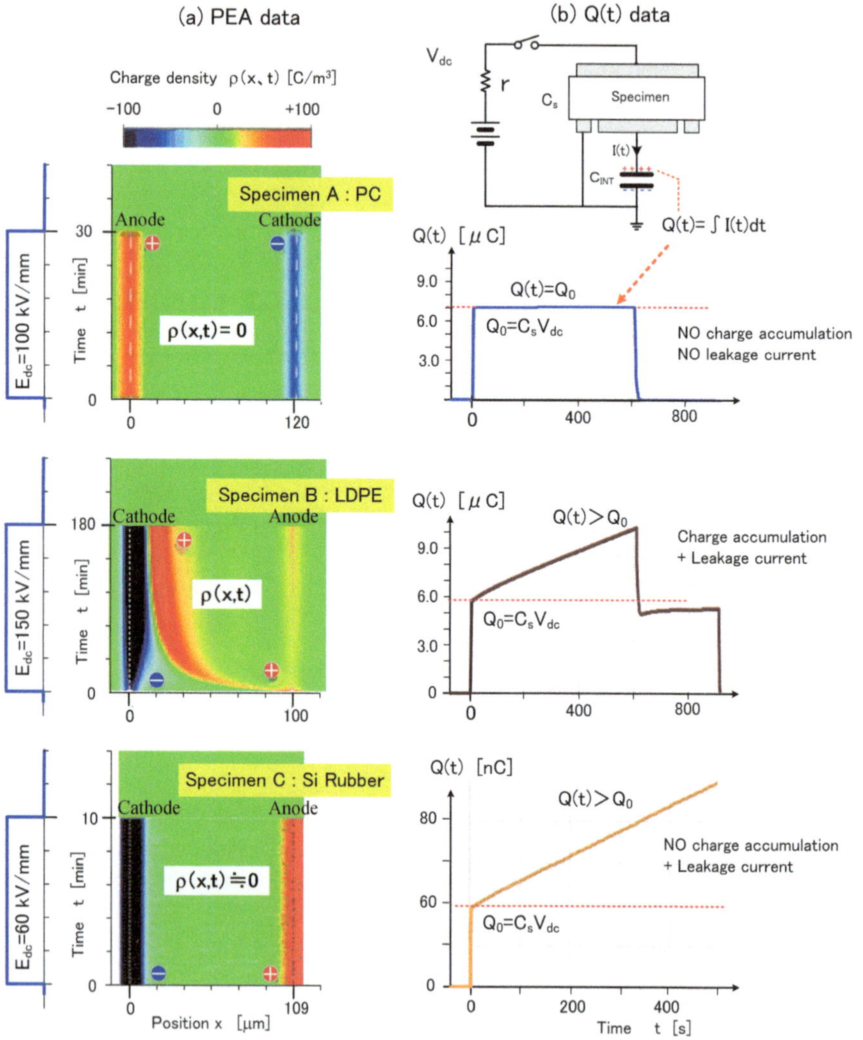

Fig. 1.2 Measured results by the PEA and Q(t) methods

From the results of both the PEA and Q(t) methods, it can be determined that there is no charge accumulation and leakage current. ⇒ PC (polycarbonate) is an excellent DC insulation material.

Specimen B: LDPE (low-density polyethylene).

PEA data: Positive charge (red) is injected from the anode and moves toward the cathode. As the positive charge (red) approaches the cathode, a negative charge injection from the cathode is observed. The electric field between the cathode and the positive charge exceeds the applied electric field strength, and the electric field in the sample is not uniform, gradually close to the breakdown value.

Q(t) data: Q(t) data increases from the initial charge amount Q_0, and $Q(t) > Q_0$. Charge accumulation is observed. From the results of both PEA and $Q(t)$ data, it can be evaluated that there is charge accumulation and the leakage current is large. LDPE (low-density polyethylene) is not a good DC insulating material.

Specimen C: Si Rubber.

PEA data: Anode charge (red) and cathode charge (blue) are observed, and there is no charge accumulation.

$\rho(x, t) = 0$ (green) inside the sample.

Q(t) data: Q(t) data increases from the initial charge Q_0, and $Q(t) > Q_0$. Then charge accumulation is observed.

This Specimen C (Si Rubber) gets contradictory evaluation results, i.e. PEA data shows "a good DC insulating material" and Q(t) data shows "a bad DC insulating material". Which is correct?

The PEA method cannot measure shallow trapped charges. On the other hand, the Q(t) data can measure all charges in shallow and deep traps. Thus, the Q(t) method provides a correct result. No charge accumulation is observed, but the leakage current is large. Hence, Specimen C (Si Rubber) is not a good DC insulating material.

(b) **Summary**

The PEA measurement method measures the charge accumulation distribution and the internal electric field distribution. However, it is difficult to measure shallow trapped charges. The Q(t) measurement method can evaluate the total charge accumulation from shallow to deep trapped charges by the relationship of $Q(t) = Q_0$ and $Q(t) > Q_0$. However, the accumulated charge distribution and the internal electric field cannot be measured. To correctly evaluate the accumulation of space charge, it is necessary to use "PEA method and Q(t) method together".

1.4 Explanation of Current Terms

In order to avoid confusion about the names of currents, this section summarizes the terms of involved current.

(1) **Electric current measurement** Fig. 1.3a

When a DC voltage V_{dc} is applied to the dielectric sample, the pico-ammeter (pA) can measure the current characteristic $I(t)$ indicated by the red line in Fig. 1.3a. Immediately after the application of the voltage V_{dc}, a steep charging current flows, followed by an absorption current I_{abs} (t) (space charge forming current I_{spac} (t)), and subsequently a leakage current $I_{leak}(t)$ (conduction current $I_{cond}(t)$).

(a) Absorption current and space charge forming current
The textbook generally states that the absorption current $I_{abs}(t)$ flows due to the orientation of dipole polarization. The orientation polarization of the polymer insulating material can be completed within a short measurement time (about 1 s), but the current in the figure is completed in a time period longer than 1 s. Therefore, this absorption current is included in this charging current. According to the real results of space charge measurement, space charges are formed during about 1 to 1000 s, and the current $I_{spac}(t)$ is observed. Therefore, the space charge forming current $I_{spac}(t)$ is adopted here instead of using the absorption current $I_{abs}(t)$.

(b) Conduction current and leakage current
It is extremely difficult to define the conduction current $I_{cond}(t)$. In the case of a particularly excellent insulating material, it takes a long time (1 h or more) for the charge injected from the electrode to reach the counter electrode and the equilibrium current state. Hence, it is difficult to determine the stable current

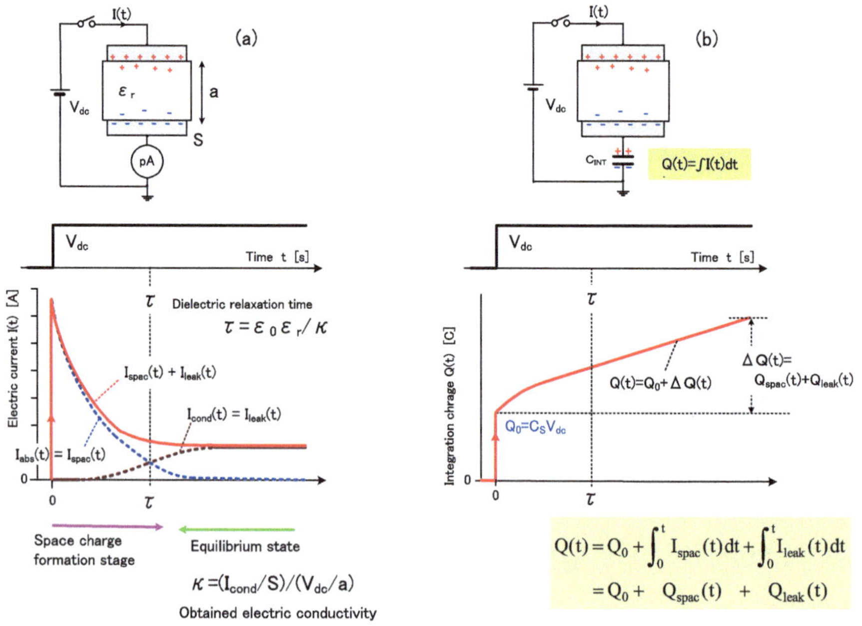

Fig. 1.3 The measured components by the Q(t) method

state. Therefore, here, the conduction current $I_{\mathrm{cond}}(t)$ in the equilibrium state is not adopted, but the term of the leakage current $I_{\mathrm{leak}}(t)$ that can be used before the equilibrium state is adopted.

(c) Very important note: relationship between measurement time and conductivity For example, assume that a conductivity $\kappa = 10^{-16}$ S/m is obtained from the experimental data of voltage and current characteristics. Then we can obtain the dielectric relaxation time $\tau = \varepsilon / \kappa = 2 \times 10^5$ s = 56 h from this value $\kappa = 10^{-16}$ S/m. A stable current cannot be obtained without taking more than 56-hour measurement.

However, in reality, the value of the conductivity $\kappa = 10^{-16}$ S/m is calculated by the current value under the measurement time of about 10 min.

According to the observed results of space charge measurement, in the case of an excellent insulating material, the charge injected from the electrode is slowly migrating in the sample within a measurement time of about 10 min, and it has not yet reached an equilibrium state. Therefore, the carriers with a value of $\kappa = 10^{-16}$ S/m is currently moving through the insulating material.

At present, there is a demand for "evaluation of DC insulating materials" for excellent insulating materials, and the Q(t) method described below is useful. Finally, as described in Sect. 3.4, the PEA method and the Q (t) method are used together.

(2) **Current integration measurement**: **Q(t) method** Fig. 1.3b

In this current integration measurement, an integrating capacitor C_{INT} is inserted instead of the pico-ammeter (pA), and the charge amount $Q(t) = \int I(t)\, dt$ obtained by integrating the total current through the dielectric sample is measured. The Q(t) components are shown by Eq. (1.1).

$$
\begin{aligned}
Q(t) &= Q_0 + \int_0^t I_{spac}(t)\, dt + \int_0^t I_{leak}(t)\, dt \\
&= Q_0 + \quad Q_{spac}(t) \quad + \quad Q_{leak}(t)
\end{aligned}
\tag{1.1}
$$

As shown in Fig. 1.3b, the Q(t) characteristic shows that the initial charge Q_0 rises immediately after the application of the voltage V_{dc}. Here, the first term of Eq. (1.1) is the initial charge $Q_0 = C_{\mathrm{s}} V_{\mathrm{dc}}$. C_{s} is the capacitance of the sample. Thereafter, the Q(t) characteristic gradually increases due to the space charge forming current $I_{\mathrm{spac}}(t)$. The second term ($Q_{\mathrm{spac}}(t)$) of Eq. (1.1) represents the time integration of the space charge forming current $I_{\mathrm{spac}}(t)$. Further, the Q(t) characteristic increases almost linearly and shifts to the leakage current $I_{\mathrm{leak}}(t)$. The third term ($Q_{\mathrm{leak}}(t)$) represents the integration of the leakage current $I_{\mathrm{leak}}(t)$.

(3) Feature of Q(t) method

① DC permittivity ε_{r}(dc) can be evaluated. The charge amount of the initial charge is $Q_0 = C_{\mathrm{s}} V_{\mathrm{dc}}$, where the capacitance of the sample is $C_{\mathrm{s}} = \varepsilon_0\, \varepsilon_{\mathrm{r}}(\mathrm{dc})\, S/a$. Since Q_0,

V_{dc}, the sample thickness a and the area S are known, $\varepsilon_r(\mathrm{dc})$ can be evaluated from this relation.

② If the measurement result is $Q(t) = Q_0$, it is determined that there is no charge accumulation and leakage current since $Q_{spac}(t) = Q_{leak}(t) = 0$.

③ If the measurement result is $Q(t) > Q_0$, it is determined that there is charge accumulation and leakage current. Since the value of Q_0 has been measured, the charge accumulation can be evaluated by comparing the magnitudes of $Q(t)$ and Q_0. This is a unique feature of the Q(t) method.

Separation of Q_{spac}(t) and Q_{leak}(t) is difficult by the Q(t) method alone. This separation can be realized by using the PEA method and the Q(t) method together as shown in Sect. 3.4.

Appendix 1.1 Pico-ammeter Measurement

(1) **Measurement of infinitesimal electric current**

The measurement result of electric current in Fig. 1.4 shows an example for measuring a very small current. Immediately after the application of the DC voltage, the instantaneous charging current I_{inst} (about 10^{-3} to 10^{-8} A) for charging the capacitance of the dielectric sample flows, and then the absorption current I_{abs} appears (about 10^{-8} to 10^{-11} A). Finally, the leakage current I_{leak} (about 10^{-11} to 10^{-12} A) in the equilibrium state flows. As described above, the picoammeter measures a very small amount of current. In practice, it is important to understand the measurement principle of the picoammeter and to learn the basic precautions of the measurement method for accurate and stable measurements [4].

(2) **Electronic circuit of pico-ammeter**

The leakage current in dielectric insulating materials typically ranges from 10^{-9} to 10^{-13} A. An electronic measurement technology that can measure such a small current has been established [4]. The basic electronic circuits of the picoammeter are classified as shown in Fig. 1.5.

(a) Shunt method
(b) Feedback method
(c) Integration method

The output voltages V_{PA} of these methods are proportional to the input current I_X to be measured. fA in the figure represents the Femto ampea 10^{-15}A.

(3) **Precaution for measuring small current**

The sample is inserted between the high-voltage power supply and the pico-ammeter. The low-voltage electrode contacting with the sample needs to be sufficiently insulated. That is, it should be paid attention to correctly measure the small current I_x as follows.

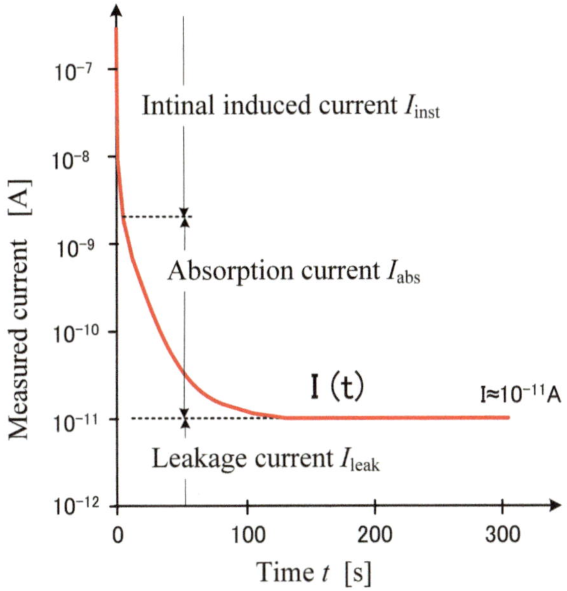

Fig. 1.4 The current components passing through the sample

(a) In the shunt method, I_x have to flow through the resistance R_S.
(b) In the feedback method, I_x have to flow through the resistance R_f.
(c) In the integration method, I_x have to flow through the C_f.

Specifically, it should be noted that the measured current I_x cannot flow through the resistor R_0. This resistor is the resistance of the insulator supporting the low-voltage electrode. Therefore, sufficient consideration should be given to the insulating properties of the supporting insulator. However, if the humidity is high, the surface resistance of the supporting insulator can decrease, and I_x often flows through the resistor R_0.

Appendix 1.2 Precautions for Guard Ring Electrode and Pair Diode

The basis concept for correctly measuring the small current is to make sure $I_X = I_{xm}$, and the measurement currents I_X and I_{xm} are shown in the figure.

(1) **Case 1: Guard ring electrode**

The leakage current of the insulating sample is extremely small ($I_X = 10^{-9}$ to 10^{-13} A). Generally, a sample is placed on the main electrode and the guard ring electrode, and a high-voltage electrode is placed on the sample (Fig. 1.6a). The pico-ammeter should measure the leakage current passing through the sample bulk only. Actually,

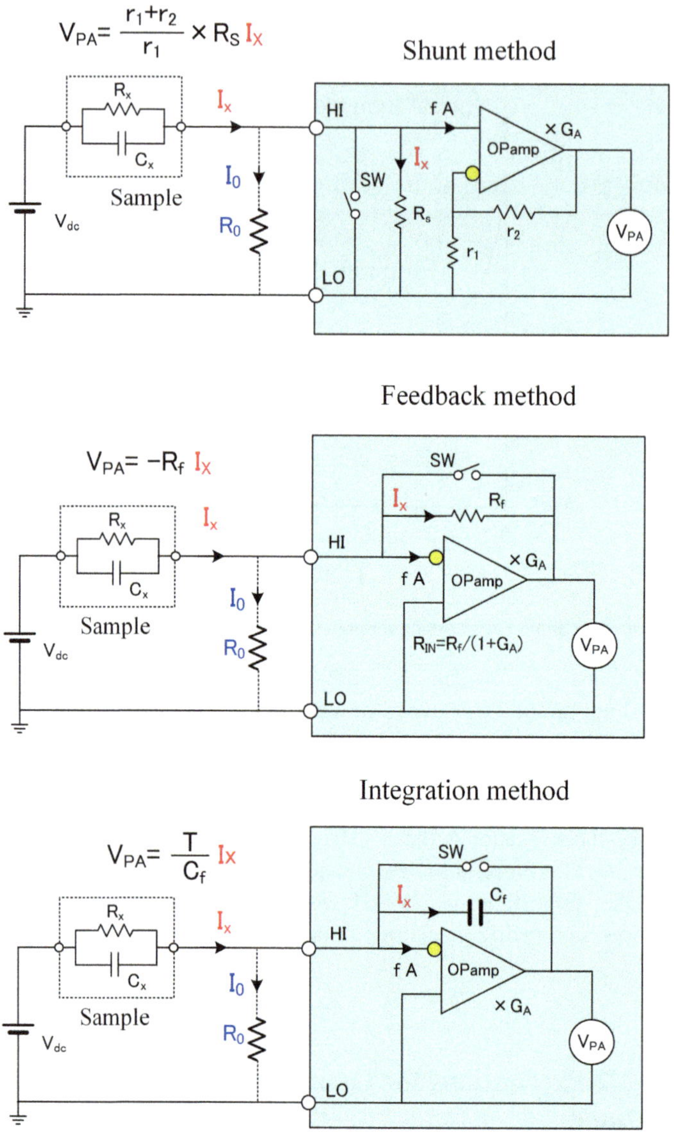

Fig. 1.5 The basic electronic circuits of the pico-ammeter

the surface of the sample may contain dirt, humidity and moisture. Then, the pico-ammeter may measure the current (I_{surf}) flowing on the surface instead of the sample bulk. Therefore, it is necessary to remove this surface current. For this purpose, guard ring electrodes are provided to prevent surface currents. However, when the humidity is high or the surface of the supporting insulator is dirty, the current I_1 flows through

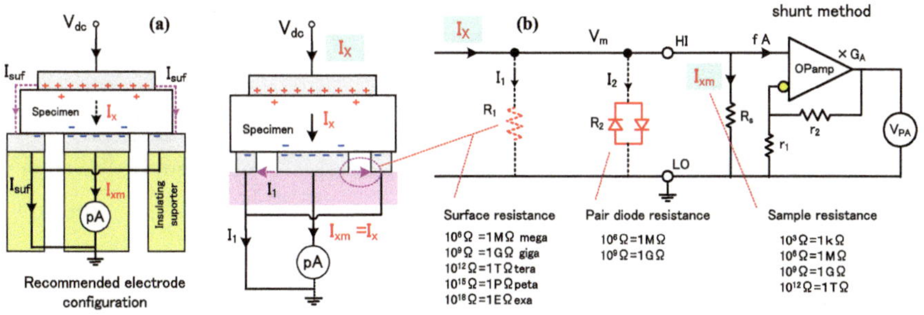

Fig. 1.6 The equivalent circuit of the measurement system

the resistance (*R*) between the main electrode and the guard ring electrode, as shown in Fig. 1.6a.

In order to satisfy the condition of $I_X = I_{xm}$, $I_1 << I_{xm}$ must be satisfied. That is, it is necessary to satisfy the condition of $R_S << R$. In particular, when $R_S = 1\ T\Omega$ $(= 10^{12}\ \Omega)$ in ultra-small current measurement, be careful of surface contamination of the supporting insulator so that $R_S = 10^{12}\ \Omega << R$.

(2) **Case 2: Pair diode**

A pair diode may be connected in parallel to the input terminal to protect the picoammeter. In this case, it is necessary that the input terminal voltage is about $V_m \fallingdotseq 0.5$ V and the internal resistance *R* of the diode satisfies the condition of $R_S << R$. In particular, in the case of $R_S = 1\ T\Omega$ $(= 10^{12}\ \Omega)$ in ultra-small current measurement, it is necessary to confirm whether $R_S = 10^{12}\ \Omega << R$. Since the internal resistance of the diode usually ranges from $10^6\ \Omega$ to $10^9\ \Omega$, it is difficult to satisfy this condition. Therefore, it should be careful.

Appendix 1.3 Noise Comparison Between pA Method and Q(t) Method

(1) **Noise source** Fig. 1.7a
In the modern information and communication society using mobile phones with radio waves and broadcast radio waves, the "displacement current due to spatial noise" of artificial electromagnetic fields is filling up around us. Thus, it is occupied by artificial electromagnetic noise in around space. The metal plate electrode (Fig. 1.7a) connected to the pico-ammeter works as an antenna for receiving the artificial electromagnetic noise. Hence, electrical engineers always have to face many challenges to avoid this.
(2) **Noise appeared in pico-ammeter measurement** Fig. 1.7b
LDPE (Low-density polyethylene): The average value of the measured current *I*(*t*) is about 4 nA at an applied electric field of 50 kV/mm, and its noise component is dominant.

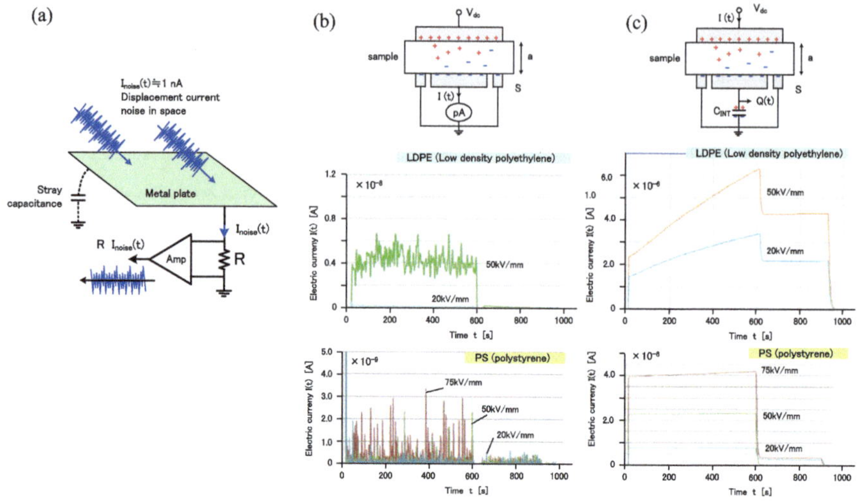

Fig. 1.7 Comparison of the pico-ammeter and Q(t) measurement results

PS (Polystyrene): With an applied electric field of 20 to 75 kV/mm, the measured current $I(t)$ is less than 0.1 nA- 100 pA, it makes the measurement very difficult. The noise component is dominant. We have to evaluate the average current value by the visual measurement.

(3) **Noise in Q(t) method** Fig. 1.7c
LDPE (Low density polyethylene): Q(t) data has no noise component under an applied electric field of 20 and 50 kV/mm. Since Q(t) method reduces noise by integrating and averaging positive and negative noise voltages, smooth Q(t) data can be obtained. From the result of $Q(\mathrm{t}) > Q_0$, it can be determined that the charge is accumulated in the LDPE sample.

PS (Polystyrene): The Q(t) characteristic is $Q(t) = Q_0$ in the applied electric field range of 20 to 75 kV/mm. From this result, it can be determined that no charge is accumulated in the PS sample.

(4) **Q(t) method is strong against noise.**
The Q(t) method for evaluating DC insulating materials is extremely excellent. By comparing the measurement results of the pA method and the Q(t) method in Fig. 1.7b and c, anyone can realize that the measurement results of the Q(t) method can be clearly evaluated for DC insulation. The pA method can read the current value directly, but it contains a large amount of noise. Of course, the current value can be calculated from the slope $\Delta Q(t)/\Delta t$ (time derivative) of Q(t) data. Moreover, there is no noise component. Q(t) method is extremely useful for DC insulation evaluation.

Appendix 1.4 Measurement of Electric Field Intensity $E(0, t)$ at Electrode Surface

The details of the measuring device for electric fields $E(a, t)$ and $E(0, t)$ on electrode surfaces described in Sect. 1.1 are introduced. This is to evaluate the charge injection from the electrode from the time characteristics of $E(a, t)$ and $E(0, t)$ [5].

(1) **Measurement equipment** Fig. 1.8a

When the space charge $\rho(x)$ is accumulated by the application of DC high voltage V_{dc}, the electric field on the electrode surfaces becomes $E(a, t)$ and $E(0, t)$. As a result, the electrode surface charge is given by the following equation.

Anode surface charge: $\sigma(a, t) = \varepsilon E(a, t)$.

Cathode surface charge: $\sigma(0, t) = \varepsilon E(0, t)$.

A high-frequency AC electric field $e(t)$ is superimposed on the charged sample. The following vibration force acts on each electrode surface.

Anode surface: $f(a, t) = \sigma(a, t)e(t) = \varepsilon E(a, t)e(t)$.

Cathode surface: $f(0, t) = \sigma(0, t)e(t) = \varepsilon E(0, t)e(t)$.

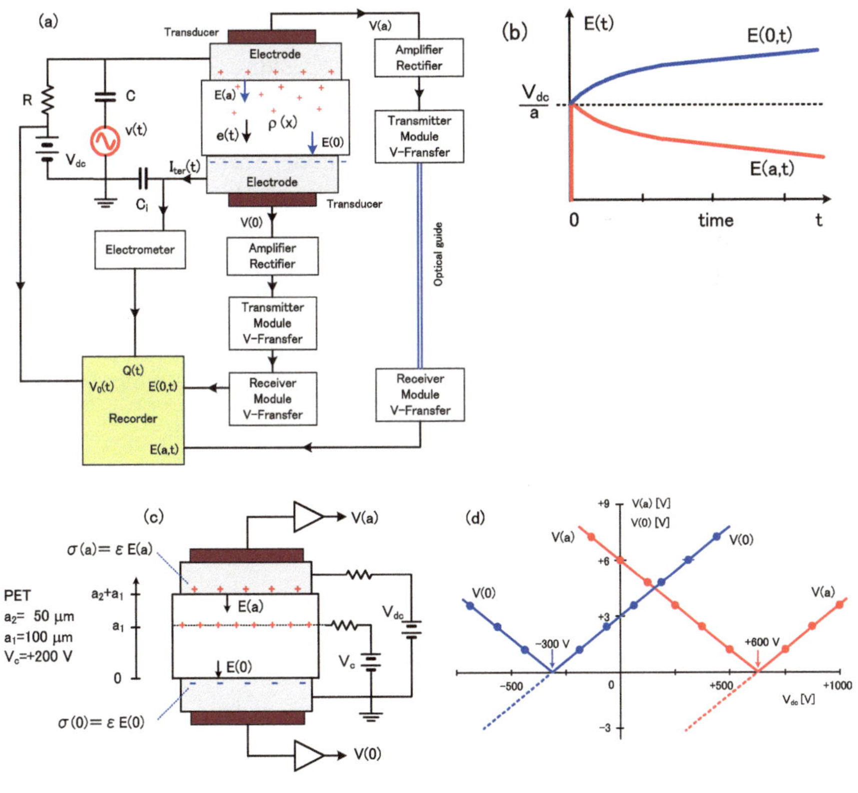

Fig. 1.8 The measurement method of electric field intensity

An oscillating wave is generated from the electrode surface by this oscillating force. This vibration wave is measured with a piezoelectric device stuck on the back of both electrodes.

In this case, the frequency and width are designed so that standing waves and resonance occur at the interface between the electrode and the piezoelectric element. The signal voltages $V(a)$ and $V(0)$ from the piezoelectric element are given by the following equations. The electrode surface electric fields $E(a, t)$ and $E(0, t)$ can be measured.

Anode surface: $V(a) = KE\ (a, t)$.

Cathode surface: $V(0) = KE(0, t)$ where K is a constant

(2) **Measurement results** Fig. 1.8b

$E(a, t)$ decreases over time.

→ Positive charge is injected from the anode.

$E(0, t)$ increases with time.

→ Positive charge approaches the cathode.

This method can evaluate charge injection from the electrode. However, information on the charge distribution in the sample cannot be obtained.

(3) **Operation check of measuring system** Figure 1.8c and d

In order to confirm that this measuring system is operating correctly, a sample of double-layer PET (thickness $a_1 = 100\ \mu\text{m}$ and $a_2 = 50\ \mu\text{m}$) is prepared. The interface of the double-layer PET sample is provided with a deposition electrode. DC voltage V_c (= + 200 V) is applied to this deposition electrode. Then, the DC voltage V_{dc} (–1000 V to + 1000 V) is applied to the two electrodes, and the signal voltages $V(a)$ and $V(0)$ are measured.

The measurement results are shown in Fig. 1.8d.

The condition of V_{dc} at which the signal voltage becomes $V(a) = 0$ and $V(0) = 0$ is as follows.

$$\begin{aligned} E(a) &= \frac{V_{dc}}{a_1 + a_2} - \frac{V_c}{a_2} = 0 \quad \Rightarrow \quad V_{dc} = +600\text{V} \\ E(a) &= \frac{V_{dc}}{a_1 + a_2} + \frac{V_c}{a_1} = 0 \quad \Rightarrow \quad V_{dc} = -300\text{V} \end{aligned} \tag{1.2}$$

When $V_{dc} = +\ 600$ V, $E(a) = 0$ and $V(a) = 0$.

When $V_{dc} = +\ 300$ V, $E(0) = 0$ and $V(0) = 0$.

From this, it can be confirmed that this measuring system is operating properly.

References

1. T. Takada, T. Sakai, Y. Toriyama, Evaluation of electric charge distribution in polymer films, *Electrical Engineering in Japan*, 92(6), 537–544 (1972).

2. T. Takada, T. Sakai, Y. Toriyama, Measurement of space charge quantity in dielectric film by maxwell stress method, *The Transactions of the Institute of Electrical Engineers of Japan.A*, 99(10), 451–457 (1979)
3. H. Ren, Y. Tanaka, H. Miyake, Q. Li, H. Gao, C. Li, Z. Wang, Effect of sinusoidal waveform and frequency on space charge characteristics of polyimide, *High Voltage*, 6(5), 760–769 (2021)
4. Naoshi Suzuki, Principle of low-current measurement and actual picoammeter, *Applied Physics*, 70(7), 868–871 (2001)
5. T. Takada, T. Sakai, Measurement of electric fields at a dielectric/electrode interface using an acoustic transducer technique, IEEE Transactions on Electrical Insulation, EI-18(6), 619–628 (1983)

Chapter 2
Fundamentals of Q(t) Measurement

2.1 Q(t) Meter Circuit Configuration

Q(t) meter produced by A and D Co. can be used on high voltage side in the measurement system, as shown in Fig. 2.1. Therefore, the measured data of Q(t) meter is also with a transmitting antenna. Figure 2.2a shows the basic configuration of its electronic circuit diagram, and Fig. 2.2b shows a photograph from the internal view.

(1) **Integration capacitor C_{INT}**

A capacitor C_{INT} for integrating the current $I(t)$ is shown in Fig. 2.2a. This current $I(t)$ enters from Terminal 1 and flows to Terminal 2. The value of the capacitance C_{INT} is selected based on the relationship between the capacitance C_S of the sample and the applied voltage V_{dc}. The selection method is described in Sect. 2.3.

(2) **Impedance conversion device**

This device can converts the charge amount $Q(t)$ charged in the integration capacitor C_{INT} into a voltage $V_Q(t)$. An OP amp with a high-input impedance ($R_{IN} = 10^{15}\,\Omega$) is used. Then, the discharge time constant of the integration capacitor with $C_{INT} =$ 1.0 nF is $\tau = R_{IN}C_{INT} = 10^9$ s $\fallingdotseq$ 300 h. Normally, an integration capacitor with C_{INT} = 1.0 to 10 μF is used, and thus there is no worry about leakage of the measured charge in the device even for a long measurement time (12 h). Section 2.4 introduces the reason. The actual leakage of the integrated charge is often caused by surface leakage of the wiring board or the connection terminal. The measurement system requires careful attention.

(3) **AD converter**

The analog output voltage from the low-impedance OP amp that has undergone impedance conversion is converted to a digital signal by an AD converter (16 bit = $2^{16} = 65{,}536$) .

T. Takada et al., *Electric Charge Accumulation in Dielectrics: Measurement and Analysis*,
https://doi.org/10.1007/978-981-19-6156-4_2

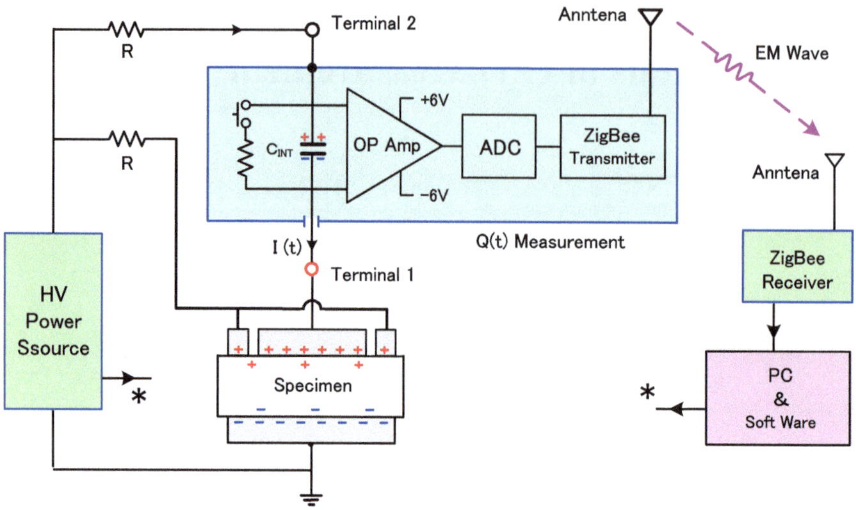

Fig. 2.1 The whole measurement circuit of the Q(t) system

(4) **Transmitter ZigBee and transmitting antenna**

The Q(t) data converted to a digital signal is transmitted to the PC placed at the ground level by ZigBee Transmitter. Therefore, ZigBee Receiver is also installed on the ground side and connected to PC.

2.2 Q(t) Meter Measurement on High Voltage Side

Sections 1.4 presents that the Q(t) method has superior features to the conventional micro-current measurement method. This section introduces the fact that a Q(t) measuring instrument can be installed and used on the high voltage side as shown in Fig. 2.3a.

Background: First, the background on why the Q (t) method was developed is introduced. A large number of wires is laid for atomic reactor control. There has been a need to evaluate how level the aged insulation layers of nuclear reactor wires exposed to high-energy radiation have deteriorated for many years. When diagnosing deterioration of the wires, it is necessary to insert a microammeter between the outer conductor of the cable and the ground. In reality, there is not enough insulation level between the outer conductor and ground. Therefore, if a pico-ammeter with high input impedance is inserted between the outer conductor and the ground level, the degraded current can not be measured due to noise, and accurate measurement is difficult. The details of this problem are described in Sect. 6.6.

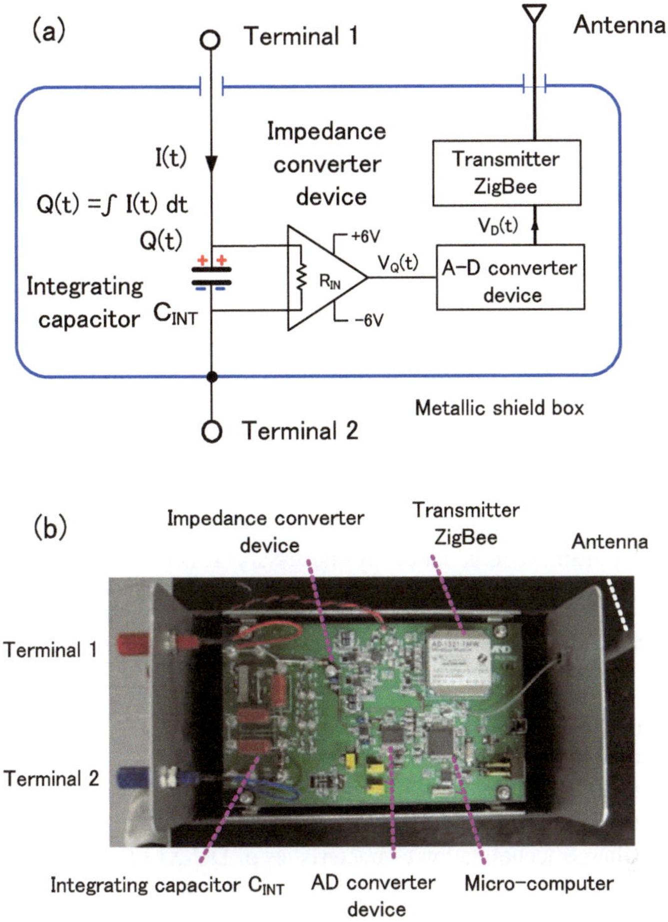

Fig. 2.2 The internal system of the Q(t) method

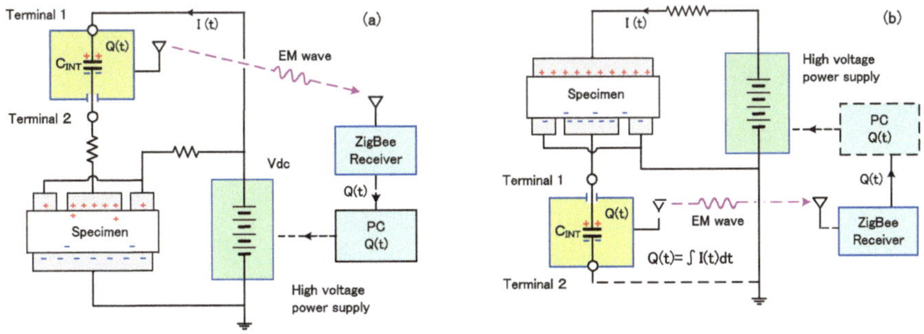

Fig. 2.3 The two connecting methods of the Q(t) system

(1) Installation of Q(t) instrument on the high voltage side

It is proposed that a new measuring instrument should be installed on the high voltage side to reduce the noise. In the background, wireless transmission technologies (ZigBee Transmitter) for digital signals and small batteries have been developed, and their applications are possible. Figure 2.3a shows a measurement system in which a Q(t) measuring instrument is installed on the high voltage side. The guard electrode is used as the high voltage electrode of the sample. The main measurement electrode is connected to a high voltage power supply via a Q(t) measuring instrument. The measured digital signal is transmitted to the PC on the ground side via the ZigBee Transmitter.

(2) No difference in connections of high voltage side and low voltage side

Figure 2.3a shows the case where the Q(t) measuring instrument is installed on the high voltage side, and Fig. 2.3b shows the case where the Q(t) measuring instrument is installed on the low voltage side. In the case of the high voltage one, wireless transmission of signals is a necessary condition, but in the case of the low voltage side, wireless transmission is not always necessary. It is used just because it has a wireless transmission function. See Sect. 4.3.

(3) Application example of Q(t) meter on high voltage side

The following are measurement examples applied to insulation diagnosis of charge accumulation by Q(t) meter on the high voltage side. Details are given in each section.
Section 5.1: Q(t) data of inorganic materials.
Section 6.1: Coaxial cable.
Section 6.2: Coaxial cable with temperature rise.
Section 6.3: Charge accumulation characteristics of DC-XLPE cable.
Section 6.4: Insulation diagnosis after gamma irradiation.
Section 6.5: Diagnosis of CV cable after accelerated water-tree deterioration.
Section 6.7: Double-layer dielectric interface.
Section 6.8: Evaluation of charge accumulation in power devices.
Section 6.9: Electric tree.

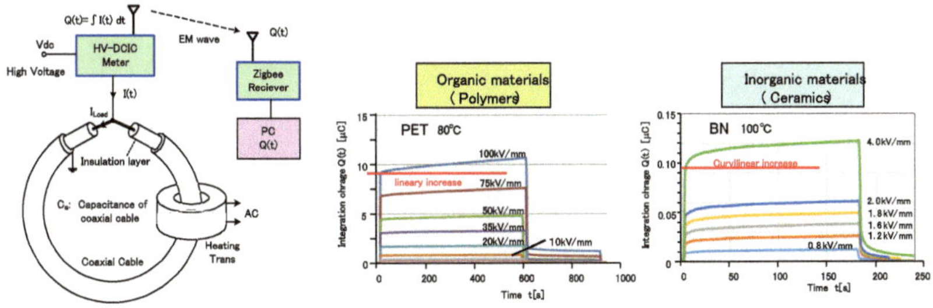

Section 6-2: Coaxial cable with temperature rise

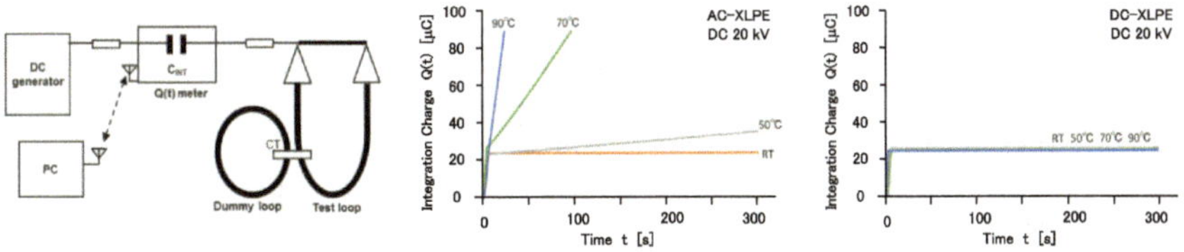

Section 6-3: Charge accumulation characteristics of DC-XLPE cable

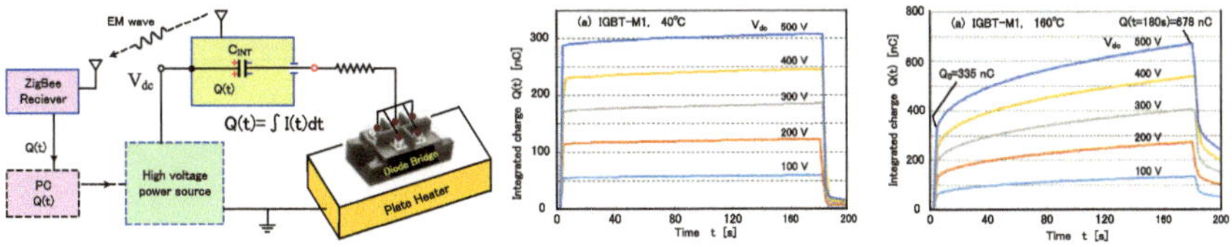

Section 6-8: Evaluation of charge accumulation in power devices

(4) Q(t) system design on high voltage side

Figure 2.4a shows the whole system when the Q(t) set is installed on the high voltage side.

① Q(t) measurement is in high voltage, ⑫ It is necessary to install on acrylic insulation shelf.

② HV DC power supply, ③ and ④ Digital conytoller are stored in the lower shelf.

⑪ Place the flat specimen on the ground side. ⑩ Measuring electrodes are placed on the high voltage side. The high voltage is supplied to ⑪ the flat plate sample through the protection resistance ($R = 1.0\ \mathrm{M\Omega}$).

As shown in Fig. 2.4b, the sample is heated by a Plate Heater and covered with a heat insulation box. This system can raise the sample temperature from room temperature (RT) to 250 °C. Therefore, the Q(t) system can apply the evaluation of electric charge accumulation in heat resistant resins and ceramic materials. In addition, it is also used to evaluate the charge accumulation of IGBT-M modules of finished power devices.

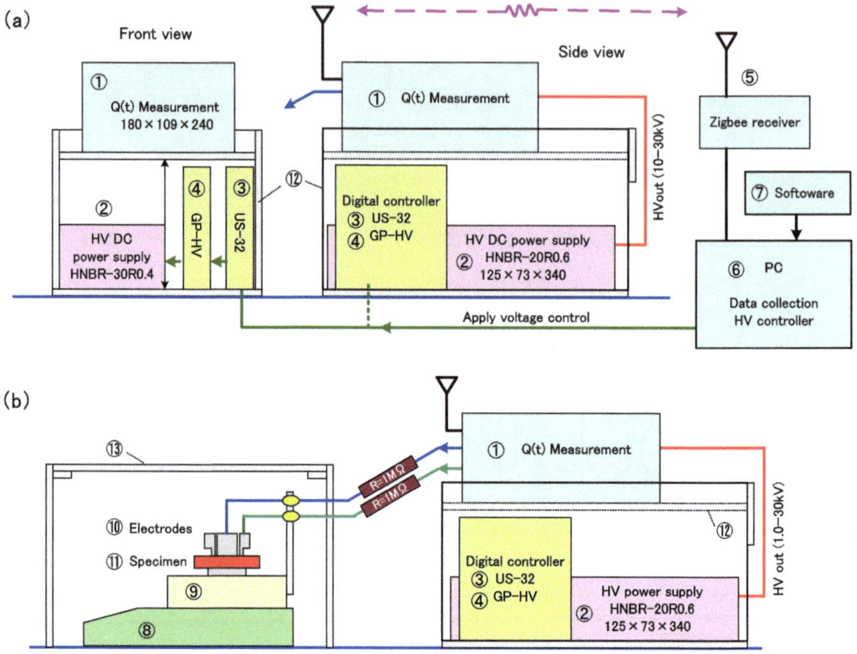

① Q(t) Measurement (180×109×240) A&D Co. AD9832
② HV DC Power Supply (125×73×340) Matsusada Precision, HMBR-30R0 (±30kV, 0.4mA)
③ Digital controller Matsusada Precision, US-32
④ Digital controller Matsusada Precision, GP-HV
⑤ ZigBee receiver A&D Co. AD9832
⑥ PC Data collection HV controller A&D Co. AD9832
⑦ Software A&D Co. AD9832
⑧ Plate Heater AS ONE
⑨ Heating Electrode Self-made
⑩ Measuring Electrodes Self-made
⑪ Specimen
⑫ Insulation Box (PMMA/ 5mm) W240×H185×L360 Self-made
⑬ Thermal Box (PMMA/ 5mm) W240×H130×L270 Self-made

Fig. 2.4 The internal design of the Q(t) system

2.3 Selection of Capacitance of Integration Capacitor

(1) **Basic equation**

Use the basic expression of Eq. (2.2) to select the capacitance of the integration capacitor, where C_{INT} is the capacitance of integrating capacitor. C_{S} is the capacitance of sample. V_{INT} is the integral capacitor voltage. V_{dc} is the applied DC voltage. The amount of charge Q supplied by the measurement circuit is given by Eq. (2.1). The capacitance ratio $C_{\mathrm{INT}}/C_{\mathrm{S}}$ in Eq. (2.2) is equal to each voltage ratio. The condition of $V_{\mathrm{dc}} >> V_{\mathrm{INT}}$ is required to sufficiently apply the voltage V_{dc} to the sample C_{S}.

In order to set the measurement accuracy to 1% and 0.1%, it is necessary to set $V_{dc} = 100 \times V_{INT}$ and $V_{dc} = 1000 \times V_{INT}$. Under such conditions, it becomes the last term of Eq. (2.2).

$$Q = C_S(V_{dc} - V_{INT}) = C_{INT}V_{INT} \tag{2.1}$$

$$\frac{C_{INT}}{C_S} = \frac{V_{dc} - V_{INT}}{V_{INT}} \cong \frac{V_{dc}}{V_{INT}} \tag{2.2}$$

(2) **Range of sample capacitance and applied voltage**

Example of film sample (flat plate sample):
Polymer insulating material with a thickness of 200 μm and electrode area of $S = 8$ cm^2 has a capacitance of $C_S = 80$ pF. Therefore, the capacitance of a general film sample is $C_S = 50$ to 150 pF (right axis of the graph in Fig. 2.5). The applied test voltage is generally $V_{dc} = 40$ V to 5 kV (upper axis of the graph).
Example of cable sample:
The capacitance of CV cable is $C_S = 250$ pF/m. Therefore, the capacitance of a CV cable (10 m long) is $C_S = 2.5$ nF. The capacitance of a generally tested cable is $C_S = 0.5$ to 10 nF (right axis of the graph). Thus, the value of the capacitance of the sample is estimated. The test voltage of the cable is generally $V_{dc} = 200$ V to 30 kV (upper axis in Fig. 2.5).

(3) **Measurement range of voltage V_{INT} on the integration capacitor**

The voltage V_{INT} on the integration capacitor C_{INT} is measured by OP amp. From these products, the measured charge amount $Q(t) = C_{INT}V_{INT}$ is obtained. The integral voltage V_{INT} is designed to have a maximum of $V_{INT} = \pm 3.0$ V in consideration of linearity and margin (bottom axis of the graph). This analog voltage $\pm V_{INT}$ is converted into a digital signal by a 16-bit AD converter. Unipolar $V_{INT} = 3.0$ V is 15 bits. 15 bits = 32,768 is equal to 5 digits (5 figures). It is enough to select a minimum of 2 to 3 digits for measurement. The reason is to ensure the measurement accuracy when calculating the current from the difference between the Q(t) measurement values. From the above, the measured range of V_{INT} is from ± 20 mV to ± 3.0 V (bottom axis of the graph).

(4) **Selection of capacitance of integration capacitor.**

Example 1: sheet sample.
The capacitance of the film sample in Fig. 2.5 is $C_S = 80$ pF. The maximum applied voltage is $V_{dc} = 5$ kV. The measurement voltage is $V_{INT} = 3.0$ V. From Eq. (2.1), the value of the integration capacitor is $C_{INT} = 0.133$ μF. Actually, select $C_{INT} = 0.15$ μF.
Example 2: cable sample.
The capacitance of the cable sample in the figure is $C_S = 2.5$ nF. The maximum applied voltage is $V_{dc} = 30$ kV. The measurement voltage is $V_{INT} = 3.0$ V. From

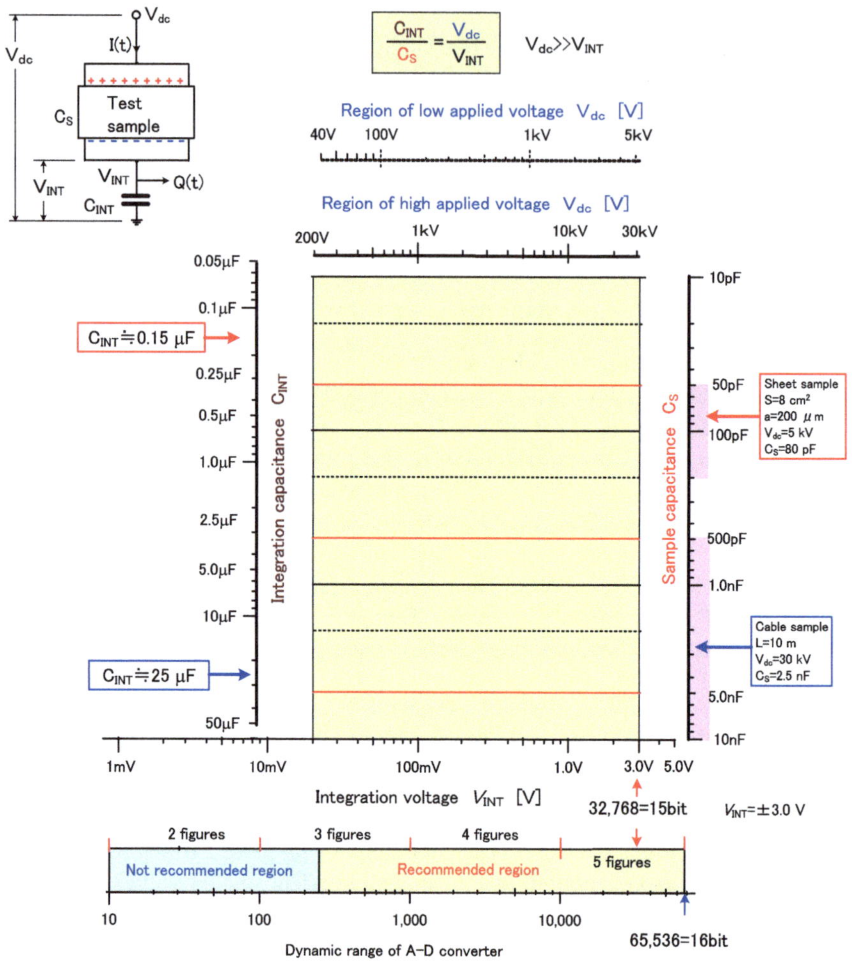

Fig. 2.5 The selection of the integration capacitance in the Q(t) system

Eq. (2.1), the value of the integration capacitor is $C_{INT} = 2.5\ \mu F$. Therefore, select $C_{INT} = 2.5\ \mu F$.

(5) **Use high-quality integration capacitors.**

When the electric charge is discharged by the leakage resistance inside the integration capacitor (C_{INT}), an error occurs in the integrated electric charge amount. Therefore, it is necessary to ensure no leakage charge in the integration capacitor itself. PP (polypropylene) capacitors or PS (polystyrene) capacitors with high quality that have this performance are used.

2.4 Stability of Q(t) System

When used for evaluating the electric charge accumulation inside DC power cables, the Q(t) measurement system is required to be able to measure stably for a long time, such as 12 h. That is, it is required that the leakage of the electric charge in the integration capacitor C_{INT} (= 1.0 μF) is small and negligible. Therefore, the following leakage charge amount is evaluated.

(1) **Evaluation of leakage charge.**

A measurement circuit for evaluating the amount of leakage charge inside the Q(t) system is shown in Fig. 2.6. Without connecting the test sample, positive voltage $V = 2.1$ V is applied to the integration capacitor C_{INT}, which isn't disconnected with the power supply. The integrated charge amount is $Q_0 = C_{INT} \times V = 210$ nC. The discharge characteristics of the charge amount Q_0 are measured for 12 h. The results are also shown in Fig. 2.6. The discharge rate of the charge after 12 h is as small as 0.6%.

Similarly, even in the case of negative voltage of − 2.5 V, the discharge rate is as small as 0.5%.

Further, the confirmation of the drift of the Q(t) system in the case of no applied voltage is shown in the lower part of Fig. 2.6. No drift is confirmed at all, and it can be evaluated as a stable measuring instrument.

Such a system with a discharge rate of 0.5 to 0.6% can be sufficiently used for evaluating the electric charge accumulation of DC power cables.

2.5 Consideration of Stray Capacitance

(1) **Connection method of Q(t) measuring instrument**

The Q(t) measuring instrument has two terminals, i.e. Terminal 1 and Terminal 2. Therefore, there are the following two connection methods.

Case A: Terminal 1 is connected to high voltage, and Terminal 2 is connected to sample electrode.

Case B: Terminal 1 is connected to sample electrode, and Terminal 2 is connected to high voltage.

(2) **Correct connection method**

When installing on the high voltage side, connect terminal 1 to the sample electrode and terminal 2 to the high voltage power supply as in Case B.

Reason

One end of the integration capacitor C_{INT} is connected to the metal case of the Q(t) measuring set. As shown in the figure of case B, there is a stray capacitance C_S

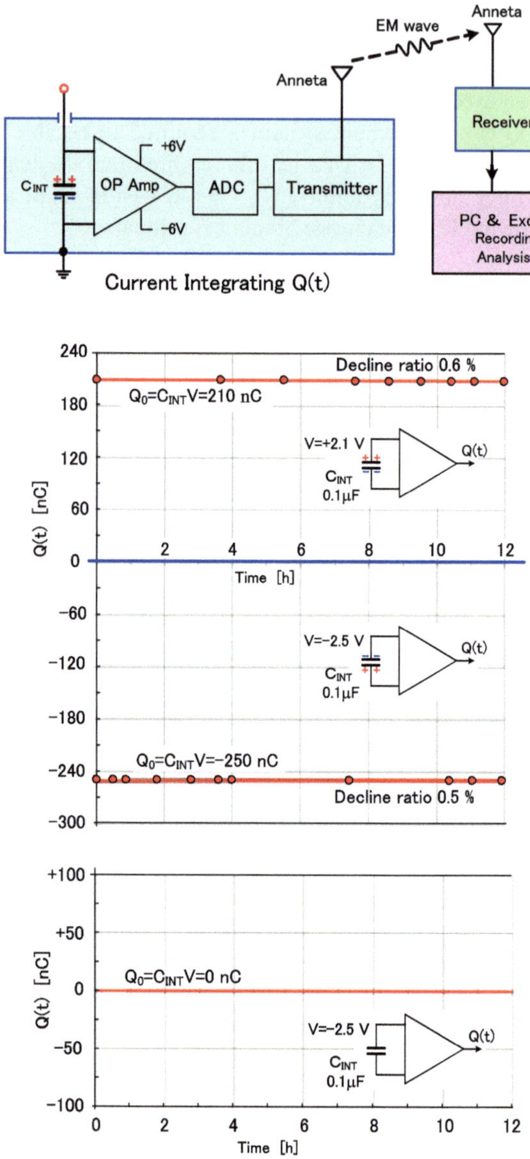

Fig. 2.6 The verification of the stability of Q(t) system

between the metal case and the floor or wall. Therefore, a displacement current flows through the space between the metal case and the floor when a voltage is applied.

Case A: The integration capacitor C_{INT} integrates both the sample charging current and the spatial displacement current.

Case B: The integration capacitor C_{INT} can only integrate the sample charging current.

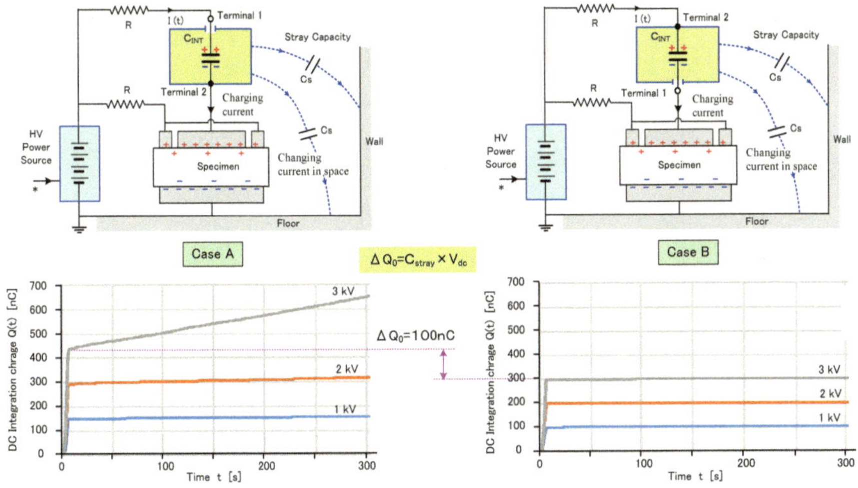

Fig. 2.7 The measured results by different connections of the Q(t) system

(3) **Experiment confirmation**

The connections of Case A and Case B and their respective Q(t) data are shown in Fig. 2.7. The initial charge amount Q_0 is larger in Case A than in Case B. Here, since the difference between the initial charge amounts is $\Delta Q_0 = 100$ nC and the applied voltage is $V_{dc} = 2500$ V, the stray capacitance is $C_{stray} = 40$ pF from the relationship of $\Delta Q_0 = C_{stray} \times V_{dc}$. When the capacitance of the sample is small while the stray value is considerably large, the presence of such a stray capacitance causes a measurement error. This is because the connection in Case A measures an extra displacement current through C_S.

In Q(t) measurement, it is necessary to consider eliminating the stray capacitance C_S.

2.6 Q(t) Method Applied Under High Temperature and High Voltage

(1) **Traditional conductivity measurement**

The sample thickness is about 200 μm, the applied electric field is about 5 kV/mm at the applied voltage of 1000 V. Temperature is up to about 80 °C. Therefore, it becomes the conductivity measurement area in Fig. 2.8.

(2) **Traditional permittivity measurement**

The sample thickness is about 100 μm and the applied electric field is about 0.1 kV/mm at the applied voltage of 10 V. The temperature is around 80 °C. Therefore, it becomes dielectric permittivity measurement area in the figure.

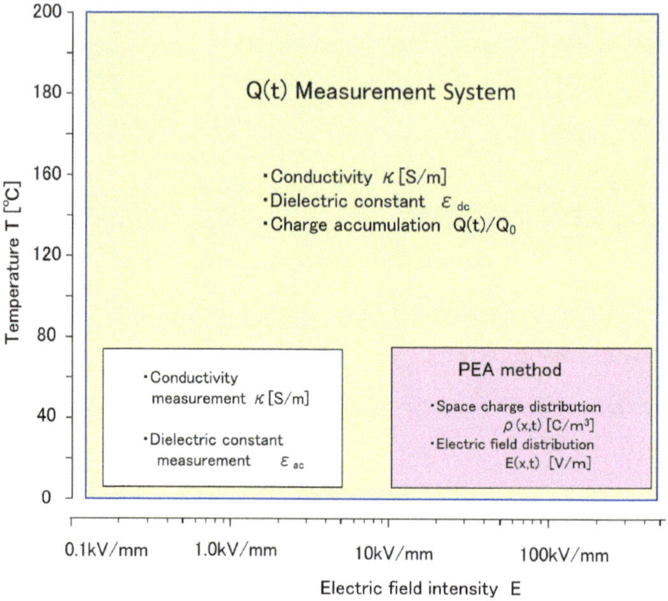

Fig. 2.8 The measurement range of the methods

(3) PEA measurement

The sample thickness is about 100 μm and the applied electric field is about 100 kV/mm at the applied voltage of 10 kV. The temperature is limited by the characteristics of the piezoelectric device for detection of sound waves and is around 80 °C. Therefore, it becomes the PEA method area in the figure.

(4) Q(t) measurement

The measurable sample is independent on the sample shape, such as (a) plate-like sample, (b) power device insulation of the finished product, (c) coaxial cable insulation of the finished product, and (d) degraded CV cable with water tree as shown in Fig. 2.9. The charge accumulation of any type samples can be evaluated. The applied electric field can range up to 500 kV/mm to breakdown electric field. The temperature range has been measured from room temperature to 200° C.

As shown in the figure, Q(t) measurement can be applied in a higher temperature and higher electric field range than other measurement methods.

PEA measurement cannot evaluate the amount of leakage charge $Q_{\mathrm{leak}}(t)$, but now it can be evaluated in combination with the Q(t) method. From the measurement results of Specimen C in Fig. 1.2, if the space charge observed in the PEA measurement has a uniform distribution (in green color), it seems at first scent that no space charge and no leakage current flows. In this case, the leakage current component $Q_{\mathrm{leak}}(t)$ can be evaluated by using the PEA method and the Q(t) method together (see Sect. 3.4) [1]. There are two main characterizations of DC insulation materials.

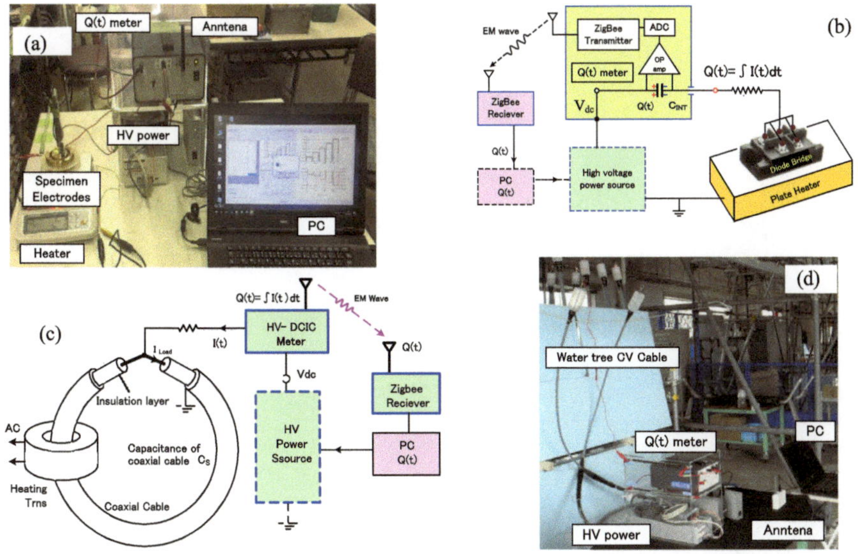

Fig. 2.9 The Q(t) systems for different power devices

- Evaluation of distortion of internal electric field $E(x, t)$ due to space charge accumulation $\rho(x, t)$. This is because the local electric field may exceed the designed electric field, leading to dielectric breakdown. It can be evaluated by PEA measurement.
- Leakage current flows and insulation is lost. The insulation can be evaluated by measuring Q(t) data based on the characteristic of $Q(\mathrm{t}) > > Q_0$.

For this reason, it is essential to use both PEA and Q(t) measurements to evaluate the characteristics of DC insulating materials and evaluate both $E(x, t)$ distribution and leakage current.

Reference

1. T. Takada, T. Hanawa, K. Ogura, Y. Tanaka, Separation of electric charge accumulation and leakage current under DC stress - data analysis by using PEA method and Q(t) method -, Paper presented at the 50th Symposium on Electrical and Electronics Insulating Materials and Application in Systems, Tokyo, Japan, 1–10, 2019.

Chapter 3
Evaluation of Charge Accumulation

3.1 Evaluation of Charge Accumulation by Charge Ratio $Q(t)/Q_0$

(1) **Q(t) characteristics of polypropylene (PP)**

Figure 3.1a shows a Q(t) measuring circuit of the current integration charge method. Figure 3.1b and c show the results of the Q(t) characteristics of the PP sample. The applied conditions include the electric field of $E = 10$–100 kV/mm, the temperature of 20 °C and 80 °C, and the measurement time of $t_{\mathrm{m}} = 600$ s. The charge accumulation properties of PP samples are significantly dependent on temperature. At the measurement time of $t_{\mathrm{m}} = 600$ s, $Q(\mathrm{t}) \fallingdotseq \mathrm{Q}_0$ at 20 °C (Fig. 3.1b), which means no charge accumulation. However, at 80 °C (Fig. 3.1c) $\mathrm{Q(t)} > \mathrm{Q}_0$, then it is determined that there is charge accumulation.

(2) **Charge ratio evaluation $Q(t)/Q_0$**

The Q(t) measurement data on vertical axis measures the time-integrated charge amount of all the current $I(t)$ in Eq. (3.1). Equation (3.1) has already been introduced in Sect. 1.4. This Q(t) data measures the sum of the initial charge amount Q_0, the accumulated space charge amount $Q_{\mathrm{spac}}(\mathrm{t})$, and the leakage current charge amount $Q_{\mathrm{leak}}(\mathrm{t})$. These three components are discriminated from the Q(t) data to evaluate the charge accumulation. The algorithm will be described in order.

$$\begin{aligned} Q(t) &= Q_0 + \int_0^t I_{spac}(t)\,dt + \int_0^t I_{leak}(t)\,dt \\ &= Q_0 + \quad Q_{spac}(t) \quad + \quad Q_{leak}(t) \end{aligned} \tag{3.1}$$

T. Takada et al., *Electric Charge Accumulation in Dielectrics: Measurement and Analysis*,
https://doi.org/10.1007/978-981-19-6156-4_3

Fig. 3.1 The measured results of PP material

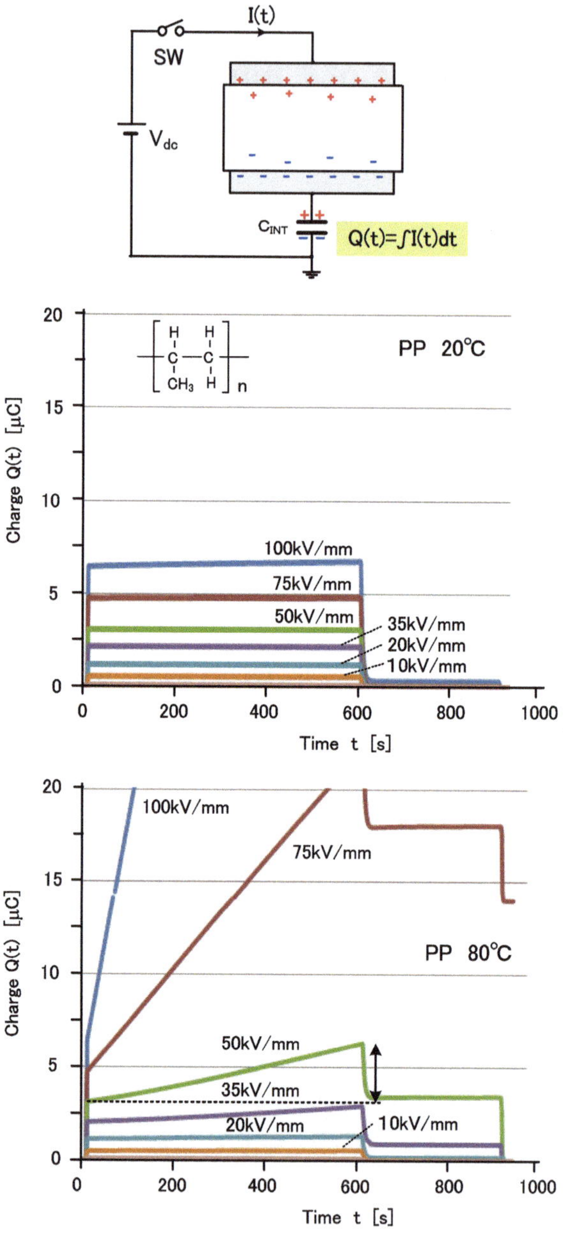

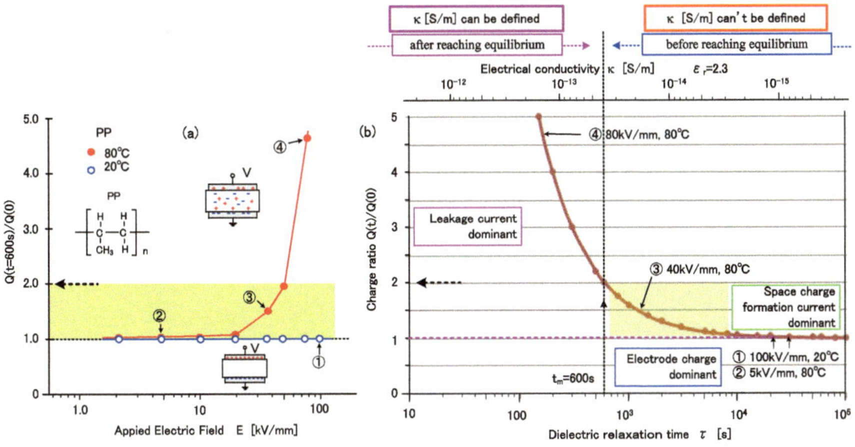

Fig. 3.2 Analysis on the charge ratio and corresponding conductivity

(3) Electric field dependence of Q(t) characteristic

In order to observe the electric field dependence of the Q(t) characteristic, Eq. (3.1) is expressed by a charge ratio $Q(t)/Q_0$ as in Eq. (3.2). The results are summarized in Fig. 3.2a.

$$\frac{Q(t)}{Q_0} = 1 + \frac{Q_{spac}(t)}{Q_0} + \frac{Q_{leak}(t)}{Q_0} \tag{3.2}$$

The Q(t) characteristic at 20 °C shows that the charge ratio is $Q(t)/Q_0 = 1$ even when a high electric field of 100 kV/mm is applied. This result means that $Q_{spac}(t) + Q_{leak}(t) = 0$, indicating there are no charge accumulation. Therefore, PP maintains excellent DC insulation at 20 °C.

The Q(t) characteristic at 80 °C starts to deviate at 10 kV/m from $Q(t)/Q_0 = 1$ and remarkably deviates at about 50 kV/mm. This result means that $Q_{spac}(t) + Q_{leak}(t) > 0$, indicating that charges are accumulated and leakage current is flowing. It can be seen that when the electric field becomes high at 80 °C, the DC insulation properties of PP is significantly reduced.

(4) Dependence of charge ratio $Q(t)/Q_0$ on measurement time t_m, relaxation time τ and conductivity κ

The following two characteristics are evaluated for DC insulating materials.

(a) Internal electric field distortion due to charge accumulation:

Evaluate whether $Q_{spac}(t)$ is distorting the electric field.

(b) Deterioration of insulation properties due to leakage current:

$Q_{\text{leak}}(t)$ evaluates the leakage current charge.

From the above, it is desired to evaluate the process of shifting from $Q_{\text{spac}}(\text{t})$ to $Q_{\text{leak}}(\text{t})$ after the start of charge accumulation of $Q(t)/Q_0 > 1$.

At the measurement time t_{m}, the charge ratio $Q(t_{\text{m}})/Q_0$ in Eq. (3.2) is derived as in Eq. (3.3). This Eq. (3.3) is from the Eq. (3.6) in the following Sect. 3.2. In order to consider this transition process, Eq. (3.3) is drawn in Fig. 3.2b. $\tau = \varepsilon/\kappa$ is the dielectric relaxation time. Where ε is the permittivity and κ is the conductivity. First, the evaluation results are described.

$$\frac{Q(t_m)}{Q_0} = 1 + \frac{t_m}{\tau} \quad where \quad \tau = \frac{\varepsilon}{\kappa} \tag{3.3}$$

Since the measurement results in Fig. 3.1b and c have a measurement time of $t_{\text{m}} = 600$ s, the calculation result is drawn in Fig. 3.2b by substituting $t_{\text{m}} = 600$ s into the Eq. (3.3). The vertical axis represents the charge ratio $Q(t_{\text{m}})/Q_0$, and the horizontal axis represents the dielectric relaxation time τ. Further, the conductivity κ is also labled on the upper horizontal axis, where the relative permittivity of $\varepsilon_{\text{r}} = 2.3$ is used.

Overview of Fig. 3.2b:

$Q(t)/Q_0 = 1$ for an excellent insulating material having a long relaxation time of $\tau >> t_{\text{m}}$. It is determined that there is no charge accumulation. When τ is short and the material shows no good insulating property, the charge ratio shifts to $Q(t_{\text{m}})/Q_0 > 1$. Just when $t_{\text{m}} = \tau$, $Q(t_{\text{m}})/Q_0 = 2$. Further, in the region of $\tau < t_{\text{m}}$, $2 < Q(t_{\text{m}})/Q_0$, and the characteristic shows that the leakage current is so large that it cannot be said that it is an insulating material.

(5) **Discrimination between space charge accumulation and leakage charge**

Observation of the Q(t) measurement results of PP:

Observe the results of the charge ratio $Q(t_{\text{m}})/Q_0$ in Fig. 3.2a which are obtained from experimental results of Fig. 3.1b and c. In case of $Q(t_{\text{m}})/Q_0 > 1$, it can be determined that the charge is accumulated, but it i s dfficult to determine which of $Q_{\text{spac}}(\text{t})$ and $Q_{\text{leak}}(\text{t})$ is dominant. Therefore, this will be discussed based on experimental results of the PEA and Q(t) methods.

① **in Fig. 3.2:** PP (20 °C) is under an high electric field of 100 kV/mm, and the result is $Q(t_{\text{m}})/Q_0 = 1.05$.

It is judged that there is no accumulated charge. This region is before the start of space charge formation.

② **in Fig. 3.2** (80 °C)**:** In the result of PP under a high temperature of 80 °C and a low electric field of 5 kV/mm,

$Q(t_{\text{m}})/Q_0 = 1.05$. It is determined that there is almost no accumulated charge.

This result is in the region of the relaxation time τ (= 10,000 s) > > the measurement time t_m (= 600 s).

③ **in Fig. 3.2** (80 °C)**:** In the result of PP under a high temperature of 80 °C and a medium electric field of 50 kV/mm, it shows $Q(t_m)/Q_0$ =1.3. In this region, space charge formation is dominant, where relaxation time τ (= 1500 s) > measurement time t_m (= 600 s).

④ **in Fig. 3.2** (80 °C)**:** In the result of PP under a high temperature of 80 °C and a high electric field of 70 kV/mm, it shows $Q(t_m)/Q_0$ =3.2. There is the process of shifting to leakage current. It is in the region of relaxation time τ (= 400 s) < measurement time t_m (= 600 s).

⑤ **in Fig. 3.2** (80 °C)**:** When the high temperature is 80 °C and the high electric field is 75 kV/mm, the result is $Q(t_m)/Q_0 = 4.5$. In this area, the leakage current is dominant. The material cannot be called insulating one. It is in the region of relaxation time τ (= 200 s) < measurement time t_m (= 600 s).

Conclusion

If the value of the charge ratio $Q(t_m)/Q_0$ is entered in the graph of Fig. 3.2a and b, the rough boundary between $Q_{spac}(t)$ and $Q_{leak}(t)$ can be determined.

$\boldsymbol{Q(t_m)/Q_0 = 1}$**:** No charge accumulation.

$\boldsymbol{1 < Q(t_m)/Q_0 < 2}$**:** Space charge accumulation is dominant until measurement time t_m = relaxation time τ.

$\boldsymbol{2 < Q(t_m)/Q_0}$**:** Leakage current charge is dominant. The relaxation time τ < the measurement time t_m.

3.2 Relationship Between Relaxation Time *t* and Measurement Time t_m

(1) Electric conduction and relaxation phenomena in Q(t) data

Equation (3.3) in Sect. 3.1 is derived based on the relaxation phenomenon model. Focusing on the time change of the Q(t) characteristic in Fig. 3.3, consider the relaxation phenomenon. Figure 3.3 shows the simulated Q(t) characteristics.

When the applied voltage V increases ($V_1 \rightarrow V_2 \rightarrow V_3$), the amount of changing charge $\Delta Q(t)$ (= $Q(t)$ - Q_0) increases. It is experimentally observed that $\Delta Q(t)$ also increases ($\Delta Q_1(t) \rightarrow \Delta Q_2(t) \rightarrow \Delta Q_3(t)$). Here, $Q_0 = C_s V$.

The characteristics in Fig. 3.3 indicates that the increase rate $\Delta Q(t)/\Delta t$ in Eq. (3.4) is proportional to the applied voltage V. That is, the increase rate $\Delta Q(t)/\Delta t$ is proportional to $Q_0 = C_s V$. Then, the proportional coefficient is set to the reciprocal $1/\tau$ of the dielectric relaxation time. $\tau = \varepsilon_0 \varepsilon_r / \kappa$, where ε_0 is the dielectric constant of

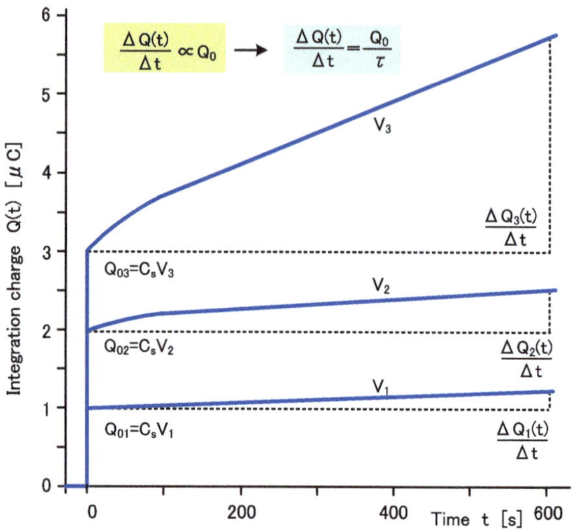

Fig. 3.3 The time change of the Q(t) characteristic

vacuum, ε_r is the relative dielectric constant of the sample, and κ is the conductivity of the sample. When the above relational expressions are arranged, they are summarized in the last term of Eq. (3.4).

$$\frac{\Delta Q(t)}{\Delta t} = \frac{1}{\tau} C_s V = \frac{1}{\tau} Q_0 \quad where \quad \tau = \frac{\varepsilon}{\kappa} \Rightarrow \frac{\Delta Q(t)}{Q_0} = \frac{\Delta t}{\tau} \Rightarrow \frac{Q(t) - Q_0}{Q_0} = \frac{\Delta t}{\tau} \tag{3.4}$$

The charge amount difference $\Delta Q(t) = Q(t) - Q_0$ is taken from Eq. (3.4), and the terms $Q_{spac}(t)$ and $Q_{leak}(t)$ are summarized on the right side as in the following Eq. (3.5). Further, from the Eqs. (3.4) and (3.5), the expression of Eq. (3.6) when $\Delta t = t_m$ at the measurement time is derived. Then Eq. (3.3) in Sect. 3.1 is gotten.

$$\frac{Q(t) - Q_0}{Q_0} = \frac{\Delta Q(t)}{Q_0} = \frac{Q_{spac}(t)}{Q_0} + \frac{Q_{leak}(t)}{Q_0} \tag{3.5}$$

$$\frac{Q(t_m)}{Q_0} = 1 + \frac{t_m}{\tau} \quad where \quad \tau = \frac{\varepsilon}{\kappa} \tag{3.6}$$

Here are some important points. Equation (3.6) is derived by the first approximation model in which the conductivity κ is constant. Actually, the trap depth is energetically and spatially distributed, then the conductivity does not have a constant value. These are the remaining research topics in future.

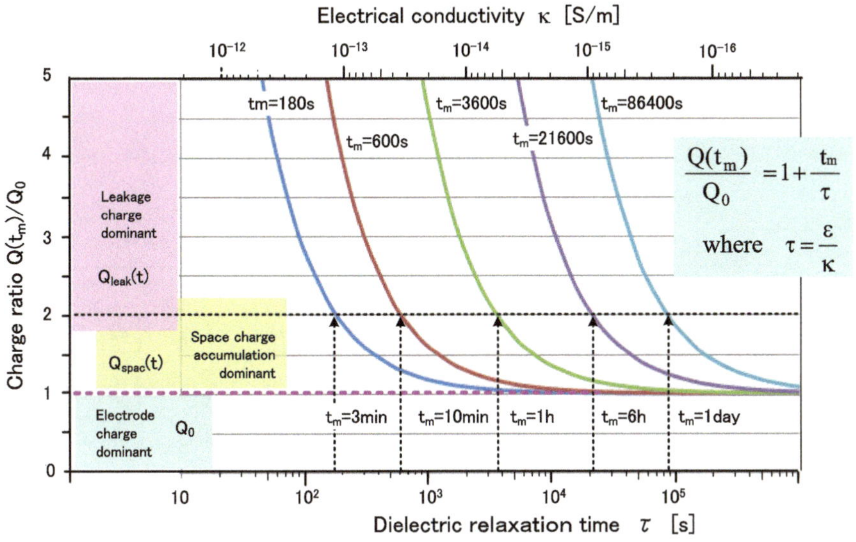

Fig. 3.4 The relationship between the charge ratio, the dielectric relaxation time and the measurement time

(2) **Relation of charge ratio $Q(t_m)/Q_0$, dielectric relaxation time τ and measurement time t_m**

Figure 3.4 shows a relationship between the charge ratio $Q(t_m)/Q_0$, the dielectric relaxation time τ and the measurement time t_m. It is the same as Fig. 3.2b in Sect. 3.1. However, the calculation results are obtained over a wide range of measurement times: $t_m = 180$ s (3 min), 600 s (10 min), 3600 s (1 h), 21,600 s (6 h), and 86,400 s (1 day). This measurement time t_m has the same scale as the relaxation time τ below the horizontal axis. Therefore, when $t_m = \tau$, the charge ratio in Eq. (3.6) is $Q(t_m)/Q_0 = 2$, and the characteristic curve is "2" on the vertical axis.

Here is an important published paper that measures leakage current over a long period of 12 h. This paper describes that DC cable requires a long-term measurement for charge accumulation and leakage current [1].

(3) Carrier movement within a time shorter than the relaxation time.

Figure 3.5c shows what happens to the equivalent electric circuit of dielectric relaxation in the time zone in which the voltage application time t_m is shorter than the relaxation time τ ($t_m < \tau$)

Immediately after voltage application (SW ON): The conductivity is defined as $\kappa = en\mu$, where en is the charge carrier density and μ is the mobility. At this stage, the charge carrier density is $en = 0$. There are no charge carriers yet. Therefore, since there is no electric resistance component R, only the capacitance component C is present. Therefore, in Fig. 3.5c (SW ON), the capacitance C is divided into

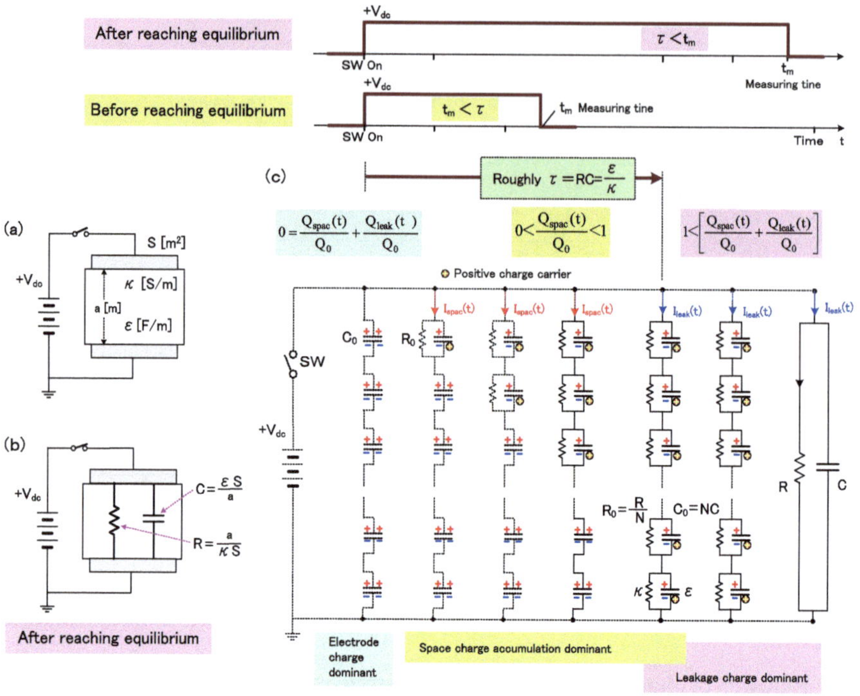

Fig. 3.5 The equivalent electric circuit of dielectric relaxation

N numbers, and a capacitor of $C_0 = NC$ is drawn in series. Furthermore, since the sample is polarized, a polarization (+ and -) is drawn.

Time period shorter than the relaxation time: The first step after voltage application is a state in which positive charge is injected from the anode. $N = 1$ step depicts the state in which positive charge moves first. The portion where the positive charge has moved has a resistance $R_0 = R/N$. The portion where the positive charge has moved can be represented by a parallel circuit of the resistor R_0 and the capacitance C_0. Then, a positive charge (red circle +) moving from the anode through the resistor R_0 is drawn. Thereafter, as in $N = 2, 3, ...$ steps, the positive charges (red circles +) sequentially charge the capacitance C_0.

Time zone longer than the relaxation time: At the Nth step, the positive charge (red circle +) reaches the counter electrode and an equilibrium state. Then, the electrical resistance R of the sample is given in the figure. At this stage, since the value of the electric resistance R is finally determined, the value of the conductivity κ can be calculated from $R = a/\kappa S$.

Summary: What has already been described in the "Summary" of Sect. 3.1 can be confirmed by the model in which charges move in the dielectric shown in Fig. 3.5c in this section. This is as follows in roughly.

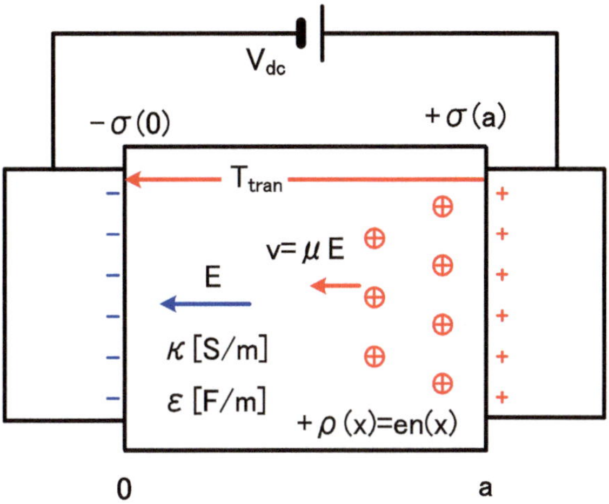

Fig. 3.6 The model for the charge migration

$t_m < \tau$ $(= \varepsilon/\kappa)$ is $1 < Q(t_m)/Q_0 < 2$ region: Space charge accumulation is dominant.

τ $(= \varepsilon/\kappa) < t_m$ is $2 < Q(t_m)/Q_0$ region: Leakage current charge is dominant.

The measurement points of the charge ratio $Q(t)/Q_0$ are entered in the graph of Fig. 3.4. Q_{spac}(t) and Q_{leak}(t) depending on whether the measurement points are in the $t_m < \tau$ region or $t_m > \tau$ region can be analysed.

(4) **Charge carrier transit time T_{tran} and dielectric relaxation time τ.**

Figure 3.6 presents a model in which carriers $\rho(x) = en(x)$ travelling through a sample (thickness a). From the transit time T_{tran}, Eq. (3.7) can be derived as follows. The meaning of Eq. (3.7) indicates that $Q_{spac}(t)$ is dominant until the measurement time t_m becomes equal to the dielectric relaxation time τ.

Equation (3.8) shows a definition of electric conductivity κ and electrode charge density σ. Equation (3.9) is a definition of transit time T_{tran} of the charge carrier, and Eq. (3.10) is the relational expression between the space charge ratio $Q_{spac}(t)/Q_0$ and $T_{tran} = t_m$.

$$\frac{Q_{spac}(t)}{Q_0} = 1 + \frac{t_m}{\tau} \tag{3.7}$$

$$\kappa = en\,\mu$$
$$\sigma = \varepsilon E \tag{3.8}$$

$$\begin{aligned} T_{tran} &= \frac{a}{v} = \frac{a}{\mu E} \Rightarrow T_{tran} = \frac{a}{\mu E} = \frac{\varepsilon}{\kappa} \times \frac{en \times aS}{\sigma S} = \frac{a}{\mu E} \\ &= \frac{\varepsilon}{\kappa} \times \frac{Q_{spac}}{Q_0} = \tau \times \frac{Q_{spac}}{Q_0} \end{aligned} \tag{3.9}$$

$$\frac{Q_{spac}(t)}{Q_0} = \frac{T_{tran}}{\tau} \Rightarrow \frac{Q_{spac}(t)}{Q_0} = \frac{T_{tran} = t_m}{\tau} \Rightarrow \frac{Q_{spac}(t)}{Q_0} = \frac{t_m}{\tau} \tag{3.10}$$

This Eq. (3.7) is applied to Fig. 3.4 in Sect. 3.2. When the measurement time t_m is equal to the dielectric relaxation time τ ($= t_m$), the charge ratio becomes $Q(t_m)/Q_0 = 2$. Judging this from Eq. (3.7), it is when the transit time of the charge carriers is $T_{tran} = t_m = \tau$, the charge carriers are also distributed throughout the sample. When the charge ratio $Q(t_m)/Q_0 = 2$, $Q_{spac}(t)/Q_0 = 1$ is obtained, and it can be determined that the space charge accumulation is dominant. Therefore, it can be determined that the leakage current is dominant in the region of $Q(t_m)/Q_0 > 2$.

(5) **Can you trust the value of conductivity $\kappa = 10^{-14}$ to 10^{-16} S/m ?**

This notation is very important. Since the relative permittivity of many polymer dielectrics is about $\varepsilon_r = 2$ to 4, $\varepsilon_r = 2.3$ is used to examine the notation. Meanwhile, for conductivity, the values of $\kappa = 10^{-13}$ to 10^{-16} S/m have been reported generally. Can we trust this value?

Conductivity, dielectric relaxation time, measurement time:
$\kappa = 10^{-16}$ S/m $\tau = \varepsilon/\kappa = 2 \times 10^5$ s = 55 h > > t_m = 10 mins = 600 s.
$\kappa = 10^{-15}$ S/m $\tau = \varepsilon/\kappa = 2 \times 10^4$ s = 5.5 h > > t_m = 10 mins = 600 s.
$\kappa = 10^{-14}$ S/m $\tau = \varepsilon/\kappa = 2 \times 10^3$ s = 34 mins > t_m = 10 mins = 600 s.
$\kappa = 10^{-13}$ S/m $\tau = \varepsilon/\kappa = 2 \times 10^2$ s = 200 s < t_m = 10 mins = 600 s.

Generally, the measurement time t_m of the V-I characteristic for obtaining the conductivity is about 10 min. To reach an equilibrium state, measurement time t_m > dielectric relaxation time τ is a necessary requirement. According to this condition, the value of the conductivity $\kappa = 10^{-14}$ to 10^{-16} S/m is at a stage where the carrier is still in the middle of the sample before reaching the equilibrium state. The only proper value is about $\kappa = 10^{-13}$ S/m. Therefore, the value of the conductivity $\kappa = 10^{-14}$ to 10^{-16} S/m cannot be trusted.

(6) **Various relaxation times**

Figure 3.7 graphically shows the potential decay of parallel circuit of capacitance (C) and resistance (R) after the voltage off. Here, a physical model of the dielectric relaxation time τ is considered. There are the following various relaxation phenomena and the relaxation time τ is defined for each of them.

Viscoelastic relaxation time: $\tau = \eta/G$ [s] where η [Ns/m^2] is viscosity, G [N/m^2] is elasticity.

Dielectric relaxation time: $\tau = \varepsilon/\kappa$ [s] where ε [F/m] is permittivity, κ [S/m] is conductivity.

And where [F] = [C/V], [S; Siemens] = [C/sV].

Potential decay of CR circuit

Relaxation time of CR electric circuit becomes $\tau = RC = \varepsilon/\kappa$ [s]. The parallel electric circuit of a capacitor C and a resistor R will be described. The voltage V_0 is charged, and the electrode on the high voltage side is separated from the power

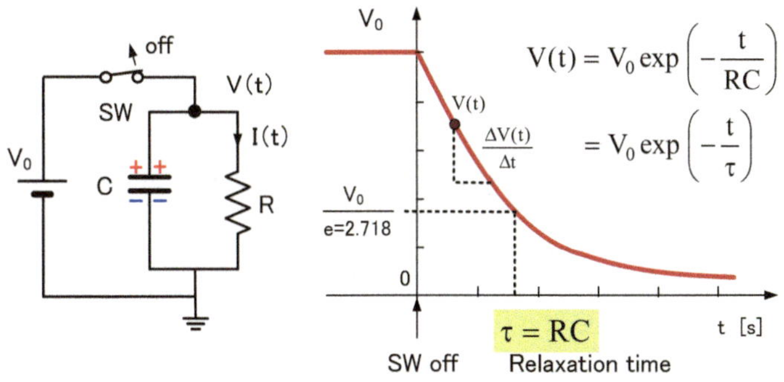

Fig. 3.7 The model of the CR circuit

supply (SW off). The potential decay $V(t)$ on the electrode is derived as in Eq. (3.11). $V(t)$ decays exponentially, and its time constant is $\tau = CR$. Here, C is the capacitance of Eq. (3.12) and R is the resistance of Eq. (3.13).

$$V(t) = V_0 \exp\left(-\frac{t}{RC}\right) = V_0 \exp\left(-\frac{t}{\tau}\right) \tag{3.11}$$

$$C = \frac{\varepsilon S}{a} \tag{3.12}$$

$$R = \frac{a}{\kappa S} \tag{3.13}$$

After the SW is turned off, the current flowing through the resistor R due to the charge Q of the capacitance C is given by the following Eq. (3.14). When the two expressions of the Eq. (3.14) are combined, the linear differential equation of the first term of the Eq. (3.15) is obtained.

$$I(t) = \frac{\Delta Q(t)}{\Delta t} = -C\frac{\Delta V(t)}{\Delta t} \text{ and } I(t) = \frac{V(t)}{R} \tag{3.14}$$

$$\frac{\Delta V(t)}{\Delta t} = -\frac{1}{RC}V(t) \Rightarrow \frac{\Delta V(t)}{V(t)} = -\frac{1}{RC}\Delta t \Rightarrow \ln\left[\frac{\Delta V(t)}{V(t)}\right] = -\frac{t}{RC} \tag{3.15}$$

Also in this case, the voltage change amount ΔV(t)/Δt is proportional to the voltage V(t). The proportional coefficient is the reciprocal of the time constant (relaxation time) of (-1/RC). Furthermore, when the variables are separated and integrated, a solution of the logarithmic expression of the third term of the Eq. (3.15) is obtained. Furthermore, the expression in the exponential function is based on Eq. (3.11).

(7) Relaxation time of dielectric samples

Here, a model of the dielectric relaxation time τ will be specifically considered. Figure 3.7 shows an electric circuit in which a DC voltage V_{dc} is applied to a dielectric sample (dielectric constant ε, conductivity κ, thickness a, electrode area S). When the voltage application time is sufficiently long and the leakage current reaches an equilibrium state, this dielectric sample can be represented with a parallel equivalent circuit of a capacitance C and a resistance R as shown in Fig. 3.7. The capacitance C and the resistance R are given by Eqs. (3.16) and (3.17). Here, the time constant is $\tau = CR = \varepsilon/\kappa$ in Eq. (3.18), which is the dielectric relaxation time. The relaxation time $\tau = \varepsilon/\kappa$ does not depend on the shape of the dielectric sample, but is determined by the values of ε and κ, which are the eigenvalues of the dielectric material. Therefore, $\tau = \varepsilon/\kappa$ is referred to dielectric relaxation time.

$$C = \frac{\varepsilon\ S}{a} = \frac{\varepsilon\ S}{n\ \Delta x} = \frac{C_o}{n} \tag{3.16}$$

$$R = \frac{a}{\kappa\ S} = \frac{n\ \Delta x}{\kappa\ S} = n\ R_0 \tag{3.17}$$

$$\tau = RC = R_0 C_0 = \frac{\varepsilon}{\kappa} \tag{3.18}$$

3.3 Space Charge Formation and Q(t) Simulation

The Q(t) characteristics (Fig. 3.1b and c) in Sect. 3.1 rise up to the initial value Q_0 after the application of a DC voltage, and then increase from the space charge accumulation $Q_{\text{spac}}(t)$ to the leakage current charge $Q_{\text{leak}}(t)$. In this section, we perform a graphical simulation of this charge transfer and observe the transition of space charge from $Q_{\text{spac}}(t)$ to $Q_{\text{leak}}(t)$.

(1) **Q(t) simulation for charge transfer.**

Measurement circuit (Fig. 3.8a): ① DC voltage $V_{\text{dc}} = 20$ kV is applied to a dielectric sample (thickness $a = 200$ μm, electrode area $S = 20$ cm^2, relative permittivity $\varepsilon_r = 2.3$). At this time, the relationship between the integrated charges $Q(t)$ supplied to the dielectric sample and the charge transfer is simulated. Immediately after the voltage application, electrode charges ($\sigma(0)$, $\sigma(a)$) are induced.

At this time, there is a relationship of $Q_0 = -\sigma(0)S = +\sigma(a)S = C_s V_{\text{dc}}$. Thereafter, the space charge $\rho(z, t)$ moves from the anode to the cathode as time passes, and finally reaches the cathode. Then, the leakage current $I_{\text{leak}}(t)$ starts to flow. The space charge $\rho(z)$ and the electric field distribution $E(x)$ i n his process can be obtained by solving Poisson's equation. The following Q(t) simulation is performed using $E(x)$ and $\rho(z)$ [2].

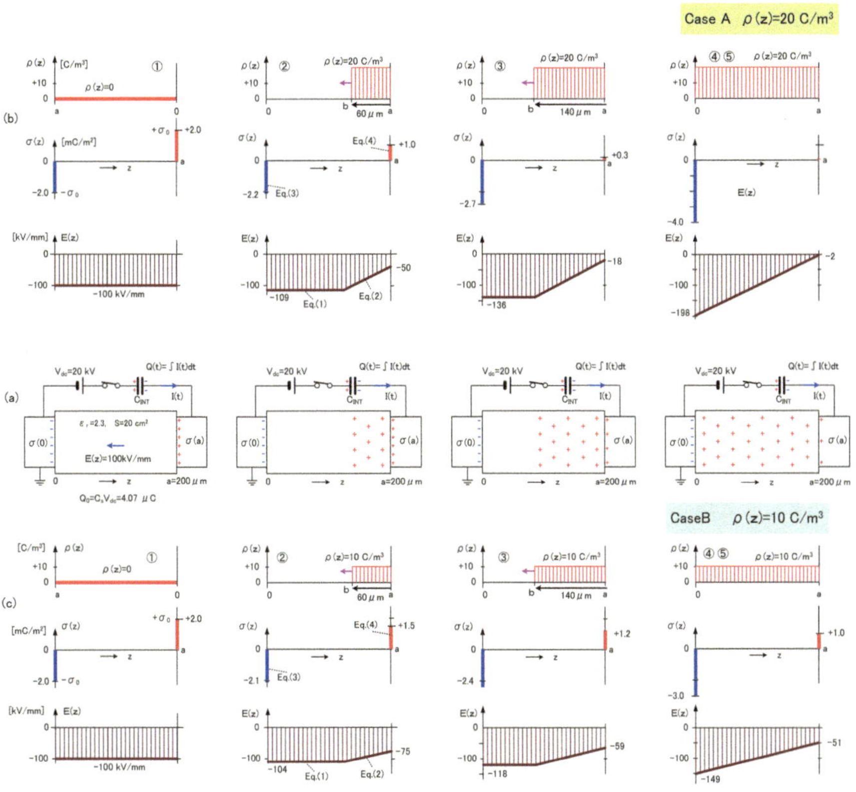

Fig. 3.8 The model of the assumed two cases

Case A $\rho(z) = 20$ **C/m³** (Fig. 3.8b).

① Immediately after the voltage application, electrode charge $\sigma_0 = 2.0$ mC/m^2 is induced, and electric field distribution is uniform at $E = 100$ kV/mm. No space charge has yet been formed, $\rho(z) = 0$.

② $\rho(z)$ is injected from the anode. After a certain period of time, it moves to $(a\text{-}b) =$ 60 μm in the sample.

This b is the position of the front of the $\rho(z)$ movement. At that time, the positive electrode charge decreases to $\sigma(a) = +1.0$ mC/m^2, and the negative electrode charge increases to $\sigma(0) = -2.2$ mC/m^2. The electric field in the $\rho(z) = 0$ region is constant at $E = -109$ kV/mm, and E in the $\rho(z) = 20$ C/m^3 region is a linear distribution.

③ When + $\rho(z)$ moves to 140 μm, the electrode charge and electric field distribution described in ③ change and advance.

④ When + $\rho(z)$ reaches the cathode, the negative electrode charge increases to $\sigma(0)$ $= -4.0$ mC/m^2 and the positive electrode charge becomes $\sigma(a) = 0$. The electric field distribution becomes a linear distribution, and the surface electric field on the positive electrode becomes *E(a)*= 0.

In Case A, we selects $\rho(z) = 20$ C/m^3 so that *E(a)*= 0 just at the stage of ④. After $+\ \rho(z)$ reaches the cathode in the measurement time t_{m} at ④, an equilibrium leakage current flows.

Case B $\rho(z)$ = 10 C/m³ (Fig. 3.8c)

Space charge $\rho(z) = 10$ C/m^3 is a half of that in Case A, and the patterns ② and ③ of Case B are exactly the same as that of Case A. The difference is that when $+\ \rho(z)$ reaches the cathode at ④, the surface electric field on the positive electrode does not reach *E(a)*=0 and remains at *E(a)*= 51 kV/mm. Thereafter, a leakage current in an equilibrium state flows to ⑤.

(2) **Poisson's equation**

Poisson's equation of Eq. (3.19) showing the relationship between charge density $\rho(x)$, electric field distribution $E(x)$, and $V(z)$ (potential distribution) is described. In Fig. 3.8, a simulation of charge transfer is performed by drawing the electrode surface charges $\sigma(0)$, $\sigma(a)$ and the electric field distribution $E(x)$ based on the value of the charge density $\rho(x)$. The relationship between $\rho(z)$ (charge distribution) and $E(z)$ (electric field distribution) can be obtained by solving Poisson's equation of Eq. (3.19). This relationship states that the second-order spatial derivative of $V(z)$ (potential distribution) and the first-order spatial derivative of $E(x)$ are equal to $\rho(z)/\varepsilon$. The first-order spatial derivative of $V(z)$ is -$E(x)$.

$$
\begin{aligned}
-\frac{\partial^2 V(z)}{\partial z^2} &= \frac{\partial E(z)}{\partial z} = \frac{\rho(z)}{\varepsilon} \\
-\frac{\partial V(z)}{\partial z} &= E(z)
\end{aligned}
\tag{3.19}
$$

When this Poisson's equation is solved based on the boundary conditions (V_{dc}, a, b), the electric field distributions $E_1(z)$ and $E_2(z)$ of the Eqs. (3.20) and (3.21) are derived. Further, the electrode charge densities $\sigma(0) = \varepsilon E(0)$ and $\sigma(a) = \varepsilon E(a)$ of the Eqs. (3.22) and (3.23) are derived.

For more guidance on this, see Takada Text Vol. II, Part 4, Chap. 2, Sect. 2.6.

$$
E_1(z) = -\frac{V_{dc}}{a} - \frac{\rho}{\varepsilon}\frac{(a-b)^2}{2a}
\tag{3.20}
$$

$$
E_2(z) = -\frac{V_{dc}}{a} - \frac{\rho}{\varepsilon}\frac{(a-b)^2}{2a} + \frac{\rho}{\varepsilon}(z-b)
\tag{3.21}
$$

$$
\sigma(0) = \varepsilon E_1(0) = \varepsilon\left[\frac{V_{dc}}{a} - \frac{\rho}{\varepsilon}\frac{(a-b)^2}{2a}\right]
\tag{3.22}
$$

$$\sigma(0) = \varepsilon E_2(a) = \varepsilon\left[-\frac{V_{dc}}{a} + \frac{\rho}{\varepsilon}\frac{(a+b)(a-b)}{2a}\right] \quad (3.23)$$

Figure 3.8 shows the results of $E_1(z)$ and $E_2(z)$ with $\sigma(0)$ and $\sigma(a)$ drawn by substituting the numerical conditions into Eqs. (3.20) to (3.23).

(3) **Explanation of Q(t) simulation** (Fig. 3.9).

Using the simulation results of Fig. 3.8, the simulation of the Q(t) characteristic of Eq. (3.24) is shown in Fig. 3.9.

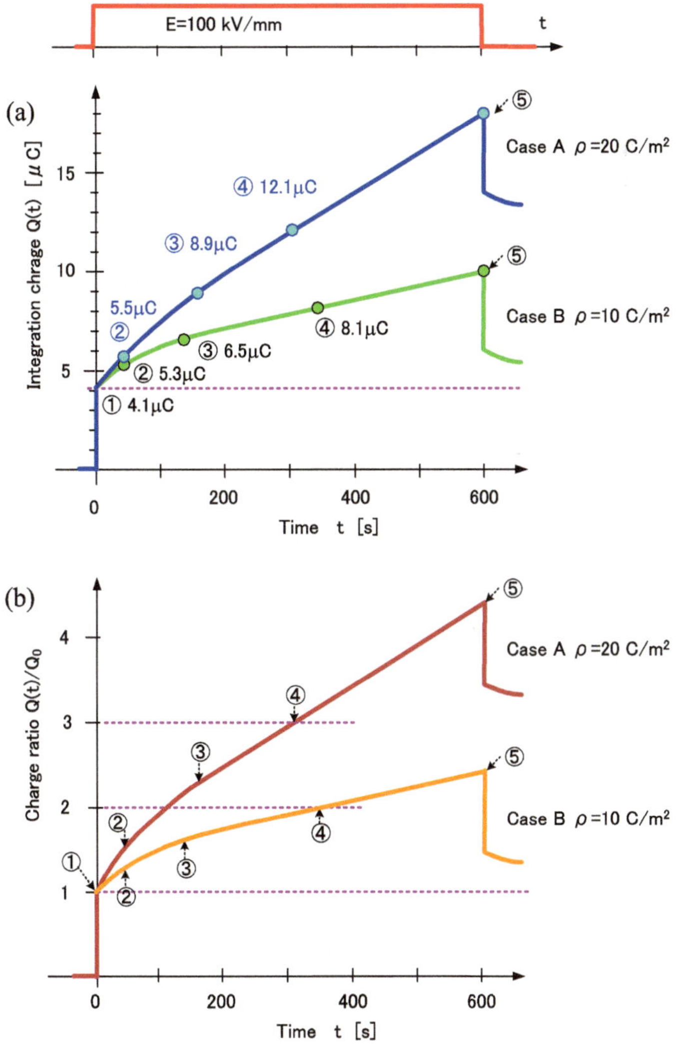

Fig. 3.9 The Q(t) results for the two cases

$$Q(t) = Q_0 + \int_0^t I_{spac}(t)\,dt + \int_0^t I_{leak}(t)\,dt = Q_0 + \quad Q_{spac}(t) \quad + \quad Q_{leak}(t) \tag{3.24}$$

Case A $\rho(z) = 20$ C/m³

① Immediately after the voltage is applied,
the electrode charge is $Q_0 = C_s V_{dc} = 4.1$ μC.

⑤ Determine the value that $Q(t = 600$ s) is 4.4 times Q_0. This 4.4 times is selected so that the current is flowing even after $\rho(z)$ fills the sample, and the charge amount becomes $Q(t = 600$ s). Then, from $Q_0 = C_s V_{dc} = 4.1$ μC in ①, $Q(t = 600$ s) = 18.0 μC in ⑤ is determined. Draw a $Q(t)$ curve (blue) between ① and ⑤.

② $Q(t) = + \rho(z)\,(a\text{-}b) = 5.5$ μC.

③ Similarly, $Q(t) = + \rho(z)\,(a\text{-}b) = 8.9$ μC.

④ $Q(t) = + \rho(z)\,a = 12.1$ μC.

Draw these values on a $Q(t)$ curve. At this stage, the front of the space charge reaches the cathode, and $\rho(z) = 20$ C/m^3 is uniformly distributed in the sample. Thereafter, in the section ④ → ⑤, a leakage current flows with uniform space charge $\rho(z)$.

Case B $\rho(z) = 10$ C/m³

Draw the $Q(t)$ curve (light green) with the same procedure as in Case A. In the section ④ → ⑤, $\rho(z) = 10$ C/m^3 is uniformly distributed in the sample, and leakage current flows.

(4) **Simulation of charge ratio $Q(t)/Q_0$** (Fig. 3.9b).

The problem in Sect. 3.1 is that it is difficult to determine where the boundary between Q_{spac}(t) and Q_{leak}(t) is in the area of $Q(t)/Q_0 > 1$. In order to specifically examine this, the Q(t) simulation is redrawn to the charge ratio $Q(t)/Q_0$ in Fig. 3.9b.

Case A of $\rho(z) = 20$ C/m³. At this time, the ratio becomes $Q(t)/Q_0 = 3$. This is the case where the space charge is accumulated in the entire sample under the application of an electric field and the accumulated charge amount is the maximum. Therefore, the electric field on the anode surface is $E = 0$, and if space charge of $\rho(z) = 20$ C/m^3 or more is injected, the direction of the electric field is reversed and the charge injection does not appear. It indicates that this is the limit of charge injection. See ④ in Fig. 3.9b.

Case B of $\rho(z) = 10$ C/m³. The space charge is uniform in the sample, but the electric field on the electrode surface is still finite ($E \neq 0$). Even in this state, the amount of leakage charge from the cathode and the amount from the anode are balanced. Now the ratio becomes $Q(t)/Q_0 = 2$. This is half of the maximum accumulated charge.

Table 3.1 Relationship with space charge density $\rho(z, t)$ in ① → ④ → ⑤ process

	Case A $\rho(z) = 20$ C/m^3	Case B $\rho(z) = 10$ C/m^3
① Only electrode charge and no charge accumulation	$Q(t)/Q_0 = 1$	$Q(t)/Q_0 = 1$
① → ④ Space charge formation process	$1 < Q(t)/Q_0 < 3$	$1 < Q(t)/Q_0 < 2$
④ → ⑤ Leakage current charge	$3 < Q(t)/Q_0$	$2 < Q(t)/Q_0$

(5) **Overall consideration.**

In Fig. 3.9b, ① → ④ indicates a space charge formation process, and ④ → ⑤ indicates a leakage current process. Table 3.1 summarizes the formation of the space charge density $\rho(z, t)$ in Case A and Case B by dividing them into ① → ⑤ processes.

It can be seen that the boundary between both processes depends on the space charge density $\rho(z, t)$. The result of Fig. 3.9 is a case where the space charge density distribution $\rho(z, t)$ in the sample is calculated from the beginning as known. On the other hand, $\rho(z, t)$ is unknown from the Q(t) measurement results in Fig. 3.1b and c of Sect. 3.1.

The problem that the boundary between both processes depends on the space charge density $\rho(z, t)$ is examined from the PEA method and the Q(t) method in Sect. 3.4. We are trying to separate $Q_{spac}(t)$ and $Q_{leak}(t)$.

3.4 Combination of PEA Measurement and Q(t) Measurement

In Sect. 1.3, it is stated that PEA measurement does not measure all the trapped space charge because it is difficult to measure shallow trapped charges. This shallow trapped charge governs the leakage current. In this section, we separate trapped space charge $Q_{spac}(t)$ and leakage charge $Q_{leak}(t)$ from the combined use of PEA data and Q(t) data, and consider the charge accumulation process and the transition process to leakage current [3].

(1) **Time characteristics of PEA data and Q(t) data**

Figure 3.10 shows (a) PEA data and (b) Q(t) data of XLPE (120 μm) measured under the same conditions (60 kV/mm). The relationship between (a) PEA and (b) Q(t) data is observed based on Eq. (3.25).

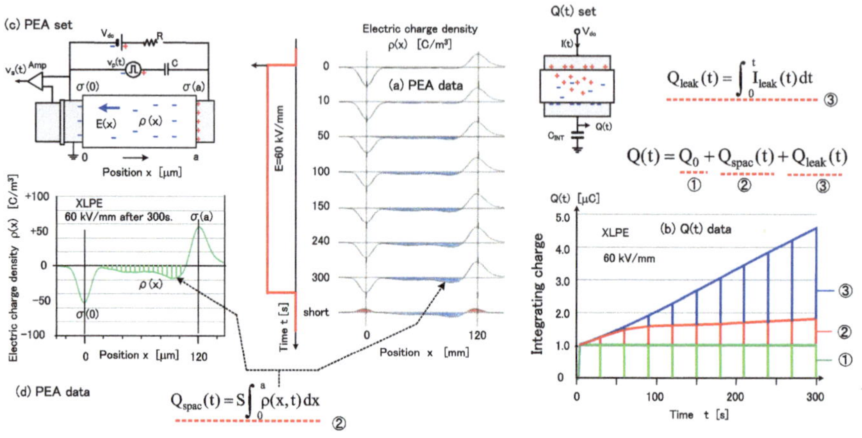

Fig. 3.10 Joint application of the two methods

$$Q(t) = Q_0 + \int_0^t I_{spac}(t)\,dt + \int_0^t I_{leak}(t)\,dt = Q_0 + Q_{spac}(t) + Q_{leak}(t)$$

$$\text{where}\quad Q_0 = C_s V_{dc} \quad Q_{spac}(t) = S\int_0^a \rho(x,t)\,dx = \int_0^t I_{spac}(t)\,dt$$

$$Q_{leak}(t) = \int_0^t I_{leak}(t)\,dt \tag{3.25}$$

PEA data: First, look at the PEA data in Fig. 3.10a. When a DC electric field of $E = 60$ kV/mm is applied to the XLPE sample, it is observed that the space charge density $\rho(x, t)$ is accumulated (blue) in the sample over time. The data is observed by the PEA equipment shown in Fig. 3.10c.

Q(t) data: Focus on the Q(t) data in Fig. 3.10b. When a DC electric field of $E = 60$ kV/mm is applied to the XLPE sample, the initial charge $Q_0 = C_s V_{dc}$ of the first term of Eq. (3.25) is observed (green). Thereafter, the sum of integrating charge ($Q_{spac}(t)$ + $Q_{leak}(t)$) is observed, where $Q_{spac}(t)$ is the second term in Eq. (3.25) obtained by integrating the space charge forming current $I_{spac}(t)$, and $Q_{leak}(t)$ is the third term in Eq. (3.25) obtained by integrating the leakage current $I_{leak}(t)$. The Q(t) data increases with time, and this Q(t) result is expressed by Eq. (3.25).

Q_{spac}(t) data: The space charge accumulation amount $Q_{spac}(t)$ in the second term of Eq. (3.25) is obtained by integrating $\rho(x, t)$ in the thickness direction of the sample (Fig. 3.10d). Figure 3.10d shows $\rho(x, t)$ at $t_m = 300$ s, and ② $Q_{spac}(t)$ is obtained by

$\rho(x, t)$ at every time ($0 \rightarrow 300$ s). Then, ② $Q_{spac}(t)$ (red) is drawn above ① $\Delta Q_0 = C_s V_{dc}$ (green) data in Fig. 3.10b.

Q_{leak}(t) data: The leakage charge amount $Q_{leak}(t)$ can be easily obtained from $Q_{leak}(t) = Q(t)$ - Q_0 - $Q_{spac}(t)$ in Eq. (3.25). ③ $Q_{leak}(t)$ (blue) is shown in Fig. 3.10b.

As described above, the Q(t) data can be decomposed into three components, Q_0, $Q_{spac}(t)$, and $Q_{leak}(t)$.

The following describes how the Q(t) data in Fig. 3.10b is separated into the three components ①, ② and ③. Immediately after the voltage is applied, $Q_0 = C_s V_{dc}$ is induced on the electrode. Subsequently, the charge injected from the electrode forms a space charge $\rho(x, t)$ [C/m^3] while hopping and moving within the sample, and then the amount $Q_{spac}(t)$ reaches an equilibrium state. When the front of the space charge reaches the counter electrode, the leakage charge amount $Q_{leak}(t)$ of ③ appears. After reaching the equilibrium state, $\Delta Q_{leak}(t)/\Delta t$ reaches a constant value, and $Q_{leak}(t)$ increases linearly.

(2) **Electric field characteristics of PEA data and Q(t) data**

The PEA data (Fig. 3.11; color display) and Q(t) data are shown in comparison when electric fields of 10, 20, 30, 40, 60, and 80 kV/mm are applied to the XLPE sample without cross-linking residual byproducts. The separation method of ② space charge formation charge $Q_{spac}(t)$ and ③ leakage charge $Q_{leak}(t)$ of these data is the same as the one described in the previous section.

Low electric field application of 10 kV/mm: Since Q(t) data is $Q(t) = Q_0$, $Q_{spac}(t) + Q_{leak}(t) = 0$, and it is determined that there is no charge accumulation and leakage current. Similarly, in the PEA data, since the charge distribution in the sample is green ($\rho(x, t) = 0$), it is determined that there is no charge accumulation. Both data shows the same results.

Medium electric field application (30 kV/mm) and high electric field application (60 kV/mm): Since the charge distribution in the sample shown by PEA data is green, it is judged that there is no charge accumulation at first glance ($\rho(x, t) = 0$). On the other hand, since the Q(t) data is $Q(t) > Q_0$, it is determined that there is charge accumulation. Different evaluations are made between the two data. This is because the maximum color scale of the charge density is set to 100 C/m^3. If the measured charge density is only 20 C/m^3, charge accumulation cannot be confirmed. The reason for this result is to trust too much on the PEA data, and it is extremely important to observe the PEA data in comparison with the Q(t) data.

(3) **Separation of $Q_{spac}(t)$ and $Q_{leak}(t)$.**

This separation is easily obtained from Eq. (3.25). Attention is paid to the Q(t) data under the electric field application of 30 kV/mm and 60 kV/mm in the figure. Q(t) data is separated into ① Q_0 (green), ② $Q_{spac}(t)$ (red), and ③ $Q_{leak}(t)$ (blue).

① Q_0 (green) is the electrode induced charge, which is constant.

② $Q_{spac}(t)$ (red) is the accumulated charge, which increases in the first 60 s but then saturates. Finally it remains constant.

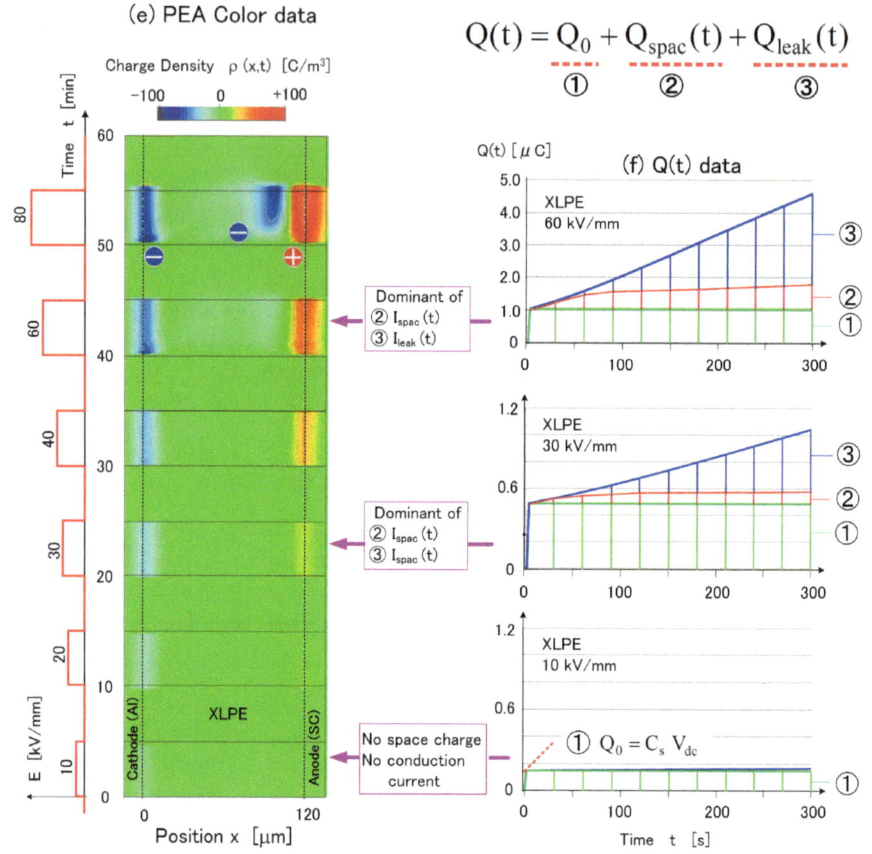

Fig. 3.11 Data analysis of the two methods

③ $Q_{\text{leak}}(t)$ (blue) is the amount of leakage charge, which appears when the tip of the moving space charge reaches the counter electrode.

Important Issues:

PEA measurement cannot evaluate the amount of leakage charge of ③ $Q_{\text{leak}}(t)$ (blue), but now it can be evaluated in combination with the Q(t) method. From the results of Fig. 3.11, if the space charge observed in the PEA measurement has a uniform distribution (green), it seems at first scent that no space charge and leakage current appears. In this case, the leakage current component $Q_{\text{leak}}(t)$ (blue) can be evaluated by using the PEA method and the Q(t) method together. There are two main characterizations of DC insulation materials.

- Evaluation of distortion of internal electric field $E(x, t)$ due to space charge accumulation $\rho(x, t)$. This is because the local electric field may exceed the designed field, leading to dielectric breakdown. It can be evaluated by PEA measurement.

- Leakage current flows and insulation is lost. The insulation can be evaluated by measuring Q(t) data with the characteristic of $Q(t) > > Q_0$.

For this reason, it is essential to use both PEA measurement and Q(t) measurement to evaluate the characteristics of DC insulating materials and evaluate both $E(x, t)$ distortion and leakage current.

References

1. H. Ghorbani, T. Christen, M. Carlen, E. Logakis, L. Herrmann, H. Hillborg, L. Petersson, J. Viertel, Long-term conductivity decrease of polyethylene and polypropylene insulation materials, *IEEE Transactions on Dielectrics and Electrical Insulation*, 24(3), 1485–1493 (2017).
2. H. Ren, T. Takada, Y. Tanaka, Q. Li, An evaluation method of observable charge trap depth for the PEA method and its complementarity with the Q(t) method, *Metrology and Measurement Systems*, 28(3), 565–580 (2021)
3. H. Ren, T. Takada, H. Uehara, S. Iwata, Q. Li, Research on charge accumulation characteristics by PEA method and Q(t) method, *IEEE Transactions on Instrumentation and Measurement*, 70, 6,004,209 (2021)

Chapter 4
Q(t) Data of Various Polymer Materials

4.1 Q(t) Characteristics of Various Polymers

(1) **Q(t) data**

The dependence of charge accumulation characteristics of various polymer materials on applied electric fields and temperatures is evaluated by Q(t) measurement. The results are introduced in the following Figs. 4.1, 4.2, 4.3, 4.4, 4.5, 4.6, and 4.7. The Q(t) characteristics can be classified into Group A, B and C as shown in Table 4.1. Figure 4.1 also shows the molecular formulas of these polymer insulating materials.

4.2 Classification of Charge Accumulation Characteristics

The Q(t) characteristics of various polymer insulating materials under different electric fields and temperatures are organized into Groups A, B, and C, as shown in Fig. 4.8.

Group A: The strong polar polymers PEN, PC, PET, PS belong here.

PTFE is not a polar molecule, but it is included here because it is difficult to accumulate the charges injected from electrodes.

Although DC-XLPE is not a polar polymer, it has polar properties due to the mixing effect of MgO nanoparticles with a high dielectric constant.

Group B: The weak polar polymers PMMA, PI, PPS belong here.

PP, AC-XLPE is not a polar polymer, but they belong here because of the overlapping effect of electron waves due to the interaction between molecules.

Group C: The non-polar molecules HDPE, LDPE belong here.

T. Takada et al., *Electric Charge Accumulation in Dielectrics: Measurement and Analysis*,
https://doi.org/10.1007/978-981-19-6156-4_4

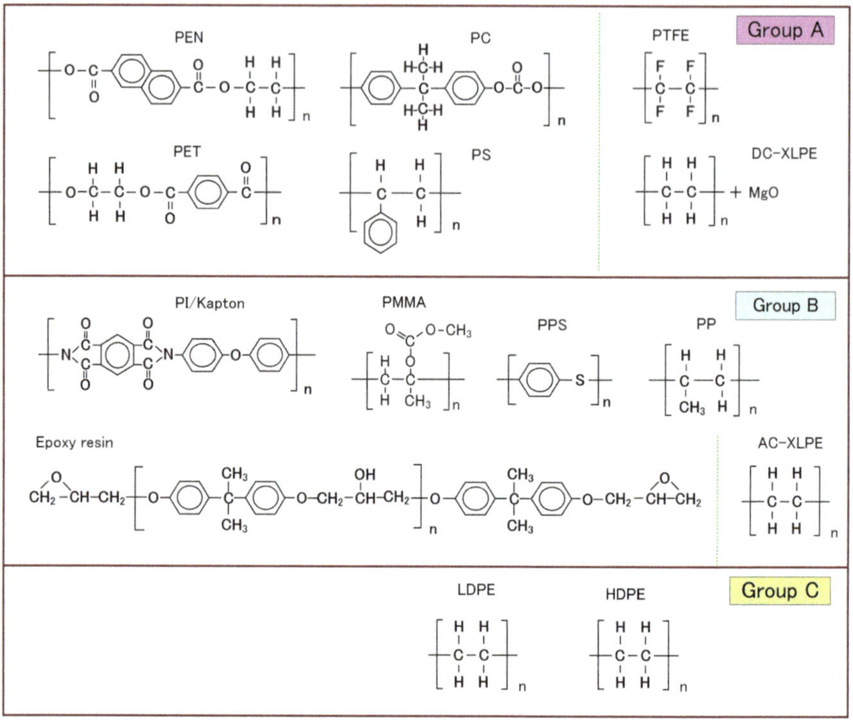

Fig. 4.1 The molecular formulas of the polymer insulating materials

4.3 Charge Accumulation Characteristics and Molecular Structure

The molecular structure that characterizes the charge accumulation properties is also shown in Fig. 4.8. An overview of the charge characteristics of different materials is described as follows.

(1) Group A: Strong polar molecules

No charge accumulates even at high electric field and high temperature.

PEN: Polyethylene naphthalate. A polar molecule with a naphthalene ester group in the main chain.
PET: Polyethylene terephthalate. A polar molecule with a benzene ring/ester group in the main chain.
PC: Polycarbonate. A polar molecule with a benzene ring/ester group in the main chain.
PS: Polystyrene. A polar molecule with a benzene ring in the side chain.
PTFE : Polytetrafluoro ethylene. The Fermi levels of the electrode and PTFE are close. Difficult to inject charge.

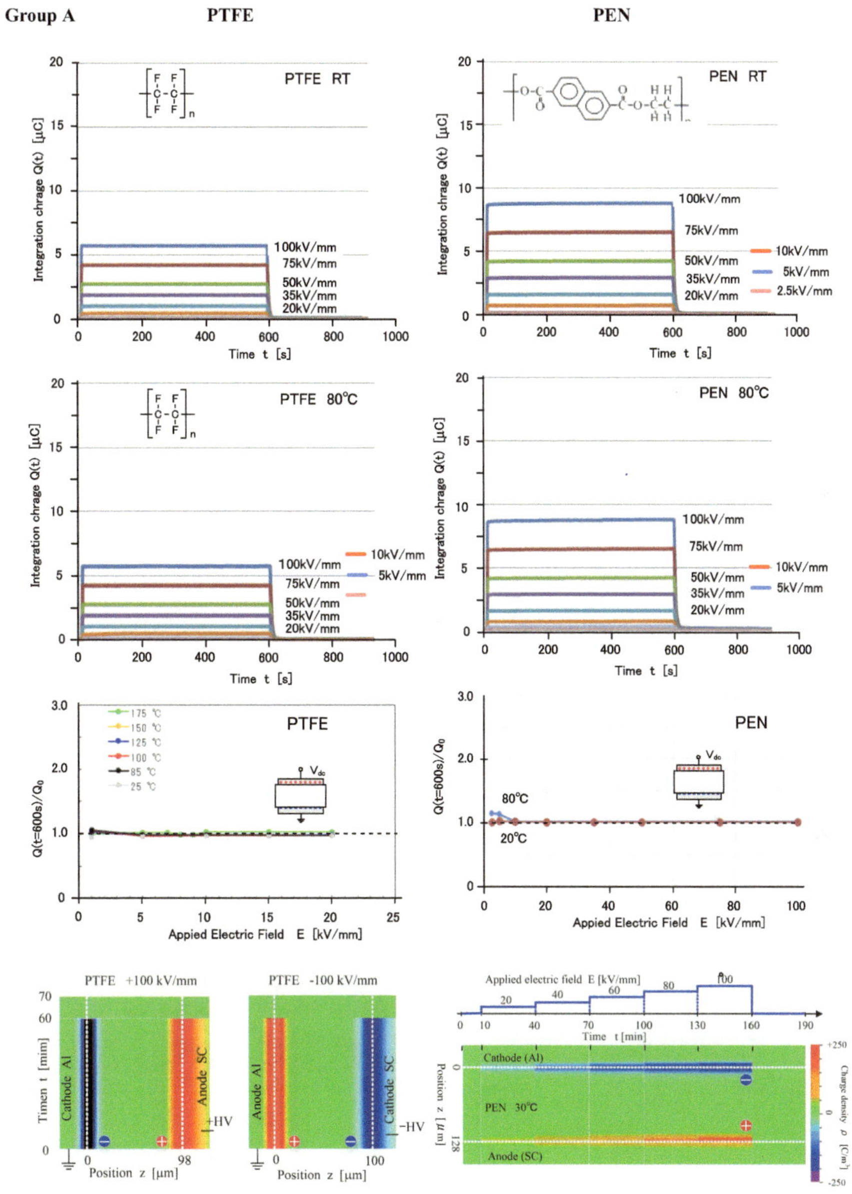

Fig. 4.2 The measured results of PTFE and PEN

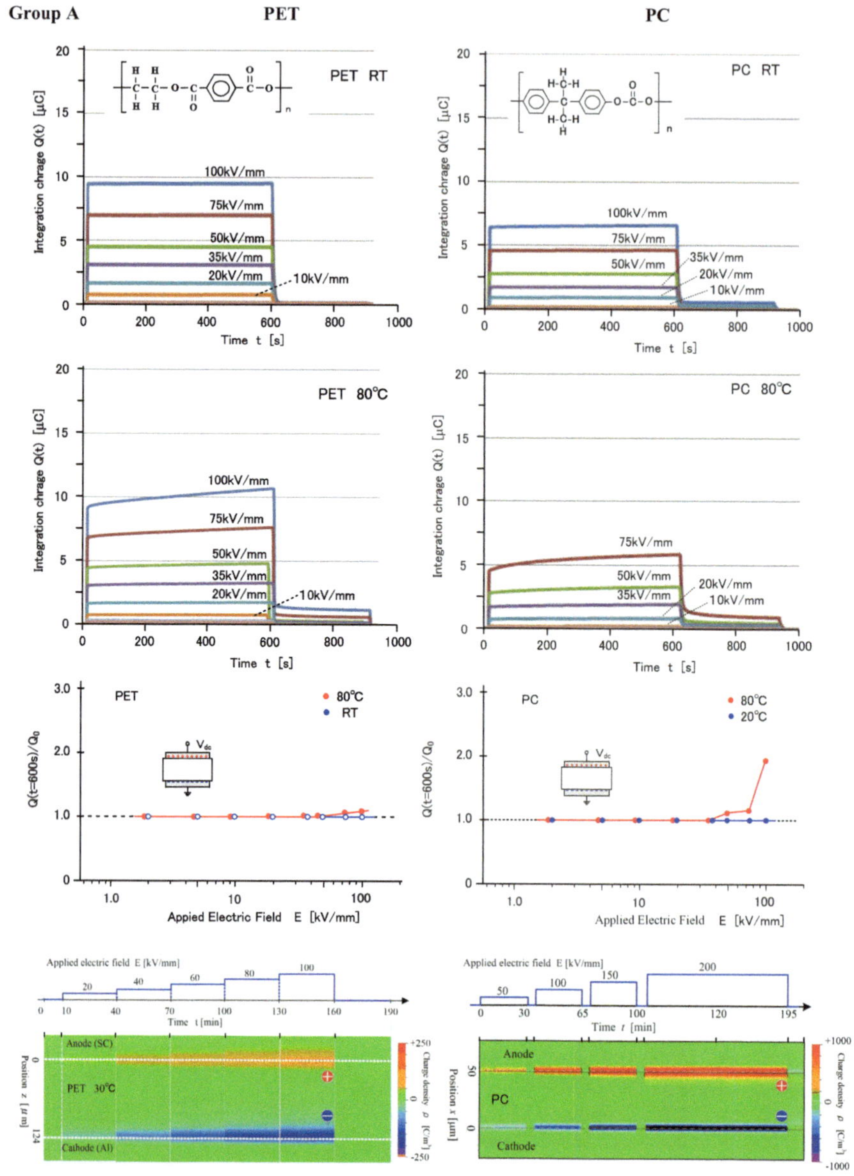

Fig. 4.3 The measured results of PET and PC

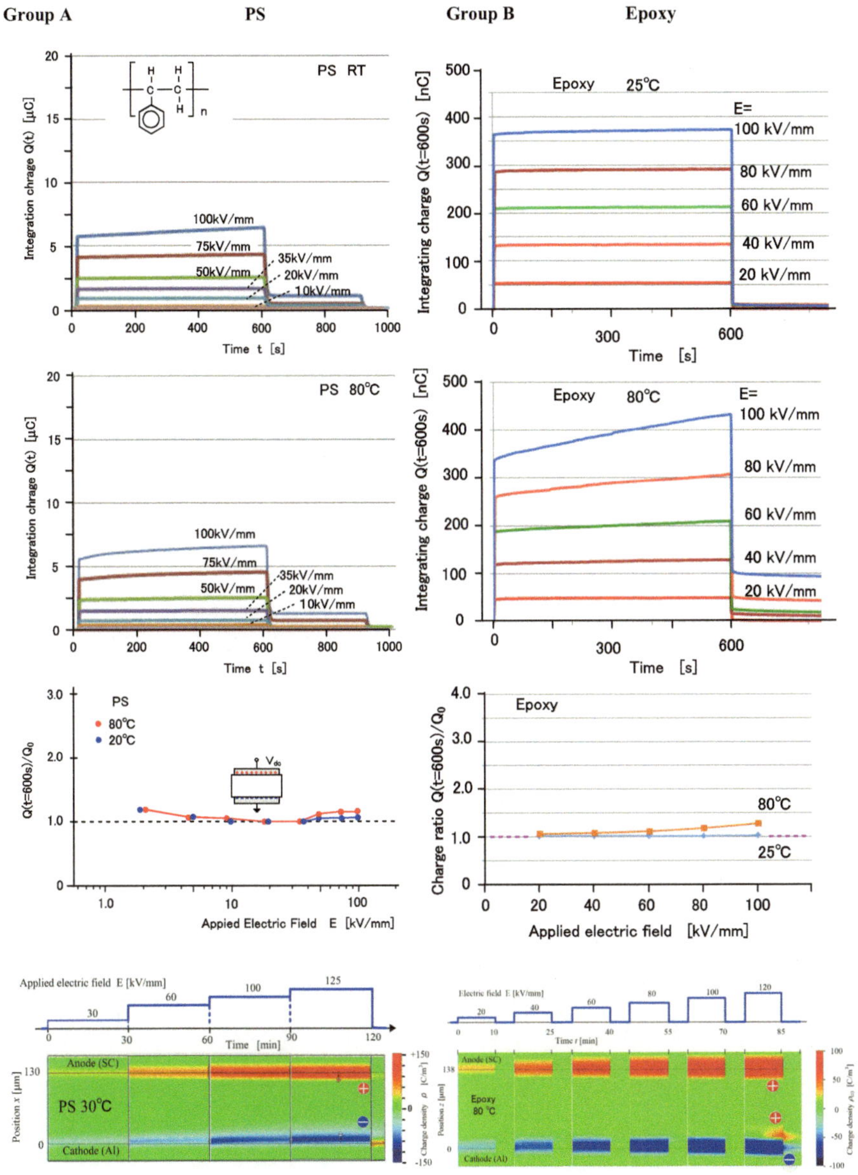

Fig. 4.4 The measured results of PS and epoxy

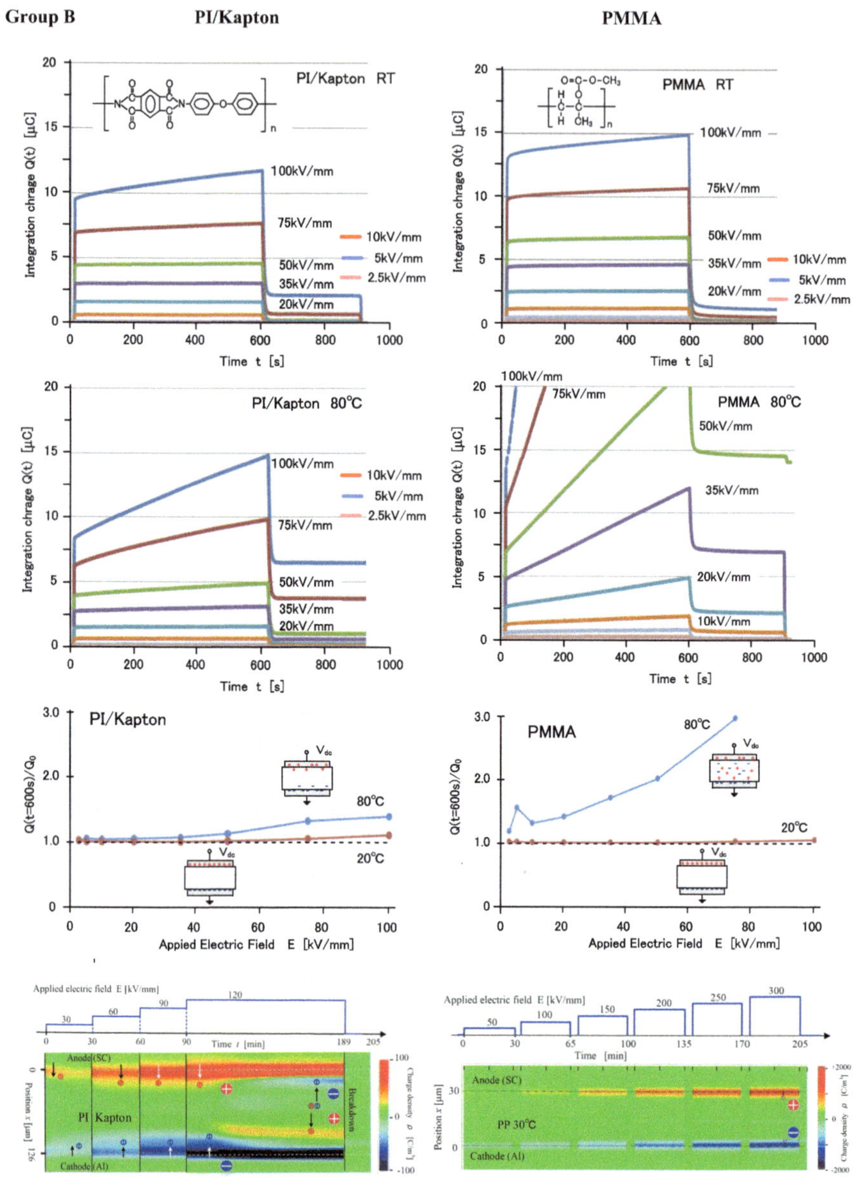

Fig. 4.5 The measured results of PI and PMMA

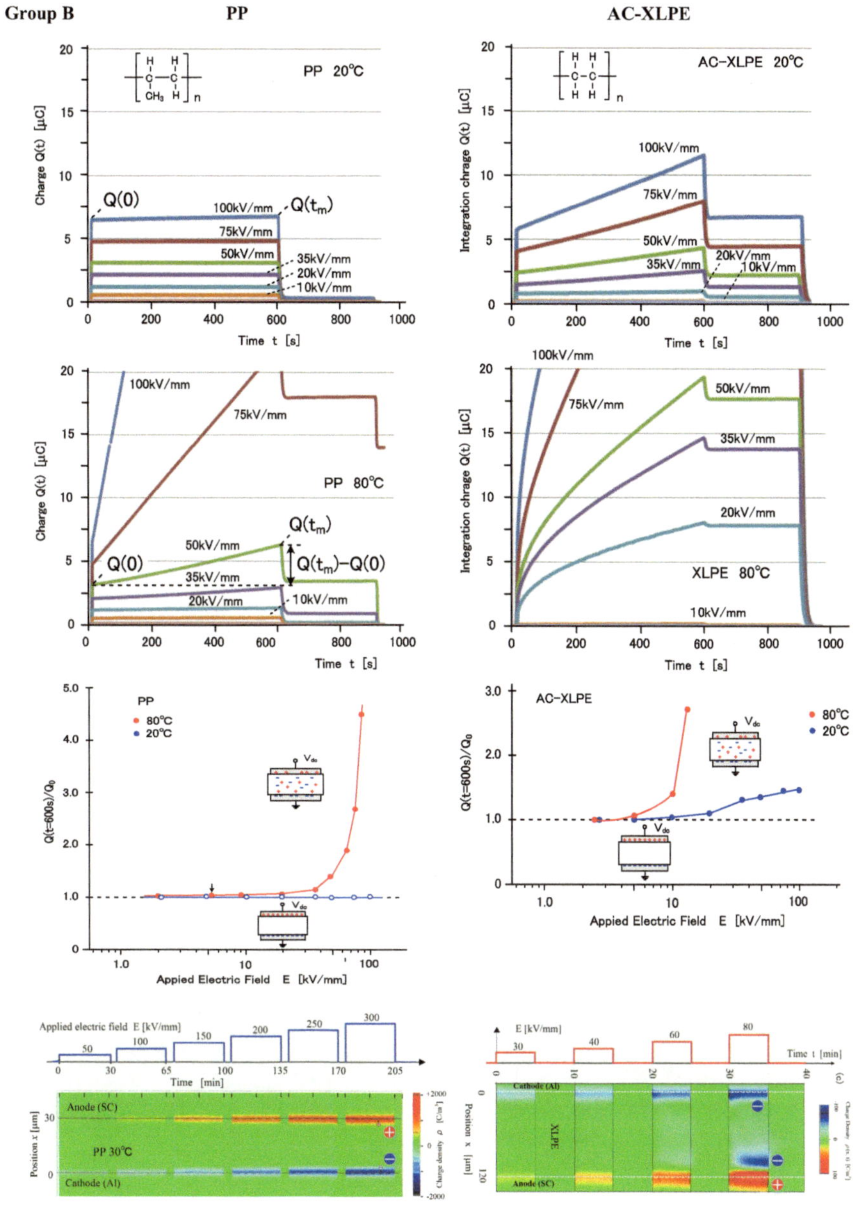

Fig. 4.6 The measured results of PP and AC-XLPE

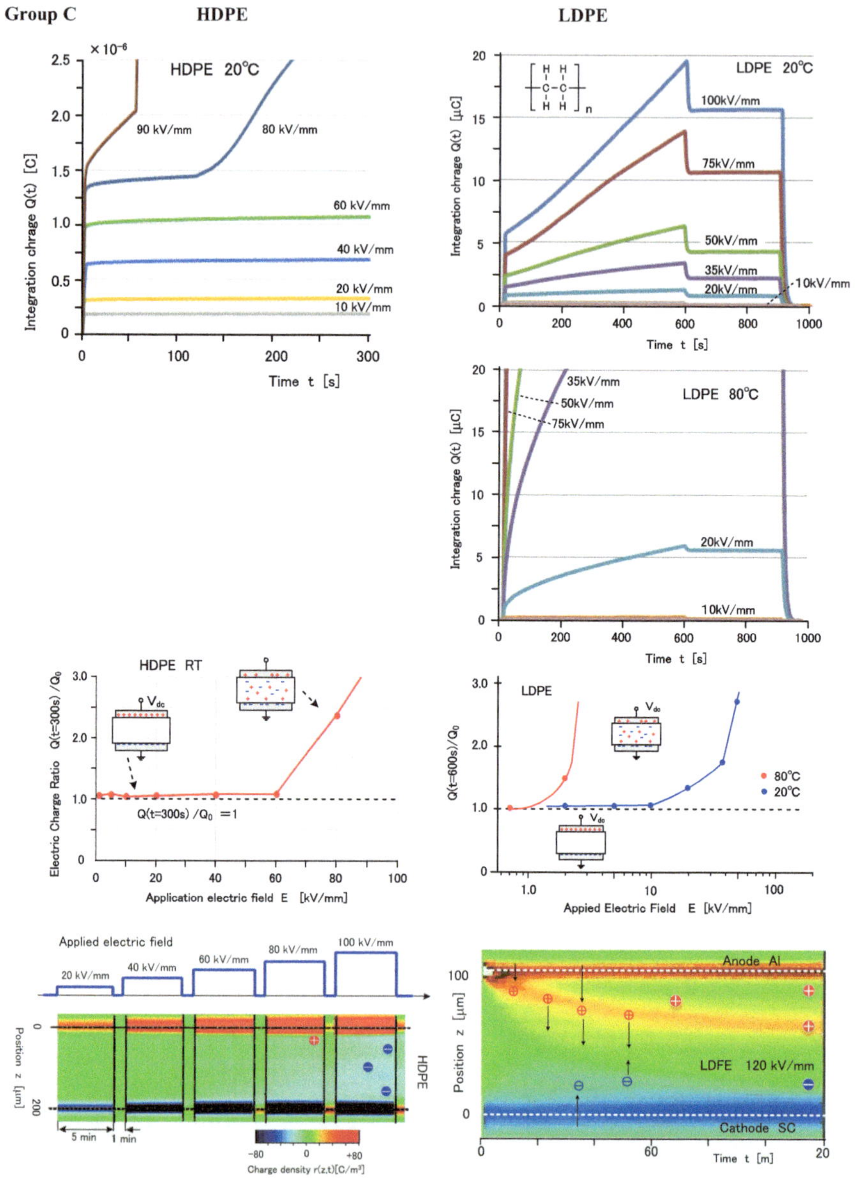

Fig. 4.7 The measured results of HDPE and LDPE

Table 4.1 The polymer insulating materials in different groups

	Polar molecular		Non-polar molecular	
Group A	PEN	Polyethylene naphthalate		
	PET	Polyethylene terephthalate		
	PC	Polycarbonate	PEFE	Polytetrafluoro ethylene
	PS	Polystyrene	DC-XLPE	DC-Cross-linked polyethylene
Group B	PI	Polyimide		
	Epoxy			
	PMMA	Polymethyl methacrylate		
	PPS	Polyphenylenesulfide		
	PP	Polypropylene		
			AC-XLPE	AC-Cross-linked polyethylene
Group C			HDPE	High density polyethylene
			LDPE	Low density polyethylene

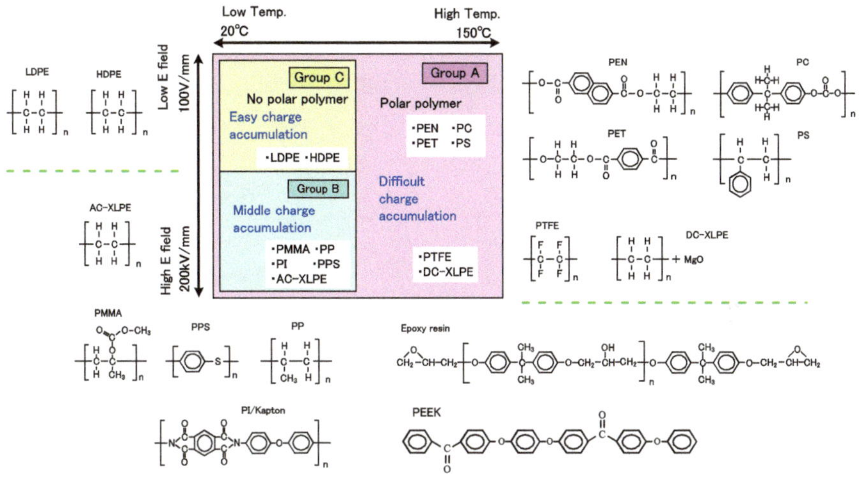

Fig. 4.8 Classification of charge accumulation characteristics

DC-XLPE: DC-cross-linked polyethylene. Deep traps are formed by adding nanocomposites (MgO) with high dielectric constants.

(2) **Group B: Weakly polar molecules**

No charge accumulates at room temperature, but charge accumulates at high temperature.

PI: Polyimide. Polar molecule with benzene ring/imide group in the main chain.

PMMA: Polymethyl methacrylate. A polar molecule with an ester group in the side chain.

PPS: Polyphenylenesulfide. A polar molecule with a benzene ring in the main chain.

PP: Polyproptlene. Non-polar molecule with methyl group (-CH_3) in the side chain.

AC-XLPE: AC-Cross-linked polyethylene. Non-polar molecule with ethylene (-C_2H_4-) polymer in the main chain.

(3) **Group C: Non-polar molecule**

Charge accumulation is observed even at room temperature, and charge accumulation with leakage current becomes more dominant at higher temperatures.

HDPE: High density polyethylene. Non-polar molecule with ethylene (-C_2H_4-) in the main chain.

LDPE: Low density polyethylene. Non-polar molecule with ethylene (-C_2H_4-) in the main chain.

Here, the types of functional groups that can create polarity are shown in Fig. 4.9.

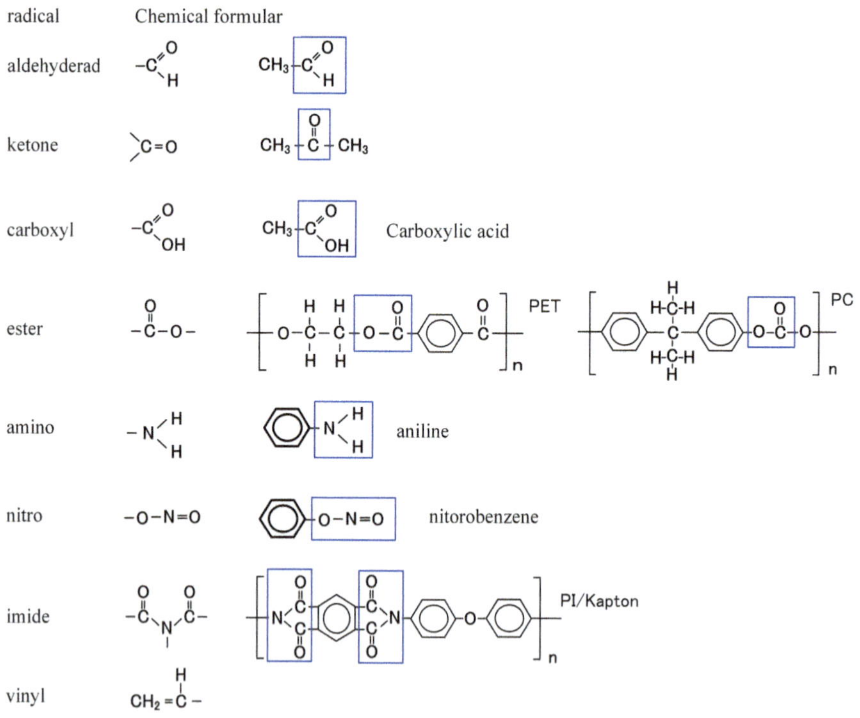

Fig. 4.9 The types of functional groups

Due to the electro-negativity of the atom, the electron orbit is shiftted between the bonded atoms, and an electric dipole is generated to become a polar molecule. This polar molecule creates a large electrostatic potential caused by the electric dipole. The electrostatic potential is described in the Sect. 4.3.

Furthermore, energy levels, trap depths, electrostatic potentials, and other properties of various polymer insulating materials can be obtained using Quantum Chemical Calculation software (Gaussian). These achievements are summarized in the text below and should be referred to. With regard to the experimental results of the Q(t) method and the PEA method, the good time has come when the properties of charge storage can be discussed from a micro scale using Quantum Chemical Calculation as shown in the following textbooks.

(a) Takada Text Vol. II, Chap. 7
(b) Takada Text Vol. II, Chaps. 12 and 13
(c) Takada Text Vol. II, Chap. 15

4.4 Relationship Between Charge Accumulation and Electrostatic Potential Distribution in Molecule

It is found that the charge accumulation characteristics depend on whether the material is a polar polymer or a non-polar polymer. To observe whether the polymer is polar or non-polar, the internal electrostatic potential distribution of the polymer is evaluated. The calculation results of the internal electrostatic potential distribution of various polymers (Group A: PEN, Group B: PP, Group C: LDPE) are shown in Fig. 4.10. The dimensions of various polymers are unified. This calculation uses Gaussian 09 software for Quantum Chemical Calculation. For details, see Takada Text Vol. III.

Group A: Strong polar molecules (Fig. 4.10): Static potential distribution of PEN pentamer, PET pentamer, PC hexamer, PEEK dimer. Similarly, the center of gravity of the orbital electron is displaced by the electronegativity of the bonding atom. In particular, the localization of the π-electron double bond and the double bond of the benzene ring generate strong electric quadrupoles. As a result, the electrostatic potential distribution of the polar polymer spreads over a range, and the overall action between the molecules becomes stronger. Further, the trap depth for charge is deeper and the material shows improved heat resistance.

Group B: Weakly polar molecules (Fig. 4.11): Static potential distribution of PI pentamer, PMMA 10-mer, PPS trimer, and PP hexamer. The electrostatic potential distribution shows an intermediate characteristic between that of Groups A and C.

Group C: Non-polar molecule (Fig. 4.12): Static potential distribution of LDPE and HDPE. Polyethylene (PE) is a compound of hydrogen H and carbon C, whose electronegativities are 2.1 and 2.5, respectively. The electron orbital center of hydrogen H shifts to the carbon C side due to the difference of the electronegativity. Therefore, hydrogen H is positively charged (orange type), and carbon C is negatively charged (blue type). As a result, an electric dipole is generated between them,

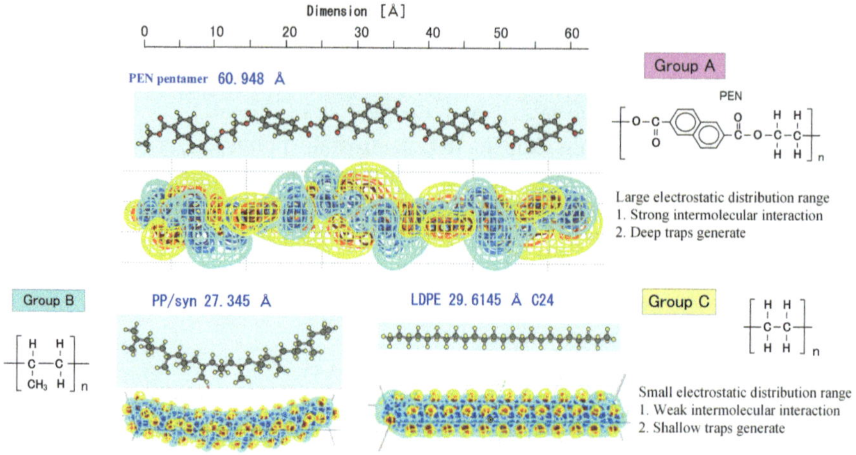

Fig. 4.10 Calculated results by Quantum Chemical Calculation

and an electrostatic potential distribution is observed. However, its corresponding distribution remains narrow.

4.5 Improvement of Charge Accumulation Characteristics of XLPE + Voltage Stabilizer

The charge accumulation characteristics of AC-XLPE become remarkable at a high temperature (80 °C) with an applied electric field of 10 kV/mm (see Sect. 4.1). The result of mixing a voltage stabilizer to improve the charge storage characteristics is introduced [1].

(1) **Q(t) measurement results**

Measured samples: XLPE, XLPE + PAC-A, XLPE + PAC-B, XLPE + PAC-C.

PAC: A voltage stabilizer for polycyclic aromatic compounds. Figure 4.14 shows the molecular structure formula.

Figure 4.13 shows the charge accumulation characteristics evaluated by the Q(t) method. The lowest figures show the result of the charge ratio $Q(t)/Q_0$. In the case of a single XLPE sample, some charge accumulation is observed at 10 kV/mm and 25 °C, and significant charge accumulation is observed at 80 °C. In contrast, in the case of XLPE + PAC-C, little charge accumulation is observed even at 25 and 80 °C. The charge storage characteristics are significantly improved by the introduction of the voltage stabilizer.

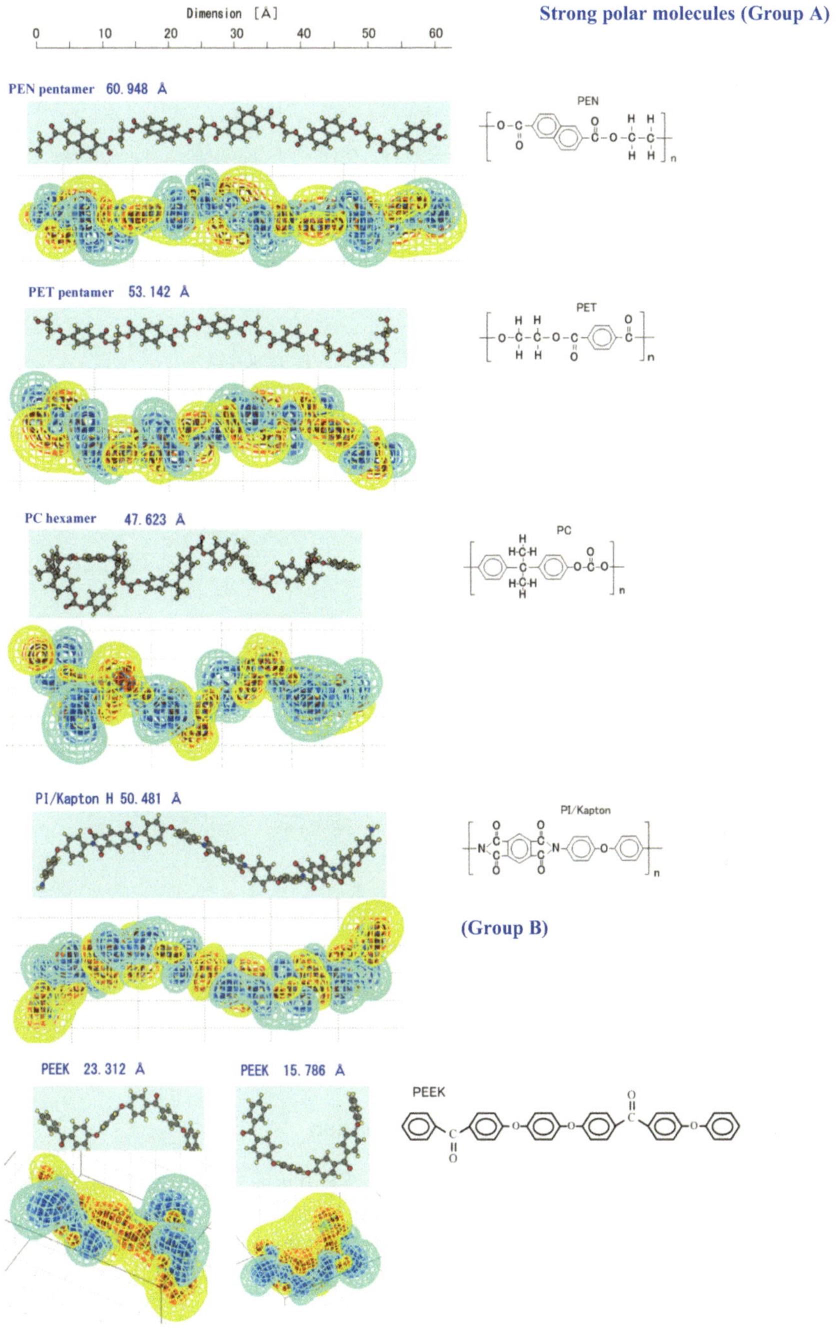

Fig. 4.11 Calculated results of strong polar molecules

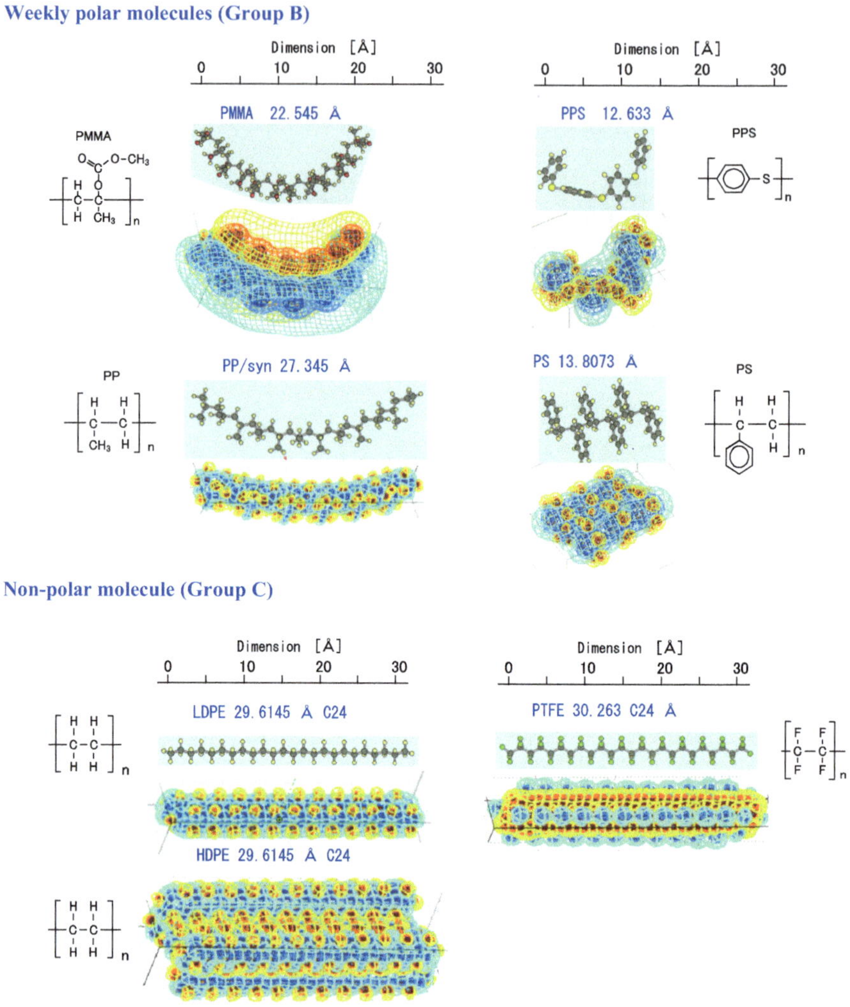

Fig. 4.12 Calculated results of the molecules in Groups C and B

(2) Analysis by Quantum Chemical Calculation

Molecular structure formula (Fig. 4.14): XLPE has a structure in which three straight-chain polyethylenes with 24 carbon atoms are cross-linked at three places with ethane, which has 2 carbon atoms. XLPE + PAC-A is obtained by mixing a voltage stabilizer PAC-A with XLPE. XLPE + PAC-B contains PAC-B and XLPE + PAC-C contains PAC-C.

Energy level (Middle of Fig. 4.14): Energy level of each molecule is simulated using Gaussian 09 software (B3LYP Hamiltonian, 6-31G basis function). The LUMO level o f XLPE is higher than the vacuum level VL, and the electron affinity is negative.

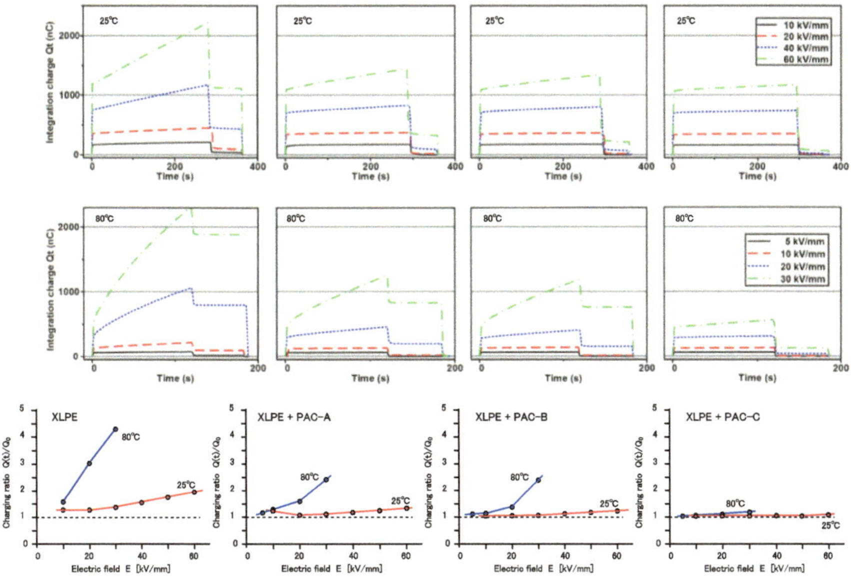

Fig. 4.13 Measured results of XLPE with different voltage stabilizers

On the other hand, the LUMO level of PAC is lower than the vacuum level VL, and the electron affinity is positive. Electrons in the vacant band (conduction level: LUMO level) are trapped in the PAC with a positive electron affinity. Similarly, holes are also trapped in the PAC with a high HOMO level. In particular, the positive traps of PAC-C are deeper than those of PAC-A and PAC-B. Therefore, the charge ratio of PAC-C is $Q(t)/Q_0 = 1$, and the charge accumulation characteristics are significantly improved.

Electrostatic potential distribution: The electrostatic potential distribution $V(x, y, z)$ can also be obtained from the calculation results by the Gaussian 09 software. The upper part of Fig. 4.14 shows the electrostatic potential distributions of the three types of voltage stabilizer molecules. Warm colors (red/brown/yellow) indicate the positive potential, and cool colors (blue/sky blue) indicate the negative potential. All the three specimens have strong positive and negative potential distributions, which spread over a wide area. The generation of this positive and negative potentials depends on the electronegativity of the atom. As a result, the electric dipole moment of the voltage stabilizer molecule is large. As described above, the gap between the LUMO and HOMO levels of a molecule with a large dipole moment becomes narrow, in other words, the energy gap becomes narrow.

Electronegativity: The electronegativity of each atom (hydrogen H, carbon C, nitrogen N, oxygen O, fluorine F) is shown in the left of Fig. 4.14. The bar color corresponds to that of the constituent atoms of the PAC. For example, in the case of C-F, fluorine F with large electronegativity (4.0) is smaller than carbon C (2.5). As the electron of carbon (2.5) is attracted by fluorine F, fluorine F is negatively charged,

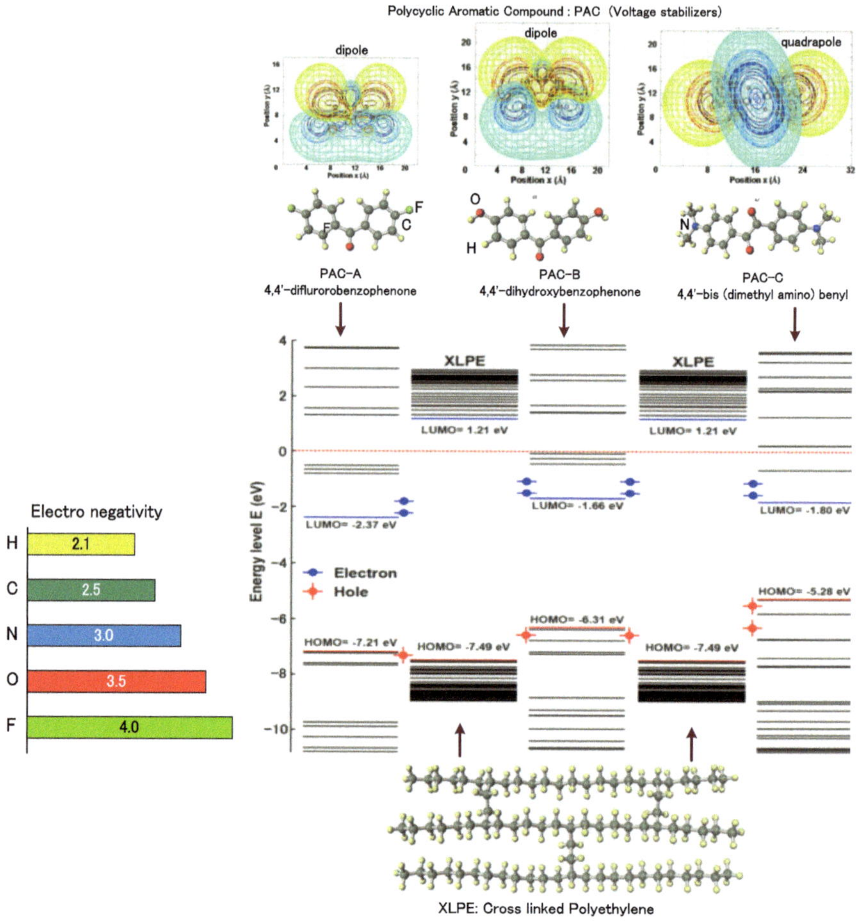

Fig. 4.14 Calculated results of XLPE with different voltage stabilizers

and carbon C is positively charged. Such coupling causes a bias in the electron distribution, resulting in a positive and negative potential distribution. In addition, the π-electron (negative) of the benzene ring with four hydrogen H (Positive) creates a quadrupole moment electrostatic potential.

Reference

1. J. Li, X. Kong, B. Du, K. Sato, S. Konishi, Y. Tanaka, H. Miyake, T. Takada, Effects of high temperature and high electric field on the space charge behaviors in epoxy resin for power module, *IEEE Transactions on Dielectrics and Electrical Insulation*, 27(3), 832–839 (2020).

Chapter 5
Charge Accumulation in Inorganic Materials

5.1 Q(t) Data of Inorganic Materials

(1) **Various inorganic materials**

Inorganic materials and ceramics are used for the insulating layer of IGBT-Modules of power devices. The functions required for inorganic materials are (a) high thermal conductivity with good heat dissipation and (b) excellent properties of low electrical conductivity and excellent electrical insulation. Generally, it is empirically known that a low thermal conductivity corresponds to good insulating properties and a high thermal conductivity corresponds to bad insulating properties. Therefore, high thermal conductivity and low electrical conductivity require completely opposite characteristics. Table 5.1 summarizes the inorganic materials that are currently being studied for the insulating layer of IGBT-M.

(2) **Q(t) measurement results**

The Q(t) characteristics of the four inorganic materials in Table 5.1 are shown in Figs. 5.1 and 5.2. The measurement conditions are as follows:

Applied electric field: $E = 1, 2, 4, 6, 8, 10$ kV/mm.
Application time of electric fields: 180 s with later short circuit.
Measurement temperature: Temp = 25, 50, 150, 200 °C. Electrode area: 7.07 cm^2.

Figures 5.1 and 5.2 summarize the data of the electric field dependence and temperature dependence of the Q(t) characteristic. The following Sect. 5.2 summarizes the electric field dependence and temperature dependence of Q(t) data.

T. Takada et al., *Electric Charge Accumulation in Dielectrics: Measurement and Analysis*,
https://doi.org/10.1007/978-981-19-6156-4_5

Table 5.1 Names of inorganic materials, thermal conductivity, and charge accumulation evaluation

Sample		Thermal conductivity (W/m·K)	Charge ratio $Q(t_m)/Q_0$ 25 °C
Si_3N_4	Silicon nitride	20–60	1.2–1.5
BN	Boron nitride	0.5	1.1–2.0
Al_2O_3	Alumina	36	1.3–3.0
AlN	Aluminum nitride	300	

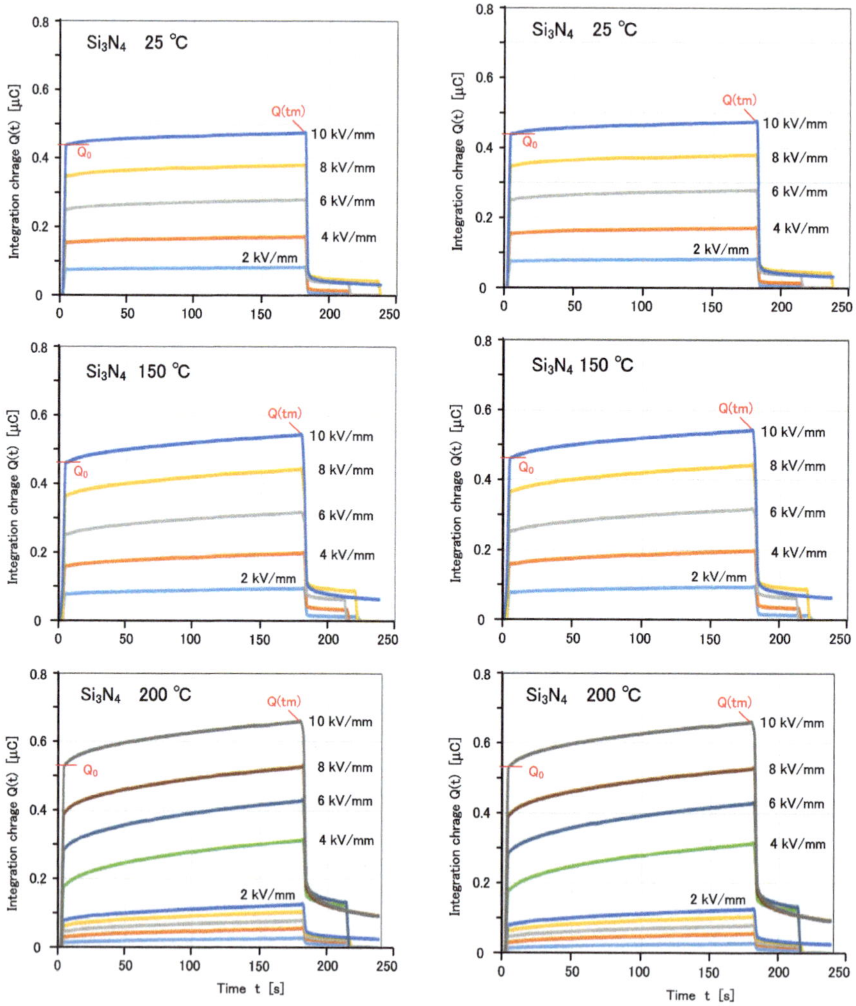

Fig. 5.1 Measured results of the Si_3N_4 material

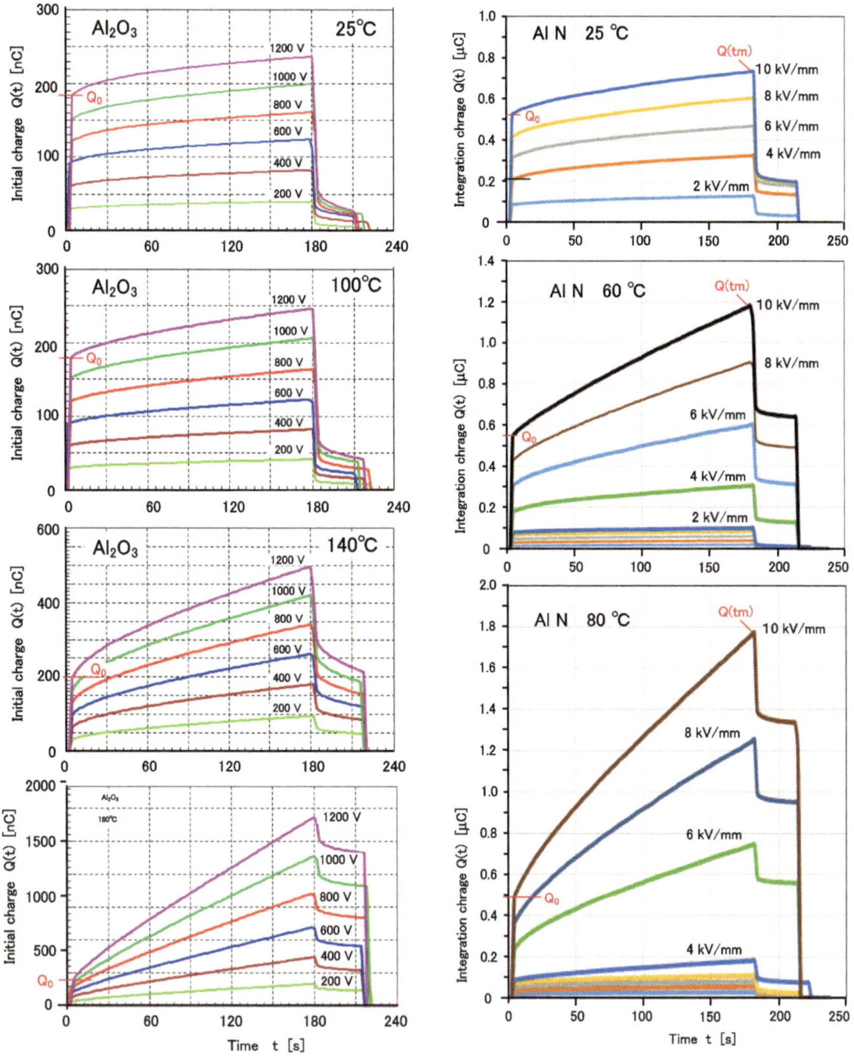

Fig. 5.2 Measured results of the two inorganic materials

5.2 Electric Field Dependence and Temperature Dependence

(1) Electric field and temperature dependence of charge accumulation

Figure 5.3 shows the electric field dependence and temperature dependence of charge accumulation. If the charge ratio $Q(t_m)/Q_0 = 1.0$, it is determined that there is no charge accumulation, and if $Q(t_m)/Q_0 > 1.0$, it is determined that there is charge

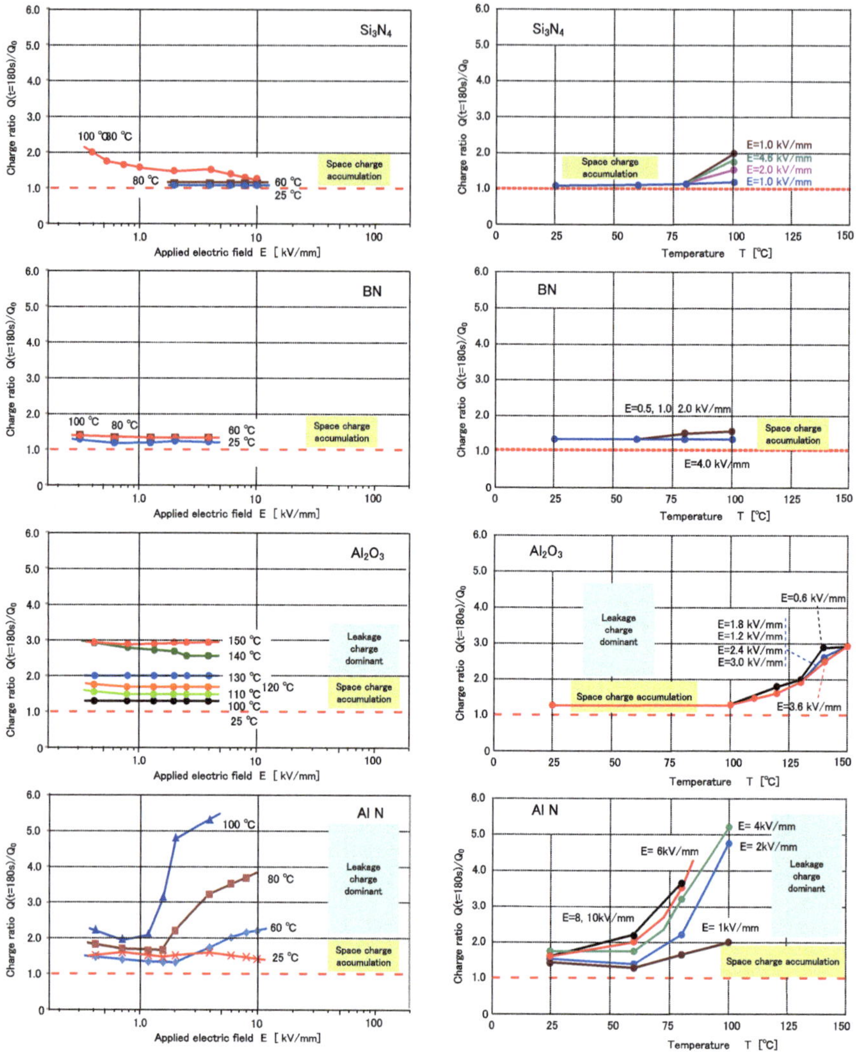

Fig. 5.3 The charge ratios of the inorganic materials

accumulation. The results are summarized in Table 5.2. It can be judged that the charge accumulation of Si_3N_4 and BN is a little, and the charge accumulation of Al_2O_3 and AlN is a lot. The corresponding physical property model is described in Sect. 5.6.

(2) **Space charge accumulation and leakage charge**

Section 5.1 shows the Q(t) data of inorganic materials. The fact that charge accumulation can be analyzed from the Q(t) data has already been described in Sect. 3.2.

Table 5.2 Names of inorganic materials and thermal conductivity, charge accumulation evaluation

	Electric field dependence 0.5–10 kV/mm	Temperature dependence 25, 50, 150, 200 °C
Si_3N_4	$Q(t_m)/Q_0 \fallingdotseq 1.1–2.0$ a little charge	Dependence appears over 75 °C
BN	$Q(t_m)/Q_0 \fallingdotseq 1.2–1.5$ a little charge	Dependence appears over 75 °C
Al_2O_3	$Q(t_m)/Q_0 \fallingdotseq 1.3–3.0$ a lot of charge	Dependence appears over 110 °C
AlN	$Q(t_m)/Q_0 \fallingdotseq 1.3–5.3$ a lot of charge	Dependence appears over 25 °C

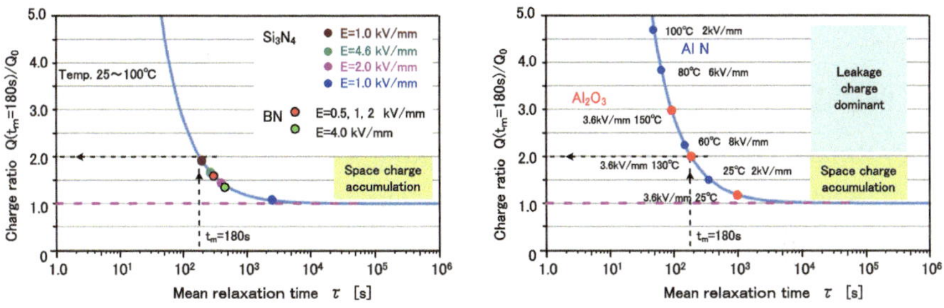

Fig. 5.4 Analysis of the charge ratios of the inorganic materials

Equation (3.3) in Sect. 3.2 is rewritten here. Equation (5.1) is arranged to calculate the data of the four types of inorganic materials, and the results are shown in Fig. 5.4. The vertical axis represents the charge ratio $Q(t_m = 180\ \text{s})/Q_0$. The horizontal axis represents the dielectric relaxation time τ, and the measurement time is $t_m = 180$ s. When the measurement time is fixed at $t_m = 180$ s, the corresponding curve is drawn. From this curve, the presence or absence of charge accumulation can be evaluated as follows.

> **$Q(t_m) = Q_0$:** It is determined that there is no charge accumulation inside the sample. Only the electrode charge is shown.
> **$2 > Q(t_m)/Q_0 > 1$:** It represents that space charge is accumulated inside the sample.
> **$Q(t_m)/Q_0 > 3$:** The charge is moving through the sample, and it is determined that the leakage current is dominant.

$$\frac{Q(t) - Q_0}{Q_0} = \frac{Q_{spac}(t)}{Q_0} + \frac{Q_{leak}(t)}{Q_0} = \frac{\Delta t}{\tau} \qquad \frac{Q(t_m)}{Q_0} = 1 + \frac{tm}{\tau}$$

$$\text{where} \quad \tau = \frac{\varepsilon}{\kappa} \tag{5.1}$$

How to analyze the figure:

Region of measurement time t_m = 180 s > dielectric relaxation time τ (blue part):
Within the measurement time t_m, free carriers can pass through the sample, and the carriers are in equilibrium.

Region of measurement time t_m = 180 s < dielectric relaxation time τ (yellow part):
Charges are migrating within the sample at the measurement time, and they are in a state before equilibrium. In other words, space charges are forming inside the sample.

(3) **Evaluation of inorganic materials**

Si_3N_4 and BN (the left of Fig. 5.4): In the temperature range of 25 to 100 °C and the electric field range of 1.0 to 4.0 kV/mm, it is judged that space charge accumulation appears inside the sample.

Al_2O_3 and AlN (the right of Fig. 5.4): Space charge accumulates in the sample at the temperature of 25 °C and the low electric field of 2.0 kV/mm. The space charge accumulation is dominant.

Leakage current appears at the temperature over 80 °C and the electric field over 3.6 kV/mm. We can judge that now the leakage charge is dominant.

5.3 DC Permittivity

DC permittivity ε_r (DC) can be obtained from the data of the Q(t) characteristic in Sect. 5.1 using Eq. (5.2). Here, ε_0 is the dielectric constant of vacuum, S is the electrode area, Q_0 is the initial charge amount, and E is the applied electric field.

$$Q_0 = \frac{\varepsilon_0\ \varepsilon_r(DC)\ S}{a} V_{dc} = \varepsilon_0\ \varepsilon_r(DC)\ S\ E \tag{5.2}$$

Figure 5.5 shows a graph of the initial charge amount Q_0 and the applied electric field E. Since the slope of the graph is $\varepsilon_0\varepsilon_r(DC)S$, ε_r(DC) can be calculated. The values are shown in the figure.

Figure 5.6 shows the temperature dependence of ε_r(DC). The DC relative permittivity of AlN and Al_2O_3 is a large value in the range of 8–10. The value of Si_3N_4 is medium with ε_r(DC) = 7–8. On the other hand, the value of BN is as small as ε_r(DC) = 3.5–4.5. Since the value of the polymer resin is about ε_r (DC) = 2–3, the DC dielectric constant of the inorganic material is larger than that of the polymer resin.

For reference, the relative dielectric constants at a high frequency of 1 MHz is also shown. The values of AlN, Al_2O_3, and Si_3N_4 are in the range from 9 to 10, and that of BN is 5.

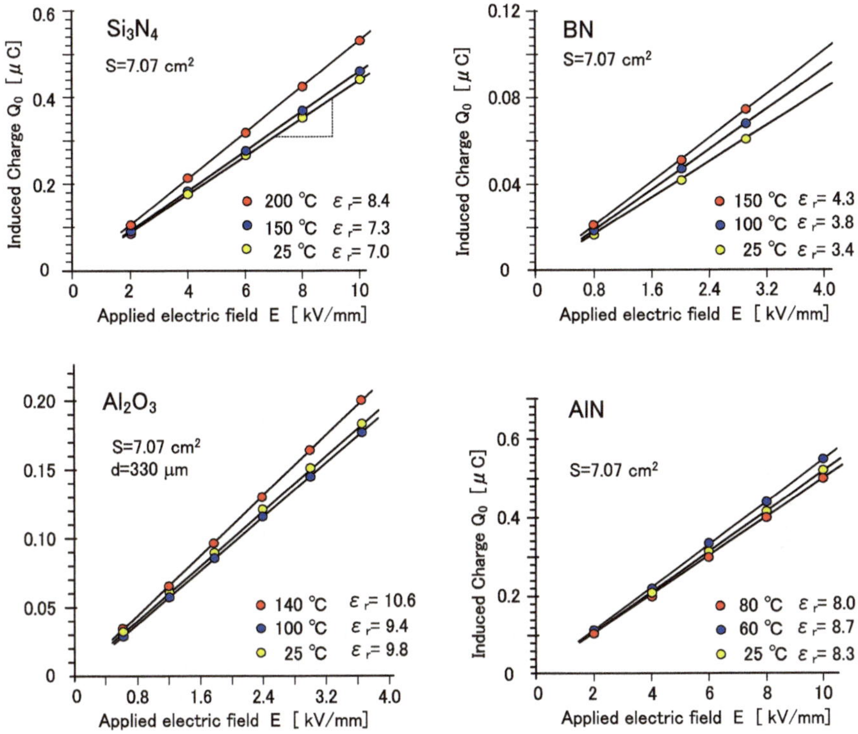

Fig. 5.5 The initial induced charge amount of the inorganic materials

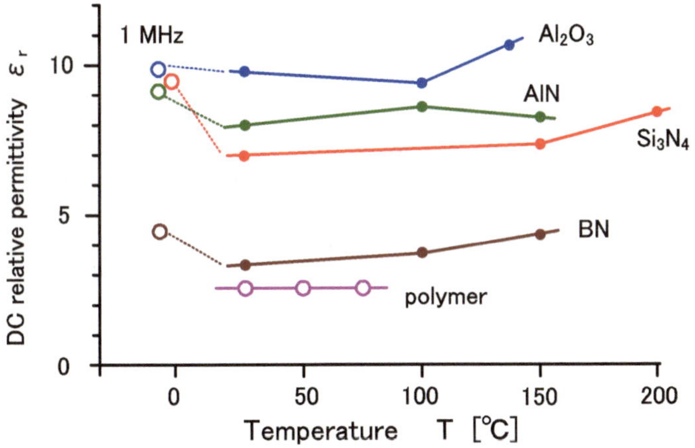

Fig. 5.6 The DC permittivity of the inorganic materials

5.4 Comparison of Q(t) Data Between Polymer Insulating Materials and Inorganic Materials

Differences are observed in the Q(t) characteristics of polymer insulating materials (PEN, PET, PI,... LDPE, etc.) and inorganic materials (BN, Si_3N_4, Al_2O_3, AlN).

(1) **Comparison of PET (polymer) and BN (ceramics)**

Figure 5.7 shows a comparison between PET and BN, which are representatives of a group in which charge accumulation is difficult shown by Q(t) characteristics. The charge ratio of PET (polymer) clearly keeps $Q(t_m = 180\text{ s})/Q_0 = 1.0$. In contrast, the charge ratio of BN (ceramics) is always larger than 1.0.

(2) **Comparison of LDPE (polymer) and AlN (ceramics)**

LDPE and AlN are also shown by the figure, which are representatives of a group in which charge accumulation is easy. Under the low electric field, the charge ratio of LDPE (polymer) clearly keeps $Q(t_m = 180\text{ s})/Q_0 = 1.0$. However, AlN (ceramics) shows $Q(t_m = 180\text{ s})/Q_0 > 1.0$ even in a low electric field, and charge injection occurs.

It is presumed that charge injection from an electrode is promoted by the polarization orientation due to the application of an electric field on an inorganic material with a large dielectric constant, and the polarization electric field traps the charge injection. This is another research topic in the future.

5.5 Multilayer Ceramic Capacitors

The capacitance of multilayer ceramic capacitor (MLCC) is always increasing for the actual application. A dielectric insulation with a high dielectric constant is usually used, such as barium titanate ($BaTiO_3$). In terms of temperature characteristics, the change rate of capacitance or temperature coefficient is specified in the temperature range of −50 to +125 °C for the classification of ceramic capacitors according to the standard. The standard evaluates the capacitance and tanδ when an AC voltage is applied. At present, MLCCs are used in DC voltage application, but charge accumulation and leakage current can result in insulation degradation and dielectric breakdown [1]. This section introduces the evaluation of charge accumulation by applying the Q(t) method, which integrates the total current flowing when a DC voltage is applied to a MLCC.

(1) **Q(t) measurement results**

MLCC sample: A commercially available capacitor shown in Table 5.3.

Q(t) measurement results in Fig. 5.8: Applied voltage is 1–8 V with an applied time of 60 s, and measurement temperatures are 30, 60, 85, 105, 125, 150, 175 °C.

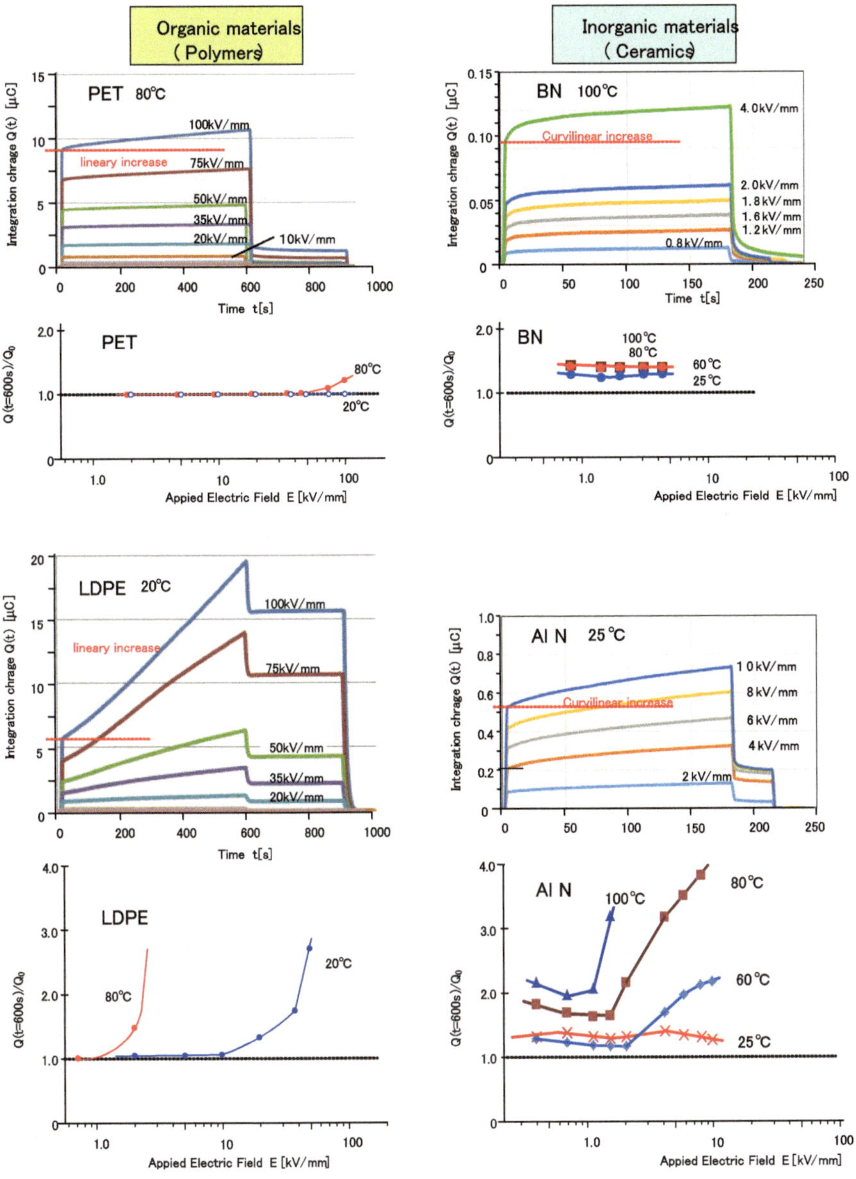

Fig. 5.7 Comparison of the two kinds of materials

Table 5.3 MLCC specifications

MLCC sample	10 μF
Capacitance (μF)	10 ± 1
Rated voltage (V)	16
tanδ (%)	5
Operating temp (°C)	−55–+85

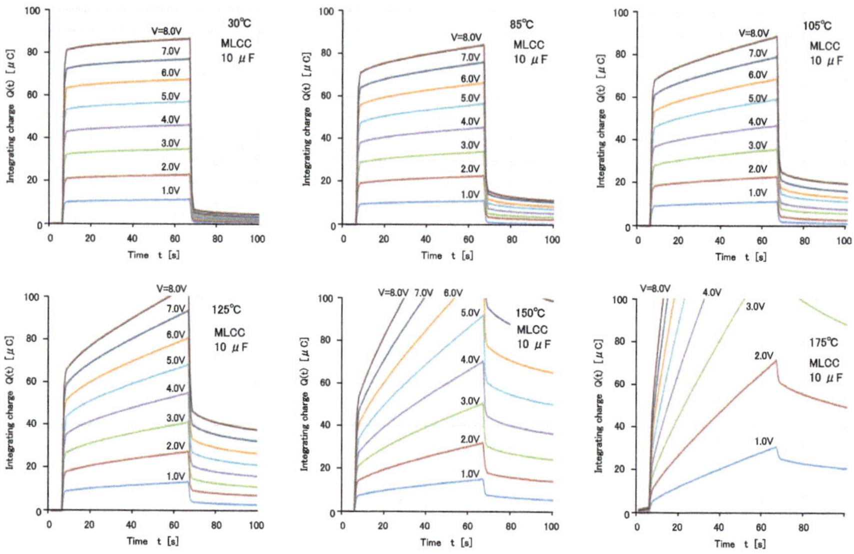

Fig. 5.8 Q(t) measurement results of the MLCC sample

The integration capacitor uses film ones with $C_{INT} = 1$ mF, which are composed of ten 100 μF capacitors in parallel. Immediately after the voltage is applied ($t = 0$), the Q(t) data quickly increases to the initial charge amount of $Q_0 = C_s V_{dc}$. Thereafter, it gradually increases to form space charge accumulation. When the temperature exceeds 125 °C, the Q(t) data sharply increases and shifts to a leakage current.

(2) **Evaluation of charge accumulation**

Figure 5.9 shows the voltage dependence of the charge ratio $Q(t)/Q_0$. At 30 °C, the charge ratio is $Q(t)/Q_0 = 1.1$, which is slightly larger than "1". In the case of polymer dielectrics (PP, PET), the charge ratio is exactly $Q(t)/Q_0 = 1.0$, which is clearly "1.0". Therefore, it is judged that there is no charge accumulation. In contrast, in the case of the dielectric ($BaTiO_3$) with a high dielectric constant, the charge ratio under the applied voltage range of 1.0–8.0 V and the temperature range of 30–105 °C is $Q(t)/Q_0$ = 1.1–1.3. Therefore, it can be judged that charge accumulation has occurred. As described above, in the case of the dielectric with a high dielectric constant, it is found that charge is accumulated even at a low voltage and a low temperature. This charge accumulation model is described in the following section.

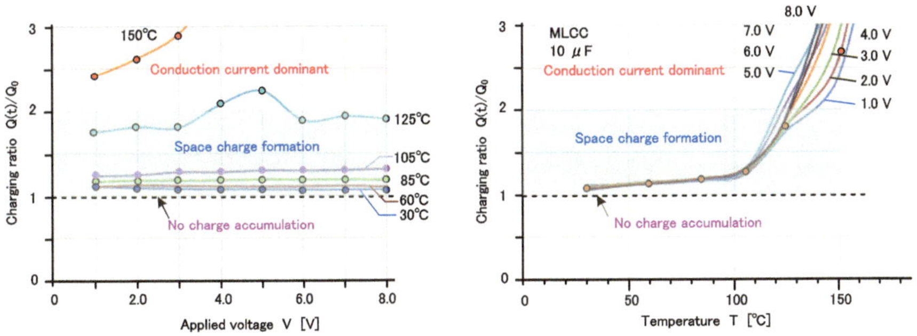

Fig. 5.9 The charge ratio of the MLCC sample

(3) Evaluation of leakage current

Section 3.1 describes that the region where the charge ratio $Q(t)/Q_0 > 2$ is dominant by conduction current (leakage current). Focusing on the results of Fig. 5.9, the range of charge ratio $Q(t)/Q_0 > 2$ can be seen at temperatures higher than 125 °C. The temperature dependence of the leakage current is stronger than that of the applied voltage.

(4) Evaluation of capacitance

Figure 5.10 shows the characteristics of the initial charge amount Q_0 and the applied voltage V. The capacitance C can be evaluated from the slope of the characteristic ($Q_0 = CV$). Since the capacitance of the MLCC is 10 μF, the value of $C = 10$ μF is drawn by a red dotted line in the figure. In the temperature range of 30–105 °C, Q_0 and V have a substantially linear relationship. The slope is the capacitance, and its value is approximately 10 μF. This value has some temperature dependence. At temperatures above 125 °C, the linear relationship between Q_0 and V is damaged. In other words, it is considered that the charge decrease $\Delta Q(V, T)$ due to the oriented polarization hindered by temperature is subtracted to the relationship of $Q_0 = CV$, as shown in Eq. (5.3).

$$Q_0 = \frac{\varepsilon_0\,(\varepsilon_r - \Delta\,\varepsilon_r(V, T))\,S}{a}V = \frac{\varepsilon_0\,\varepsilon_r\,S}{a}V - \Delta Q(V, T) \tag{5.3}$$

(5) High dielectric constant and electric charge accumulation

(a) **The basic equations for the dielectric material:** The charge density σ (C/m^2) on the electrode surface is proportional to the internal electric field strength E (V/m), and the proportional coefficient is the dielectric constant $\varepsilon = \varepsilon_0\varepsilon_r$. This relational expression is derived as in Eq. (5.4).

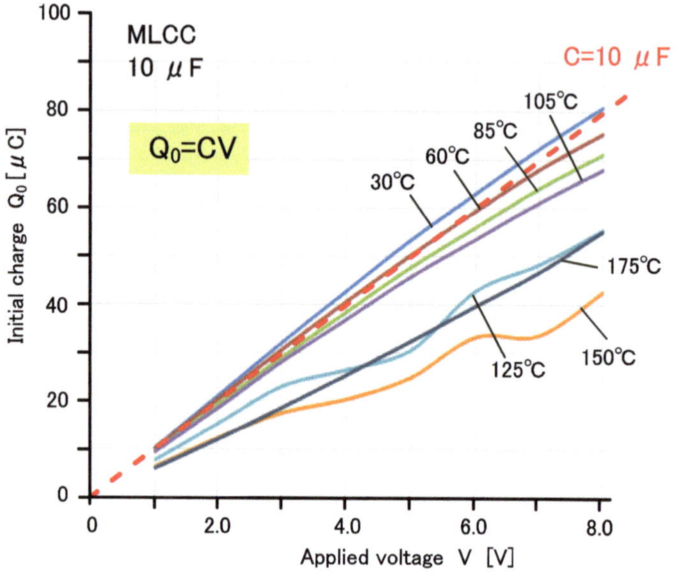

Fig. 5.10 The capacitance of the MLCC sample

$$Q = CV \quad \Rightarrow \quad Q = \frac{\varepsilon_0\,\varepsilon_r\,S}{a} \times V$$

$$\Rightarrow \quad \sigma \left[= \frac{Q}{S}\right] = \varepsilon_0\,\varepsilon_r \times E \left[= \frac{V}{a}\right] \quad \Rightarrow \quad \sigma = \varepsilon_0\,\varepsilon_r E \tag{5.4}$$

(b) **In the case of air:** When the air is between the electrodes, the charge density σ_1 is given by Eq. (5.5). The electric field is uniform with the value of $E = V/a$.

(c) **In the case of paraelectric permittivity:** A dielectric with a relative permittivity ε_r is inserted between the electrodes. The electric field in the dielectric is uniform with the value of $E = V/a$. Since there is a vacuum between the electrode surface and the dielectric, the electric field on the electrode surface is $E_2 = \varepsilon_\mathrm{r} \times E$ (= V/a), which is ε_r times larger than E, as shown in Eq. (5.6).

$$\sigma_1 = \varepsilon_0\,E \tag{5.5}$$

$$\sigma_2 = \varepsilon_0\,\varepsilon_r E \quad \varepsilon_r \approx 3 \quad \Rightarrow \quad \sigma_2 = \varepsilon_0\,E_2 \quad \Rightarrow \quad E_2 = \varepsilon_r \times E \tag{5.6}$$

(d) **In the case of high permittivity:** In the case of dielectric substance with a high relative permittivity of ε_r = 10–1000, as shown in Eq. (5.7), the electric field E_3 at the interface becomes ε_r = 10–1000 times as large as the electric field E = V/a inside the dielectric.

(e) **In the case of charge injection and high dielectric constant** ($\varepsilon_r = 100$)**:** When the applied electric field is 100 kV/mm, the interfacial electric field becomes $E_3 = 10{,}000$ kV/mm. With this ultra-high electric field, charges $\Delta Q(t)$ are injected from the electrode into the dielectric, which may neutralize with the polarization charge near the surface, and part of them are trapped. Therefore, the electric field E_3 decreases to E_4, as shown Eq. (5.8) and Fig. 5.11. As a result, the charge $\Delta Q(t)$ is further supplied from the power supply.

$$\sigma_3 = \varepsilon_0\, \varepsilon_r E \quad \varepsilon_r > 10\text{–}1000 \quad \Rightarrow \quad \sigma_3 = \varepsilon_0\, E_3 \quad \Rightarrow \quad E_3 = \varepsilon_r \times E$$
$$\Rightarrow \quad E_3 = (\,\varepsilon_r = 100\text{–}1000) \times E \tag{5.7}$$

$$E_4 = E_3 - \frac{\Delta Q(t)}{\varepsilon_0\; S} \tag{5.8}$$

$$Q(t) = Q_0 + \Delta Q(t) = \varepsilon_0\, \varepsilon_r\; E\; S + \Delta Q(t) \tag{5.9}$$

(f) **Q(t) characteristics:** Each of the Q(t) data in Fig. 5.11 is expressed by Eq. (5.9) and is sketched in Fig. 5.12. The charge amount at the initial stage becomes $Q_0 = \varepsilon_r ES$ and depends on the relative permittivity ε_r. Therefore, the Q(t) data is also drawn depending on ε_r. In the case of (d) with charge injection and high permittivity, $Q(t) = Q_0 + \Delta Q(t)$. In the case of a low temperature, the injected charge is trapped in the potential of the polarization charge. However, when the temperature rises, the trapped charge is released and $\Delta Q(t)$ continues to increase.

Classical and quantum theory: Even though the electrode and the dielectric are in contact, the interface still has some spacing, which is about 0.3–0.5 nm. The behavior of charges (electrons) in this region cannot be properly considered in the classical theory. This small region corresponds to the electron wave in the quantum theory, and the classical theory of point charges cannot be applied. In the future, we hope to consider the quantum theory to explain it.

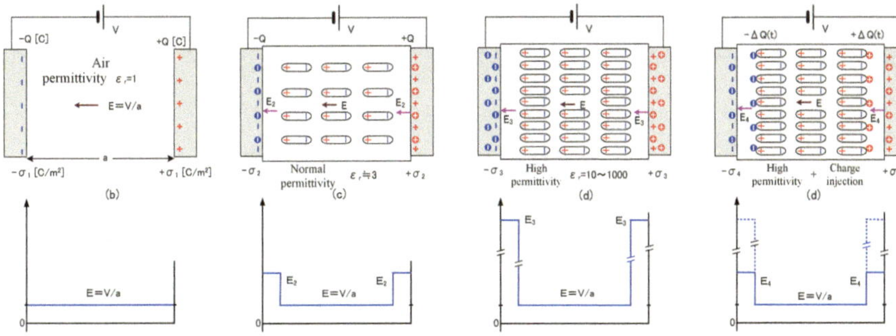

Fig. 5.11 Different sample models for analysis

Fig. 5.12 Q(t) analysis on different permittivities

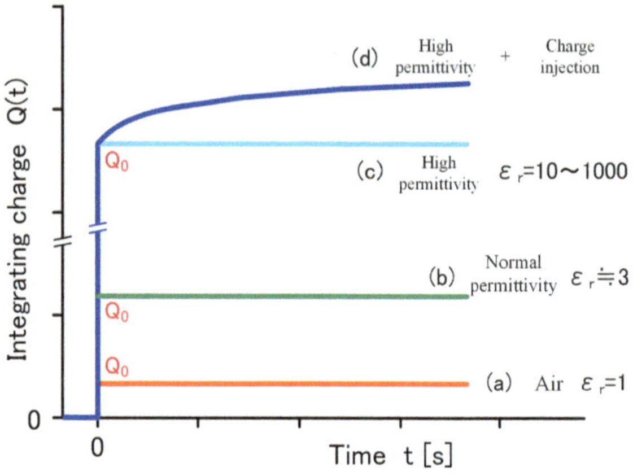

5.6 Research Subjects

(1) Comparison of DC permittivity

DC relative permittivities ε_r (DC) of the four inorganic materials obtained in Sect. 5.3 are summarized in Table 5.4. The relative permittivity of BN and Si_3N_4 is as large as the value range of 4.0–5.0, and that of Al_2O_3 and AlN is as large as the value range of 15–18. In order to consider this difference, the molecular structures of the four inorganic materials are compared in the following section.

(2) Comparison of molecular structures

The molecular structures of the inorganic materials are shown in Fig. 5.13. From this figure, it is difficult to directly explain the difference between the charge storage characteristics and the DC relative permittivity ε_r (DC). Therefore, the atoms that make up ceramic and inorganic materials are taken up and their characteristics are summarized in Fig. 5.14.

(3) Grouping of inorganic materials

Figures 5.14 and 5.15 summarizes the combinations of atom ionization in inorganic materials.

Table 5.4 DC relative permittivity ε_r(DC) of inorganic materials

Sample		DC relative permittivity	Relative permittivity at 1 MHz
BN	Silicon nitride	3.5–4.5	4.5
Si_3N_4	Boron nitride	7–8	9.6
Al_2O_3	Alumina	8–10	9.0–9.9
AlN	Aluminum nitride	8–10	8.8–9.0

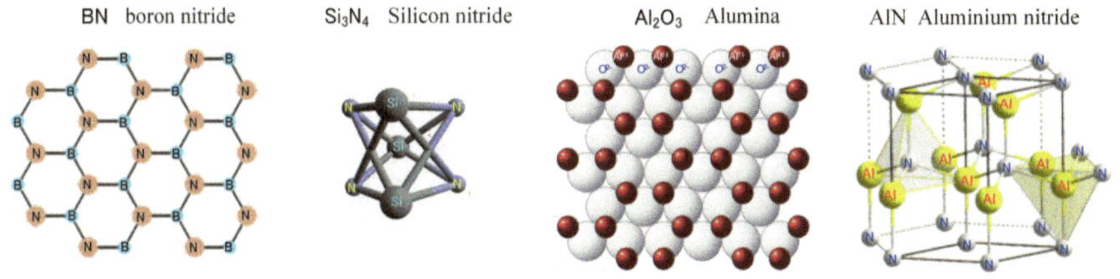

Fig. 5.13 The molecular structures of the inorganic materials

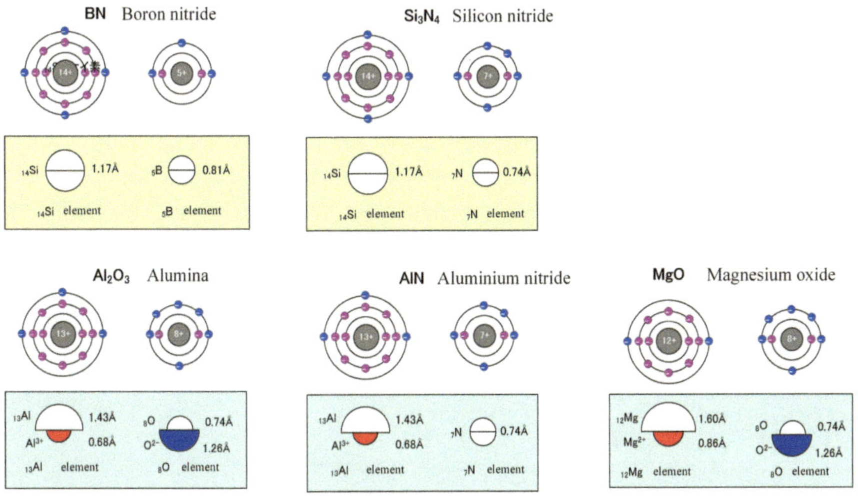

Fig. 5.14 The atoms of ceramic and inorganic materials

BN (boron nitride) which consists of neutral atoms are hardly ionized. The polarization due to the application of an electric field is small, and the DC relative permittivity is in the range from 3.5 to 4.5.

Si_3N_4 (silicon nitride) which consists of neutral atoms are not easy to ionize. Polarization due to the application of an electric field is small. The value of DC relative permittivity is medium and in the range from 7 to 8.

Al (aluminum) of Al_2O_3 (alumina) tends to be a positive ion, and O (oxygen) tends to be a negative one. It is easy to be polarized by the application of an electric field, and the DC relative permittivity is larger, which is about 8–10.

Al (aluminum) of AlN (aluminum nitride) is likely to become a positive ion, and N (nitrogen) is not easily ionized. Polarization occurs due to the application of an electric field, and the DC relative permittivity increases to the value range from 8 to 10.

The above description is an inference and a stage where an intial polarization model is presented. In the future, it is necessary to study polarization mechanism under applied electric fields by Quantum Chemical Calculation.

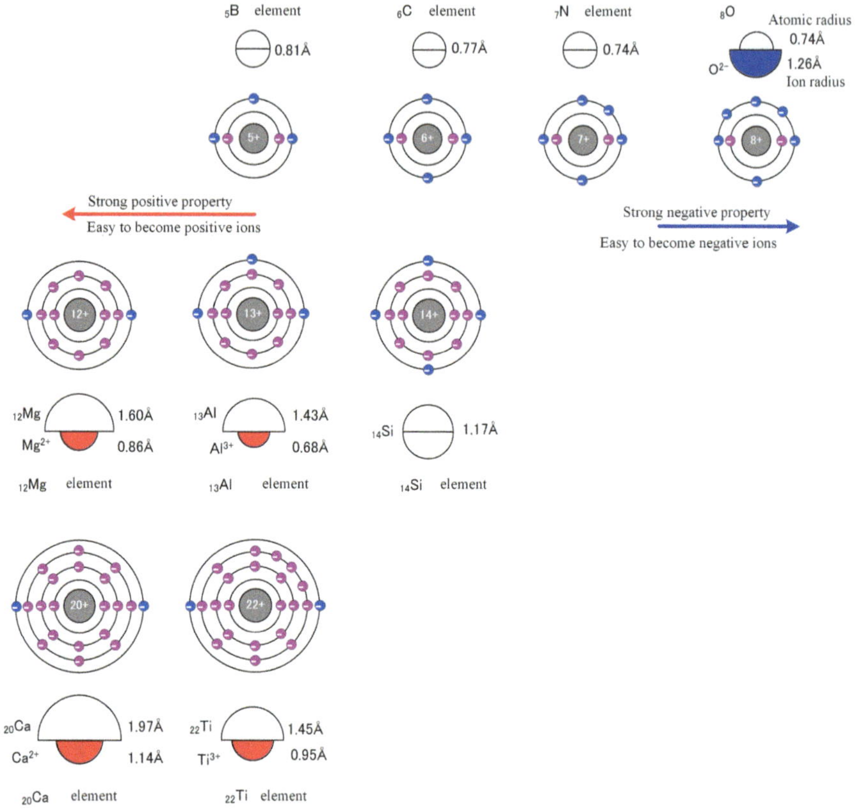

Fig. 5.15 Analysis on the atoms of the materials

Comparison of ionization radii of atoms: They are analyzed based on the groups in the Periodic Table of the Elements, including ${}_5$B (boron), ${}_6$C (carbon), ${}_7$ N (nitrogen), ${}_8$O (oxygen), ${}_{12}$ Mg (magnesium), ${}_{13}$Al (aluminum), ${}_{14}$Si (silicon), etc. Inner-shell electrons are displayed in pink, and outer-shell electrons are displayed in blue. The atom radius and ion radius are indicated. Positively ionized atoms are displayed in a red hemisphere, and negatively ionized atoms are displayed in a blue hemisphere.

One important thing in Fig. 5.15 is that atoms with a small mass number in the periodic table easily emit outer-shell electrons and are therefore easily ionized. An atom with a large mass number in the periodic table easily attracts an electron from the outside, and thus is easily ionized negatively. The ionization of the atoms of the inorganic materials in Fig. 5.14 depends on the positions of the atoms in the periodic table.

(4) **Evaluation of inorganic materials by Q(t) method**

Background: The PEA measurement method was developed to measure the charge accumulation distribution $\rho(x, t)$ (C/m^3). Until now, the PEA method has been widely used for organic materials (e.g., LDPE, PET). More recently, it has been applied to the measurement of the charge accumulation of inorganic materials (e.g. Al_2O_3, PZT). However, it has been found that it is difficult to correctly measure the accumulated charge distribution of the inorganic material from the PEA signal. Because the signals of the charge accumulation and the piezoelectric polarization are mixed. Based on this background, studies have begun on whether the Q(t) method can be used to evaluate the charge storage characteristics of inorganic materials.

Features of Q(t) data of inorganic materials: Table 5.5 summarizes the differences in charge accumulation characteristics between organic materials (e.g. LDPE, PET) and inorganic materials (e.g. Al_2O_3). The remarkable difference is that charge accumulation occurs in the inorganic material even when an ultra-low electric field is applied ($Q(t)/Q_0 \geqq 1$). In an organic material, charge accumulation occurs when an ultra-high electric field is applied ($Q(t)/Q_0 \geqq 1$). The feature of the inorganic material is that its relative permittivity ε_r is considerably larger than that of the organic material. Based on this, why the charge accumulation becomes easier when the relative dielectric constant ε_r is large is considered.

Examination of charge increase $\Delta Q(t)$: The charge accumulation in the dielectric means that the carriers are injected from the electrodes and trapped in the bulk. Therefore, in order to consider the results in Table 5.5, the relationship between the electric field strength on the electrode surface and the relative dielectric constant ε_r is considered, and the injection of electrode charges is examined.

Figure 5.16 shows the Q(t) data when a DC voltage V is applied to a dielectric with a high dielectric constant. The charge amount of the current integration is $Q(t) = Q_0 + \Delta Q(t)$, which is the sum of the initial charge amount Q_0 and the increased charge amount $\Delta Q(t)$.

Figure 5.17 shows a model of the increased amount $\Delta Q(t)$. In Fig. 5.17a, immediately after the voltage is applied, the initial charge amount $Q_0 = C_S V = (\varepsilon_0 \varepsilon_r S/a)V = \varepsilon_0 \varepsilon_r ES$. At this time, the electric field intensity on the electrode surface is $E_3 =$

Table 5.5 Characteristics of charge accumulation in inorganic materials

Specimen		Charge source	Charge carrier	Applied electric field		Relative permittivity ε
			Trap level	Low field	High field	
Organic materials	LDPE	Charge injection	Shallow	$Q(t)/Q_0 = 1$	$Q(t)/Q_0 > 3$	2–4
	PET	Charge injection	Deep	$Q(t)/Q_0 = 1$	$Q(t)/Q_0 = 1$	
Inorganic	Al_2O_3	To be studied	To be studied	$Q(t)/Q_0 \geqq 1$	$Q(t)/Q_0 > 2$	8–12–1000

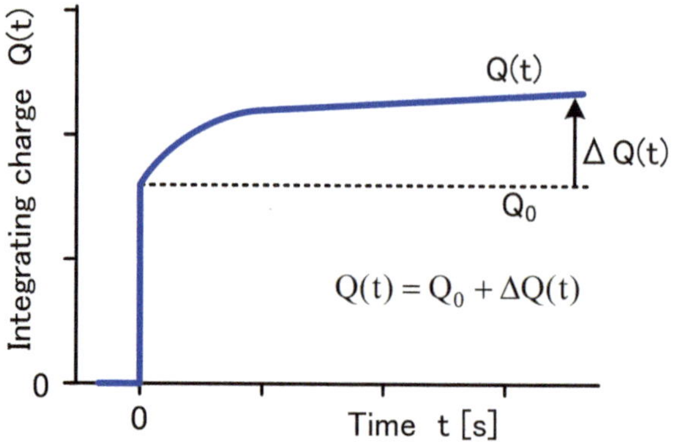

Fig. 5.16 The $\Delta Q(t)$ component

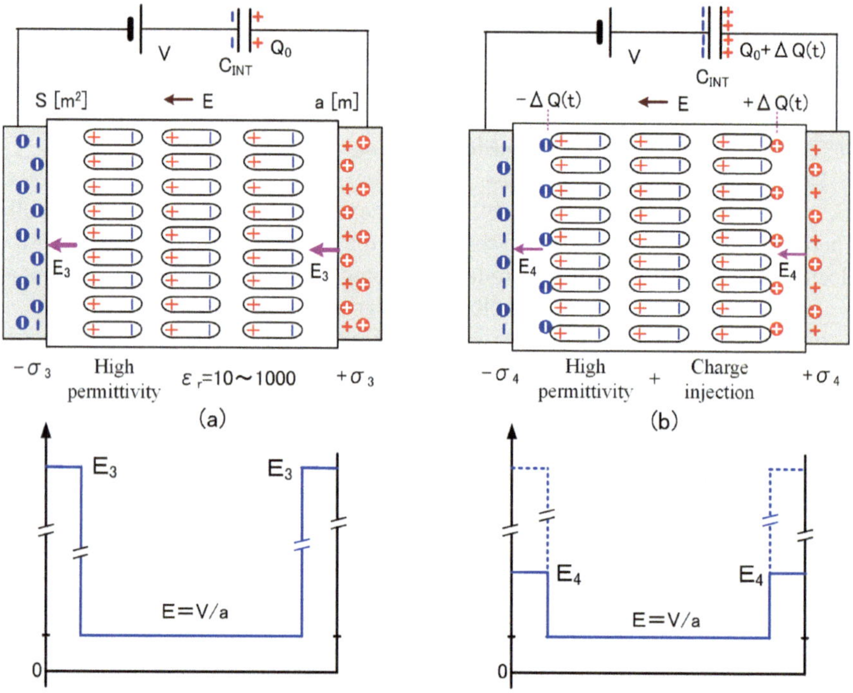

Fig. 5.17 Model analysis of the effect of $\Delta Q(t)$

$\varepsilon_r E$. For example, even when a low electric field of $E = 1.0$ kV/mm is applied to an inorganic material with $\varepsilon_r = 1000$, the electric field intensity on the electrode surface is as high as $E_3 = 1{,}000$ kV/mm. In this ultra-high electric field, a large charge injection from the electrodes occurs. As a result of the charge injection, the charge is supplied from the power supply, and the increased charge $\Delta Q(t)$ is measured. As a result, the electric field intensity on the electrode surface decreases from E_3 to E_4. Therefore, $\Delta Q(t)$ represents an electric field difference of $(E_3 - E_4)$, as shown in Eq. (5.10). Furthermore, it can be transformed to the right part of Eq. (5.10).

$$\Delta Q(t) = \varepsilon_0\ S(E_3 - E_4) = \varepsilon_0\ \varepsilon_r E\left(1 - \frac{E_4}{E_3}\right) S \tag{5.10}$$

$\varepsilon_0 \varepsilon_r E = \sigma_0 = (Q_0/S)$ of the head of the right part in Eq. (5.10) is the initial electrode charge density. The next electric field ratio E_4/E_3 indicates the change rate of the electric field on the electrode surface when the electrode charge moves into the dielectric side due to the ultra-high electric field on the electrode surface. Eventually, the factor causing the ultra-high electric field on the electrode surface is the relative permittivity ε_r in $E_3 = \varepsilon_r E$. Next, the origin of ε_r is described.

Relative permittivity: The relative permittivity ε_r is derived from the Clausious Mosottit's equation, as shown in Eqs. (5.11) and (5.12).

$$\frac{\varepsilon_r - 1}{\varepsilon_r + 2} = \frac{N\alpha}{3\ \varepsilon_0} \tag{5.11}$$

$$\frac{\varepsilon_r - 1}{\varepsilon_r + 2}\frac{M}{\rho} = \frac{N_0\,\alpha}{3\ \varepsilon_0} \equiv P_m\left(m^3\right) \tag{5.12}$$

N_0 Avogadro number
M Molecular weight (kg)
ρ Density (kg/m^3)
α Polarizability
P_m Molecular polarizability.

After all, the relative permittivity ε_r depends on the molecular polarizability P_m (m^3).

(5) Difference in polarization between organic and inorganic materials

Polarization of organic materials: The bonds between atoms are basically covalent bonds of electrons. When an electric field is applied to the organic molecule, the electron wave is shifted in the molecule and polarization occurs. In this case, the spacing between atoms does not basically extend. The polarization due to the offset of the electron wave creates a relative permittivity, and the value is in the range from 2 to 4.

Polarization of inorganic material: Figure 5.18 depicts the polarization of inorganic material (alumina; Al_2O_3). Although the molecular structure of Al_2O_3 is three-dimensional, the molecular structure in Fig. 5.18 is drawn as a two-dimensional structure for explanation.

As described in Sect. 5.6(3), when Al (aluminum) and O (oxygen) approach, aluminum is positively charged with Al^{3+} and its diameter decreases. Conversely, oxygen becomes negatively charged O^{2-} and its diameter increases. Therefore, the

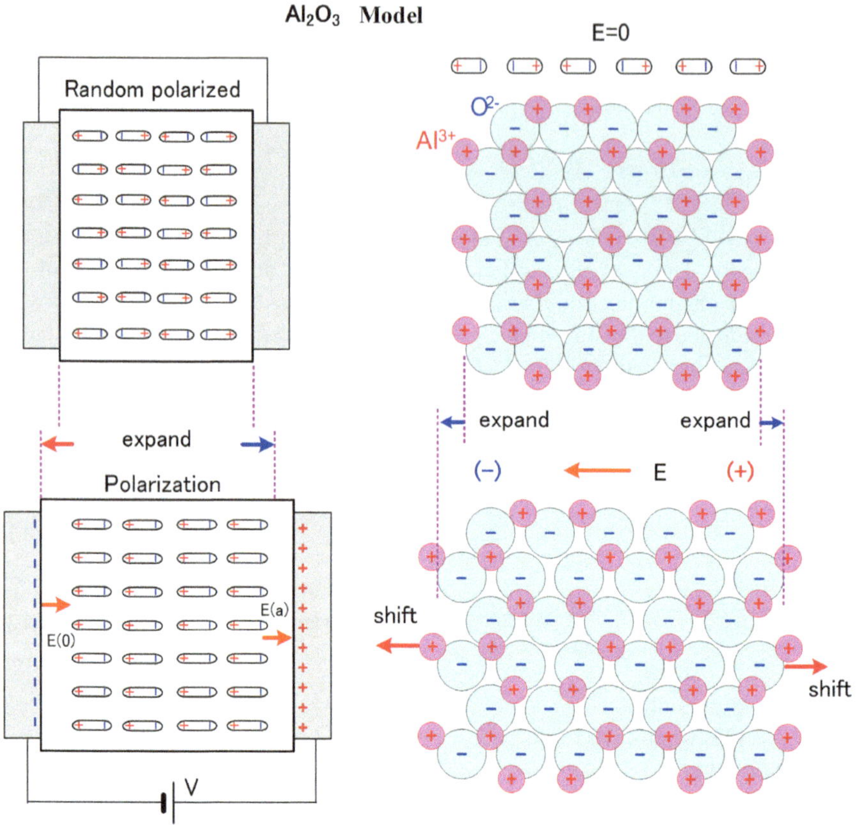

Fig. 5.18 The two-dimensional structure of Al_2O_3

structure is that positively charged aluminum Al^{3+} with a small diameter floats in the oxygen (O^{2-}) lattice under the balance of the Coulomb field.

Under this state, when no electric field is applied ($E = 0$), the dipoles of the positively charged Al^{3+} and the negatively charged O^{2-} cancel each other, and no polarization occurs.

When an electric field is applied, positively charged Al^{3+} shifts to the cathode side, and negatively charged O^{2-} shifts to the anode side, which can cause polarization. Unlike the polarization action of the organic materials in which the electron wave is biased, the polarization of the inorganic material is resulted from a shift between positive and negative charges. Here, the polarization is explained from a microscopic view and a macroscopic view. The relative permittivity of Eq. (5.12) is derived based on the following two polarizations P.

Microscopic view : $P = N\alpha E_i$ E_i is the internal electric field.

Macroscopic view : $P = \varepsilon_0(\varepsilon_r - 1)E$ E is the average field in the dielectric.

$$\frac{\varepsilon_r - 1}{\varepsilon_r + 2} = \frac{N\,\alpha}{3\,\varepsilon_0} \quad \text{Eq. (5.11) is gotten.}$$

Then, you need to have a clear understanding of the model for polarizability α.

Classical theory and quantum theory: I would like to discuss the model of the polarizability α for inorganic materials, but a t present, there isn't any proper models based on the classical theory. Therefore, the studies on the polarizability α due to the electron waves based on the quantum theory have begun. We hope to obtain accurate experimental results of the polarization phenomenon and explain the experimental results using both the classical theory and the quantum theory.

Reference

1. M. Fukuma, Y. Sekiguchi, Messurement of leakage current and capacitance of MLCC by using Q(t) meter, Paper presented at the 2020 Annual meeting of IEEJ, Tokyo, Japan, 1–4 (2020).

Chapter 6
Application to Insulation Diagnosis

6.1 Charge Accumulation Evaluation by Q(t) Method on High Voltage Side

(1) Features of Q(t) measurement on high voltage side

The charge accumulation of DC insulating materials has been evaluated by the PEA method in flat samples (sheet samples). Electrical equipment and power cables are developed and manufactured using DC insulation materials evaluated on flat samples. However, it is extremely difficult to directly evaluate the charge accumulation in the insulation system of manufactured electrical devices and power cables by the PEA method.

In contrast, the Q(t) method can easily evaluate the charge accumulation in the insulating layer of a manufactured electric device. In particular, by installing this Q(t) instrument on the high voltage side, the electric charge accumulation in the insulating material of many electrical devices and power cables can be easily evaluated. Since this charge accumulation distorts the internal electric field distribution and becomes the seeds of insulation deterioration, it is important to evaluate the charge accumulation in the actual insulating layer. Moreover, since the Q(t) measurement on high voltage side is resistant to noise, it is excellent for insulation diagnosis of installed electric equipment on site.

(2) Application of Q(t) system on coaxial cable

Figure 6.1 shows a measurement system in which Q(t) measurement is applied to a coaxial cable from high voltage side. The outer conductor of the coaxial cable is grounded, and a high voltage is applied to the inner conductor via a Q(t) measuring device. The current supplied after the application of the DC high voltage V_{dc} is time-integrated by the integration capacitor C_{INT} to measure Q(t) results. The Q(t) data measured on the high voltage side is transmitted by a radio wave to the PC on the ground side.

T. Takada et al., *Electric Charge Accumulation in Dielectrics: Measurement and Analysis*,
https://doi.org/10.1007/978-981-19-6156-4_6

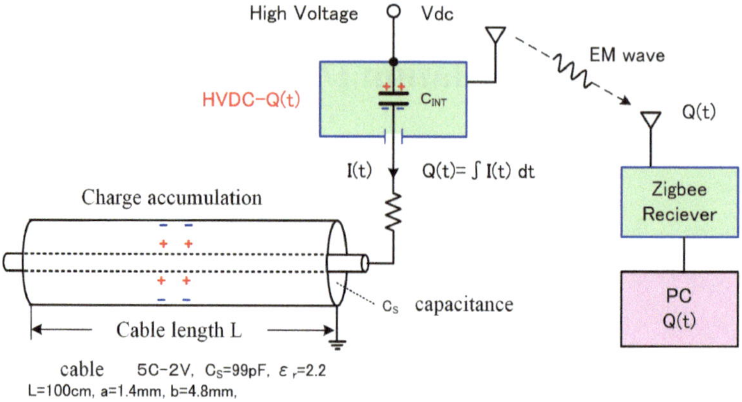

Fig. 6.1 The Q(t) system on the high voltage side

(3) **Method of evaluating charge accumulation**

Q(t) data shows no charge accumulation in Fig. 6.2:

The integrated charge amount immediately after the voltage application is $Q(t = 0) = Q_0$. This charge is the initial charge of $Q_0 = C_s V_{dc}$. Since the capacitance C_s of the coaxial cable is 100 pF and the applied voltage V_{dc} is 2 kV, the initial charge is $Q_0 = 0.200\ \mu$C.

The integrated charge amount after 300 s from the voltage application is $Q(t = 300\text{ s}) = 0.202\ \mu$C. If $Q(t = 300\text{ s}) = Q_0 = 0.200\ \mu$C, the charges stay on the electrode surface and there is no charge accumulation inside the sample. In this case,

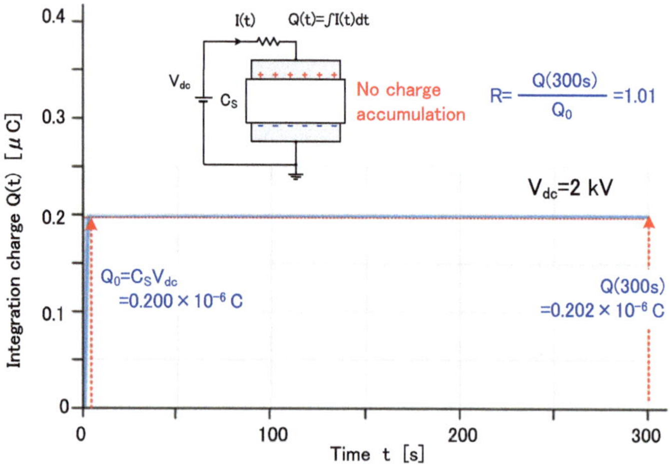

Fig. 6.2 The results under the voltage of 2 kV

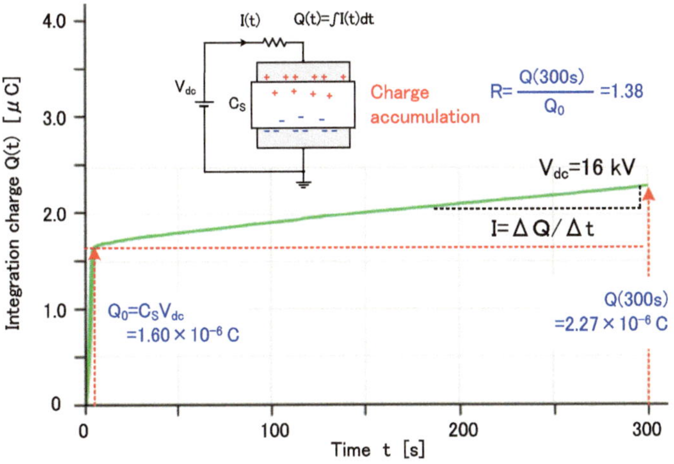

Fig. 6.3 The results under the voltage of 16 kV

the charge amount increases by 1%, and it can be determined that there is almost no charge accumulation.

Q(t) data shows charge accumulation in Fig. 6.3:

A higher voltage of $V_{dc} = 16$ kV is applied. The initial charge is increased to $Q_0 = 1.60$ μC, and the integrated charge after 300 s is increased to $Q(t = 300$ s$) = 2.27$ μC. The increased charge ratio is 1.38 (up 38%).

$$Q(t\ =\ 300\,\mathrm{s})/Q_0 = 2.27/1.60 = 1.38.$$

Since the charge ratio is $Q(t)/Q_0 > 1$, the charge clearly accumulates in the sample. Therefore, it can be judged that there is a space charge accumulation.

(4) **Q(t) results of coaxial cable**

Q(t) data in Fig. 6.4:

Different DC voltages are applied to the coaxial cable 5C-2V (length 1 m, $C_s = 99$ pF), which is in the range of 300 V to 18 kV, and the Q(t) characteristics at each voltage application are measured over 300 s. Up to 14 kV, $Q(t) \fallingdotseq Q_0$ and there is no charge accumulation. Q(t) starts to increase from the voltage of 16 kV, and charge accumulation is observed.

The Q(t) data in Fig. 6.4 is expressed by Eq. (6.1).

$$\text{Integration charge } Q(t) = Q_0 + \Delta Q(t) \tag{6.1}$$

$$\text{Charge ratio } \frac{Q(t)}{Q_0} = 1 + \frac{\Delta Q(t)}{Q_0} \tag{6.2}$$

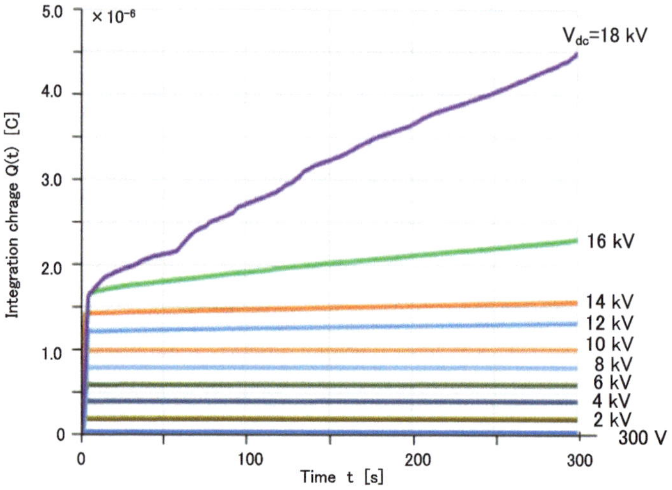

Fig. 6.4 The results under the different voltages

The initial charge amount is $Q_0 = C_s V_{dc}$, which is the product of the capacitance and the voltage. The increased charge amount $\Delta Q(t)$ is the sum of the accumulated charge amount (electrode charge injection or charge generated from impurities) and the leakage charge amount.

Observation of $\Delta Q(t)$ in Fig. 6.5:

The data in Fig. 6.4 is further processed. By setting the applied voltage V_{dc} on the horizontal axis and Q(t) data on the vertical axis, Fig. 6.5 is given. $Q(t) = Q_0$ holds until the applied voltage reaches 10 kV and there is no charge accumulation. From V_{dc} = 12 kV, the Q(t) data deviates from the straight line, and an increased amount $\Delta Q(t)$ appears. That is, charge accumulation is observed.

Observation of charge increase ratio $Q(t)/Q_0$ in Fig. 6.6:

Deriving the charge ratio $Q(t)/Q_0$ from Eq. (6.1) gives Eq. (6.2). The data in Fig. 6.4 is rearranged by plotting the charge ratio $Q(t)/Q_0$ on the vertical axis and the applied voltage V_{dc} on the horizontal axis in Fig. 6.6. Up to V_{dc} = 10 kV, the charge ratio holds $Q(t)/Q_0 = 1$, and there is no charge accumulation. $Q(t)/Q_0 > 1$ starts from V_{dc} = 12 kV, and charge accumulation is recognized.

Furthermore, when a high voltage of 18 kV is applied, the ratio becomes $Q(t)/Q_0 > 2$. The leakage current becomes dominant, and the insulating property is lost.

Summary

The Q(t) system installed on the high voltage side has the characteristic that the charge accumulation and the insulation deterioration of the insulating material with any shapes and the insulating layer in the finished electric product can be evaluated. Some of its applications are introduced in the following sections.

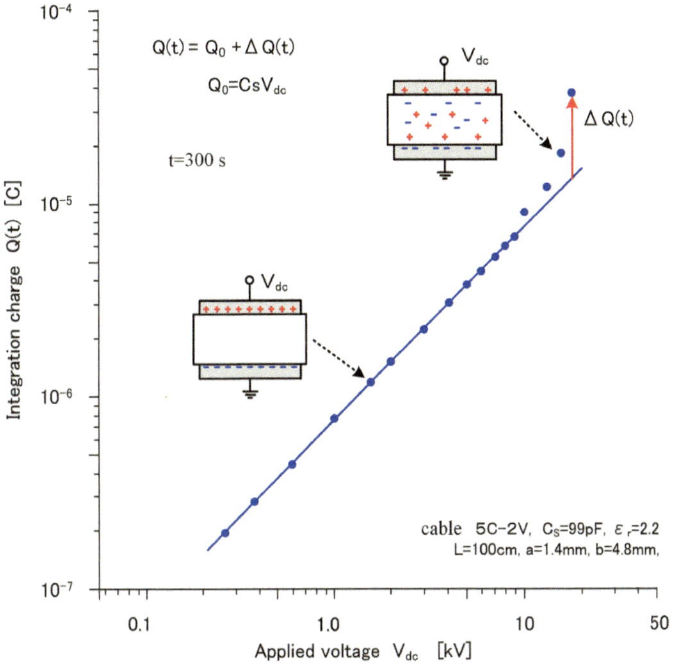

Fig. 6.5 The change of the Q(t) results

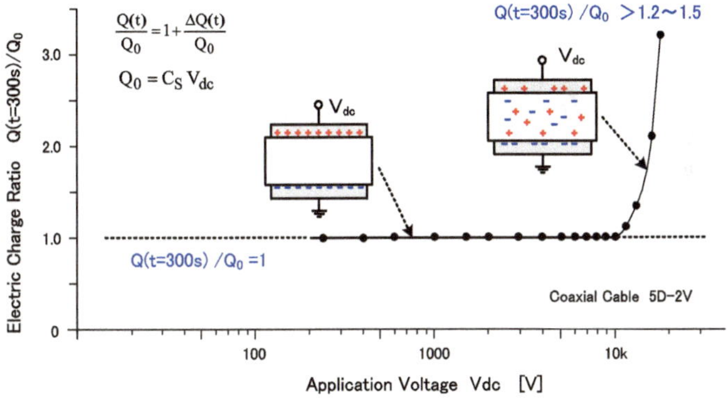

Fig. 6.6 The charge ratio results

6.2 Coaxial Cable with Temperature Rise

The first example of the application of Q(t) meter on the high voltage side is introduced [1].

(1) **Temperature rise of coaxial cable**

As shown in Fig. 6.7a, the inner conductor is heated based on the Joule heating by applying a load current with a heating current transformer. The temperature of the insulating material is increased for evaluating the charge accumulation.

This current transformer uses a disassembled Slidac. The winding wire of Slidac is wrapped on the magnetic core with $N_1 = 500$ turns, and the inner conductor of the closed-circuit cable is passing through the magnetic core to form the secondary winding ($N_2 = 1$ turn). When the current of $I_1 = 1$ A passes through the Slidac coil, the current of the one-turn coaxial cable conductor as the secondary winding is $I_2 = (N_1/N_2) \times I_1 = 500$ A. In other words, a current of 500 times can flow through the inner conductor wire. The inner conductor wire can be heated by Joule heating to 100 °C or more. Thus, the temperature of the insulating material can be increased.

(2) **Q(t) measurement data**

The Q(t) data at each temperature for inner conductor is shown.

(a) **Temperature 25 °C ($I_{\mathbf{load}}$ = 0 A) Fig. 6.7b:**

A DC voltage of 300 V to 10 kV is applied. Even at 10 kV, the $Q(t = 300\ \text{s})$ value does not change from the initial value Q_0. $Q(t = 300\ \text{s}) = Q_0$ is maintained, and it can be judged that there is no charge accumulation.

(b) **Temperature 57 °C ($I_{\mathbf{load}}$ = 57 A) Fig. 6.7c:**

Focus on the Q(t) data at a voltage of 5 kV. The $Q(t = 300\ \text{s})$ value clearly increases from the initial value Q_0. That is, it can be judged that $Q(t) > Q_0$ due to the temperature rise. Therefore, the space charge accumulation has occurred.

(c) **Temperature 75 °C ($I_{\mathbf{load}}$ = 48 A) Fig. 6.7d:**

When the voltage is increased, the Q(t) data increases from Q_0, and it increases even when a low voltage of 1000 V is applied. Eventually, dielectric breakdown has occurred.

(3) **Q(t) measurement data**

The dependence of the Q(t) data on the applied voltage and temperature is evaluated using the charge ratio $Q(t)/Q_0$. In Fig. 6.8, the horizontal axis represents the applied voltage V_{dc}, the vertical axis represents the charge ratio $Q(t)/Q_0$, and the labeled parameter represents the load current.

(1) **($I_{\mathbf{load}}$ = 0 A; Temp = 25 °C)** Up to 10 kV, charge accumulation is not observed with a charge ratio of $Q(t)/Q_0 = 1$. Above 12 kV, the current increases and breakdown occurs at 14 kV.

Fig. 6.7 The Q(t) results of the cable

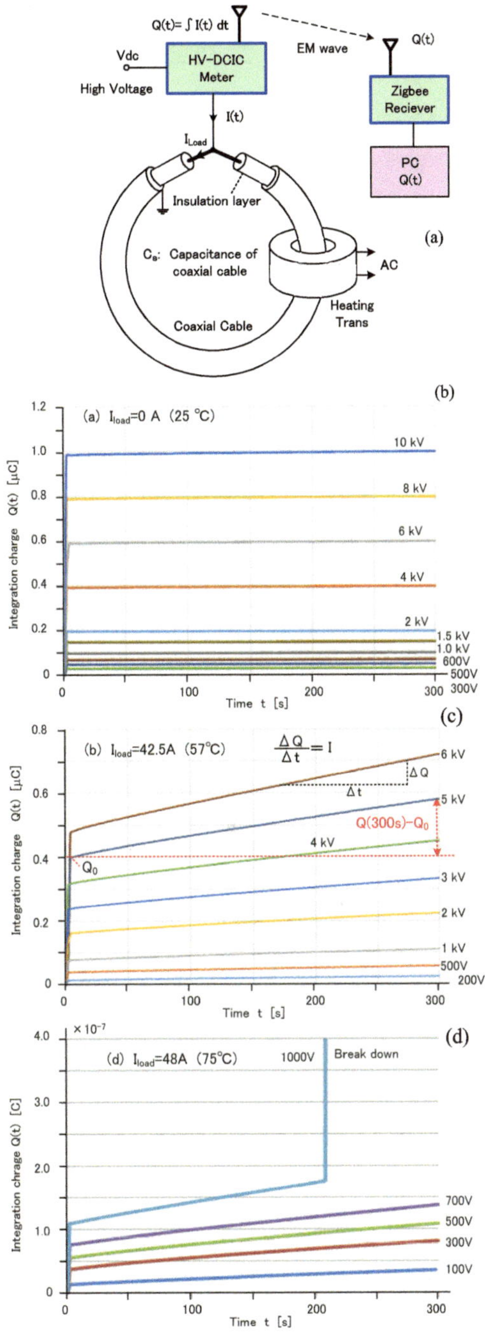

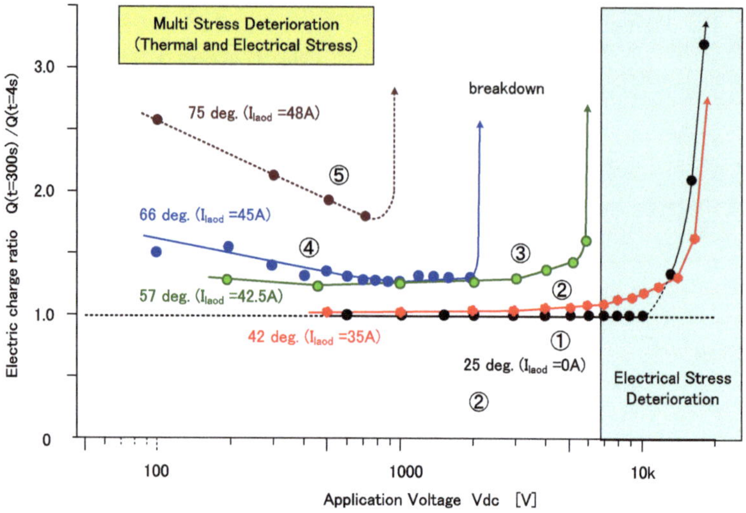

Fig. 6.8 The charge ratio results under different temperatures

(2) **(I_{load} = 35 A; Temp = 45 °C)** $Q(t)/Q_0 > 1$ above 5 kV. The charge accumulation becomes larger than (1), and electrical breakdown occurs at 12 kV.

(3) **(I_{load} = 42.5 A; Temp = 57 °C)**, (4) **(I_{load} = 45 A; Temp = 60 °C)**, (5) **(I_{load} = 48 A; Temp = 75 °C)**.
Even when a low voltage is applied, $Q(t)/Q_0 > 1$ shows charge accumulation, and breakdown occurs at 6 kV, 2 kV, and 0.8 kV, respectively.

Thus, the Q(t) measurement can easily evaluate the temperature dependence of the charge accumulation of the coaxial cable.

(4) Current $\Delta Q(t)/\Delta t$

The Q(t) data is the result of the current integration with time. Conversely, a current value can be obtained by differentiating Q(t) results with time. After calculating the current value from Q(t) data that has almost reached equilibrium, Fig. 6.9 shows the characteristics of the current value I with the applied voltage V_{dc}.

(5) Charge accumulation cannot be evaluated by current measurement

Is it possible to evaluate charge accumulation from this V_{dc}-I characteristic? It is impossible to evaluate at all whether the charge is accumulated in the insulating layer just by looking at this characteristic. On the other hand, the charge accumulation can be easily evaluated from the charge ratio $Q(t)/Q_0$ ratio in Fig. 6.8.

If the charge ratio $Q(t = 300\text{ s})/Q_0 = 1$, there is no charge accumulation.

If the charge ratio $Q(t = 300\text{ s})/Q_0 > 1$, there is charge accumulation.

The initial charge amount Q_0 is an electrode charge, and $Q_0 = C_s V_{dc}$.

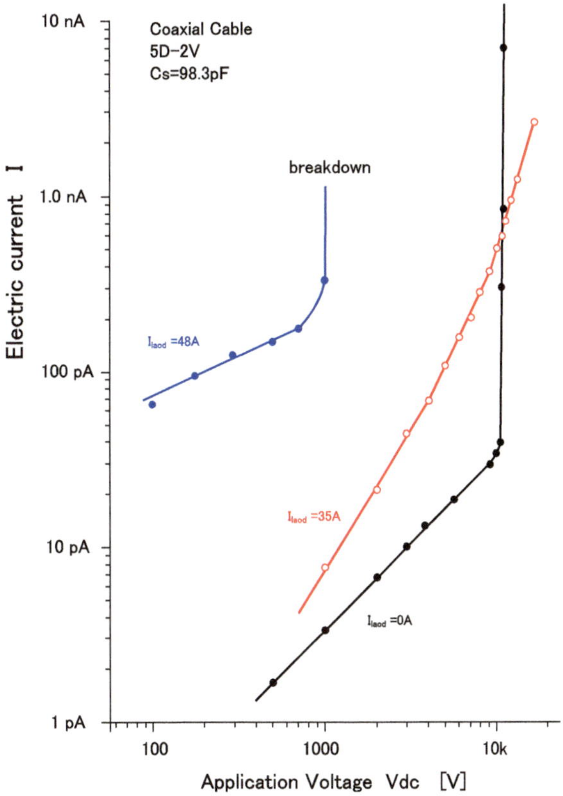

Fig. 6.9 The calculated current results

The fact that $Q(t)/Q_0 > 1$ means that more charge has been supplied than the induced charge amount, and the increment is the charge accumulated in the insulation. However, when $Q(t)/Q_0 \geqq 2$, leakage charge is included in addition to the accumulated charge. See Sect. 3.1 for details.

6.3 Charge Accumulation Characteristics of DC-XLPE Cable

(1) DC-XLPE cable

The charge accumulation characteristics of the DC-XLPE cable using the developed DC insulation materials are evaluated by the Q(t) method in a wide temperature range (RT to 90 °C).

DC-XLPE cable: The cross section of inner conductor is 2 mm^2. The thickness of insulation layer is 1.0 mm. The effective length is 5 m, and the applied DC voltage is 20 kV.

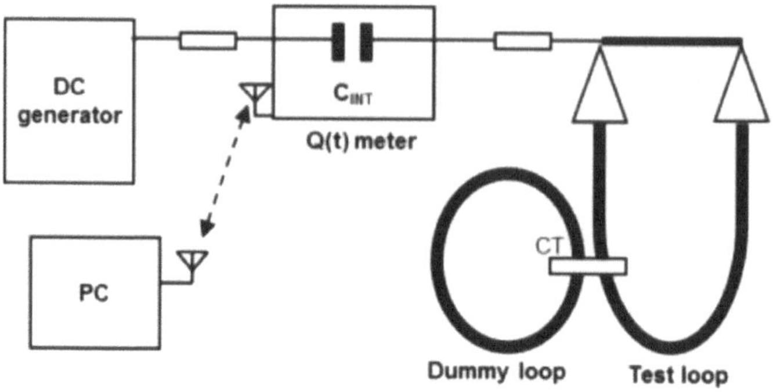

Fig. 6.10 The system for XLPE cable

Measurement system (Fig. 6.10): The test loop is the measured cable with a closed circuit in which the two ends of the DC-XLPE cable connects. The dummy loop is a similar DC-XLPE cable used for monitoring temperature. CT is a current transformer, which applies a large current to the conductors of the test loop and dummy loop to heat them.

Q(t) meter: Insert a Q(t) meter between the high voltage power supply (DC Generator) and the conductor in the measured DC-XLPE cable, and integrate the total current with time. This Q(t) data is transmitted wirelessly to the PC on the ground side [2].

(2) **Charge accumulation characteristics** (Fig. 6.11)

(a) **AC-XLPE:** There is no charge accumulation at room temperature due to $Q(t = 300\text{ s}) = Q_0$. When the temperature rises to 50 °C, $Q(t = 300\text{ s}) > Q_0$, and it can be judged that there is charge accumulation. At high temperatures of 70 and 90 °C, $Q(t = 300\text{ s}) >> Q_0$, and the leakage current becomes dominant in addition to charge accumulation.
(b) **DC-XLPE:** $Q(t = 300\text{ s}) = Q_0$ even at a high temperature of 70 and 90 °C, and it can be judged that there is no charge accumulation. The developed DC-XLPE can be thought as maintaining excellent insulation properties in a high temperature range (RT to 90 °C).

As described above, it can be said that the Q(t) measurement system can clearly evaluate the charge accumulation characteristics of the cable product.

6.4 Insulation Diagnosis After Gamma Irradiation

(1) Necessity of insulation diagnosis on electric wires in atomic reactors

A large number of electric wires are used in the power atomic reactors to supply and control power when transmitting measurement signals. These wires are exposed to the

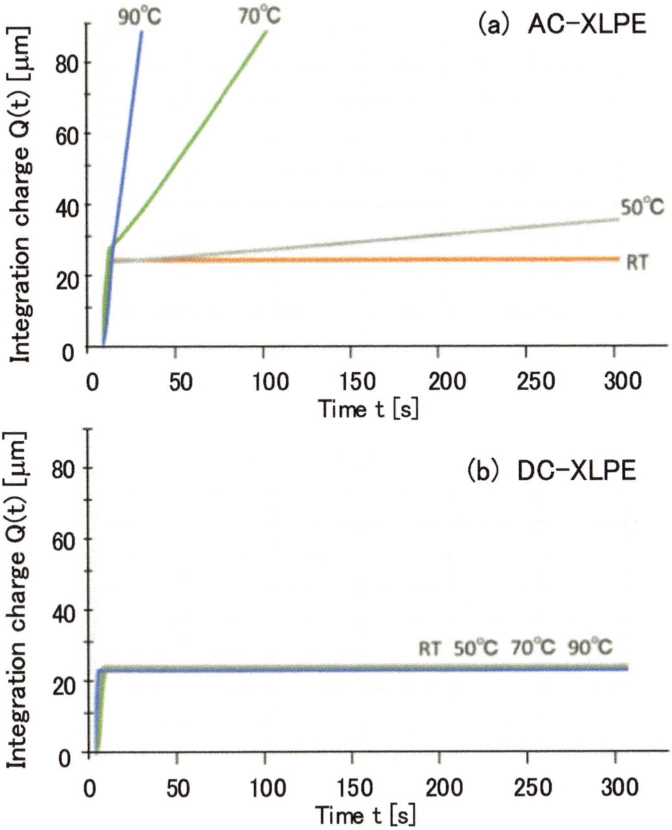

Fig. 6.11 The Q(t) results of the two cables

high-energy radiation from the nuclear reactor, and the insulation easily deteriorates. Its degradation has been diagnosed by dielectric loss (tanδ).

When high-energy radiation is applied to the insulating layer of the electric cable, the material molecules are ionized to generate electron–hole pairs. When an electric field is applied, electrons and holes move in the opposite directions and form the space charge. It is considered that the generation of this space charge is accompanied with an increase in the local electric field and the break of molecular bonds, thereby promoting insulation deterioration.

Therefore, it has been proposed to perform insulation diagnosis by evaluating the space charge characteristics based on Q(t) measurement [3].

(2) Q(t) measurement system for gamma-irradiated coaxial cable

The charge characteristics of the coaxial cable irradiated with ^{60}Co γ-rays are evaluated by the Q(t) measurement system shown in Fig. 6.12, and insulation diagnosis is performed. The outer conductor of the coaxial cable is grounded, and a high DC

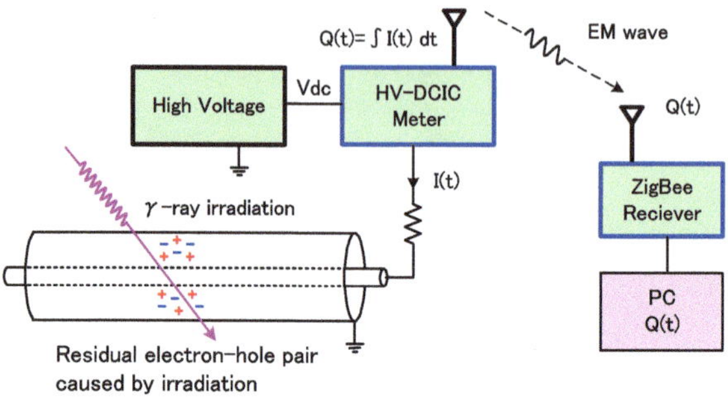

Fig. 6.12 The system for the cable after irradiation

voltage is applied to the inner conductor through a Q(t) meter. The outer conductor of the coaxial cable is laid in the reactor, and it is difficult to remove it from the ground. Therefore, the Q(t) meter is connected to the inner conductor from the high voltage side.

(3) **Q(t) measurement data** (Fig. 6.13)

Before γ-ray irradiation (Fig. 6.13a): A high DC voltage of 300 V to 10 kV is applied to the coaxial cable. Even at 10 kV, the $Q(t = 300\ \text{s})$ value does not change compared to the initial value Q_0. It can be determined that there is no charge accumulation.

After γ-ray irradiation (Fig. 6.13b–d): Focus on the Q(t) data with an applied voltage of 10 kV. The $Q(t = 300\ \text{s})$ value is clearly larger than the initial value Q_0. In other words, electron–hole pairs are generated by γ-ray irradiation, and electrons and holes move in the opposite directions by the application of an electric field, so that the space charge accumulation occurs.

6.5 Diagnosis of CV Cable After Accelerated Water-Tree Deterioration

(1) **Q(t) measurement system** (Figs. 6.14 and 6.15)

An example of applying the Q(t) method to insulation diagnosis of deteriorated CV cable with water tree is introduced [4].

Cable samples: Cable A: New, not deteriorated. Cable B: medium deteriorated. Cable C: severely deteriorated. Since each cable has a different length, each capacitance is different. The values of the respective capacitances are shown in the figure.

Q(t) measurement system: Ground the outer conductor of the cable. Install Q(t) system on the high voltage side of the inner conductor, and the system is connected to the high voltage power supply. The measured value of the integrated charge amount Q(t) is transmitted to a receiver (Zigbee) at the ground level by radio waves (EM wave). This Q(t) value is recorded on a PC and displayed as a graph. The voltage application time is 10 mins.

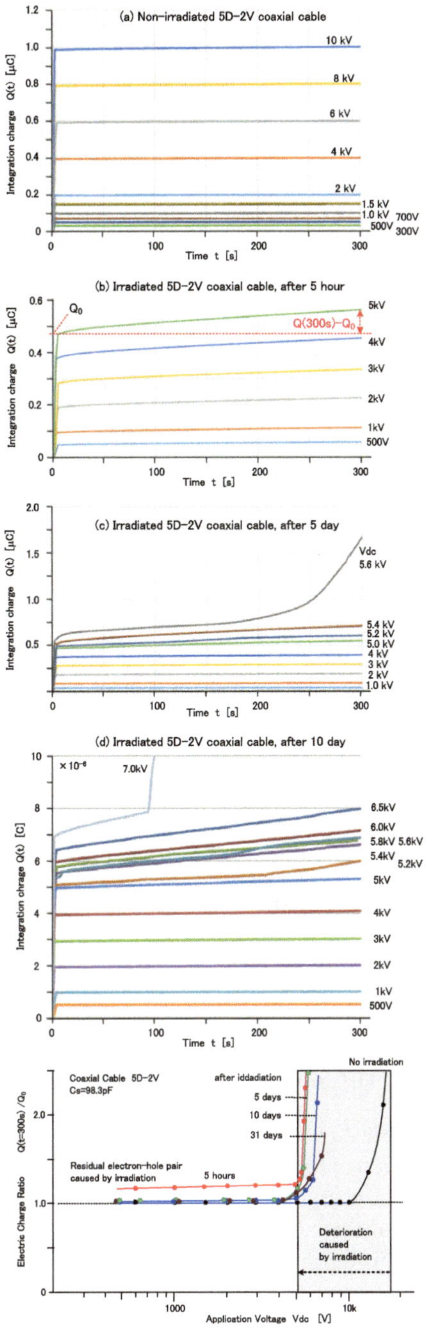

Fig. 6.13 The results under different irradiations

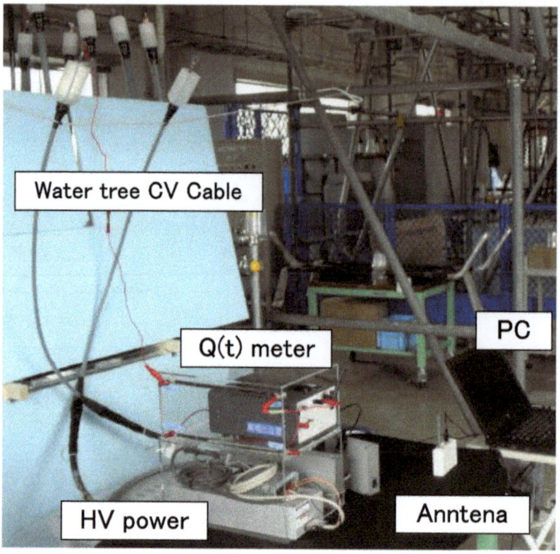

Fig. 6.14 The system for the actual cable

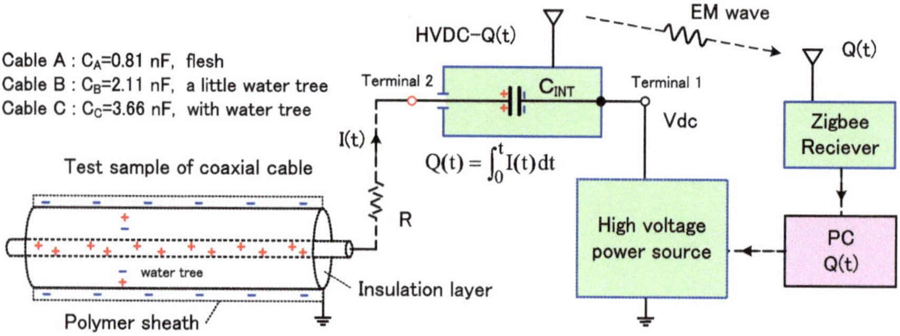

Fig. 6.15 The Q(t) system for the cable with water tree

(2) **Q(t) measurement data** (Fig. 6.16)

Cable A: Q(t) data at the applied voltages of $V_{dc} = 2.5, 5.0, 7.5, 10$ kV is shown on the left of the figure. The Q(t) value is the same as the initial one Q_0. The dependence of the charge ratio $Q(t)/Q_0$ at the applied voltages is shown on the right of the figure. It indicates $Q(t)/Q_0 = 1$ up to $V_{dc} = 10$ kV. Since there is no charge accumulation in this result, it can be determined that no water tree occurs.

Cable B: Q(t) data at the applied voltages of $V_{dc} = 10, 15, 20, 25$ kV is shown on the left of the figure. The charge ratio shows $Q(t)/Q_0$ is close to 1 up to $V_{dc} =$ 10 kV, but $Q(t)/Q_0 > 1$ at the voltage of 15 kV. That is, some charge accumulation is observed. The cable can be determined to be moderately degraded with no water tree development.

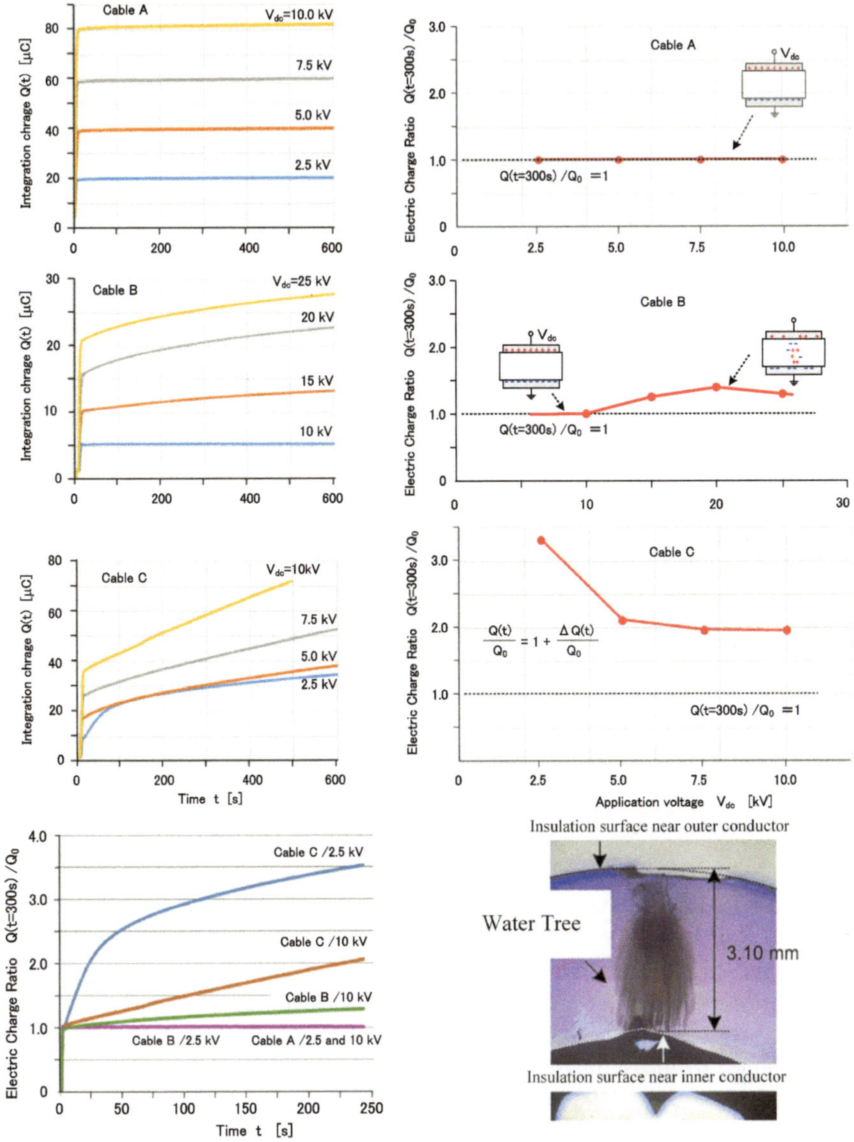

Fig. 6.16 The Q(t) results of the cable with water tree

Cable C: Q(t) data at the applied voltages of $V_{dc} = 2.5, 5.0, 7.5, 10$ kV is shown on the left of figure. At a low electric field of $V_{dc} = 2.5$ kV, $Q(t)/Q_0 > 1$, and charge accumulation and a large amount of leakage current are observed. It can be estimated that water trees have occurred and severe deterioration has progressed.

The results show some questions. At the first voltage of $V_{dc} = 2.5$ kV, $Q(t)/Q_0 = 3.3$ is large. However, when the applied voltage becomes larger, such as $V_{dc} = 5.0$,

7.5 kV, the charge ratio shows smaller with the value of 2.1 and 2.0. Is the measuring instrument defective? Such phenomenon indicates the charge is easily accumulated at low applied voltages. This will be discussed.

When a low voltage of $V_{dc} = 2.5$ kV is applied to Cable C, a large amount of charge is accumulated in the insulating layer. After 600 s, the voltage is turned off and the closed circuit is maintained for 360 s. During this time, a certain amount of electric charge remains in the insulating layer. In this state, it is considered that even if the next voltage of $V_{dc} = 5.0$ kV is applied, the movement of the electric charge is hindered by the influence of the residual accumulation. This is called a pre-effect.

(3) **Deterioration diagnosis** (Fig. 6.16)

Observation of traces of water tree: An observation photograph of water tree traces in the slice after diagnosing the severely degraded Cable C is also shown in Fig. 6.16. Apparently, a water tree is bridged in the insulating layer. Therefore, it can be said that the Q(t) measurement performs a correct deterioration diagnosis and is an easy diagnosis method.

Examination of application level of DC voltage: The sample is a CV cable of type AC 6.6 kV. The DC test voltage of 2.5 kV is about 27% of the AC peak value ($\sqrt{2} \times$ AC 6.6 kV = 9.2 kV). In this way, insulation diagnosis can be performed under the condition that the test voltage is 27% of working stress.

6.6 Q(t) Measurement is Strongly Against External Noise

(1) **Conventional insulation diagnosis method and fight against noise**

At present (2020), the DC voltage is applied to a power cable to perform the insulation diagnosis based on the magnitude of leakage current.

Figure 6.17 shows the measurement circuit. The DC voltage is obtained by the rectifier and capacitor on the secondary side of the high voltage transformer. The high voltage terminal is connected to the inner conductor of the cable sample. The low voltage terminal is grounded via a resistor ($R = 1$ kΩ). The measurement current I_m is obtained by measuring the voltage V between the two sides of the resistor (R) and calculated by $I_m = V/R$. Since a resistance of R_{leak} (about 1 MΩ) exists between the primary winding and the secondary winding of the transformer, the measurement resistance is selected as $R = 1$ kΩ. Since the measurement voltage is limited to several volts (up to 0.1 mV), the measurement current is about 0.1 μA.

Since the noise voltage at the site of the insulation is about 0.1 mV, the measurement current is actually buried in the noise, as shown in Fig. 6.18.

(2) **Q(t) measurement is strong against noise**

The Q(t) meter is installed on the high voltage side, as shown in Fig. 6.19. The Q(t) data of Cable A (fresh), Cable B (medium degraded), and Cable C (severely degraded) is obtained. No noise is observed in the data (Fig. 6.20, the upper one).

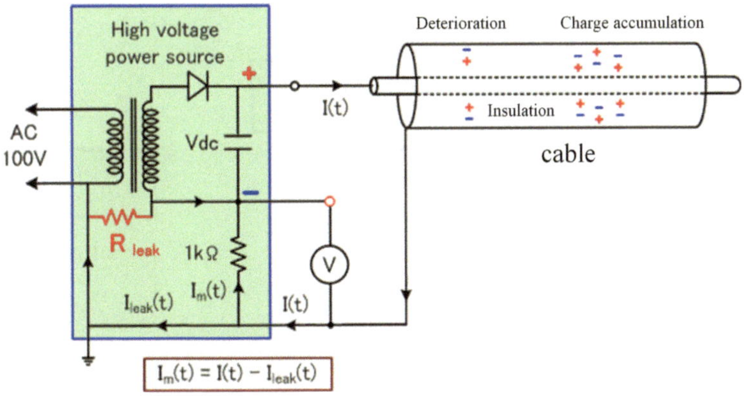

Fig. 6.17 The current measurement system

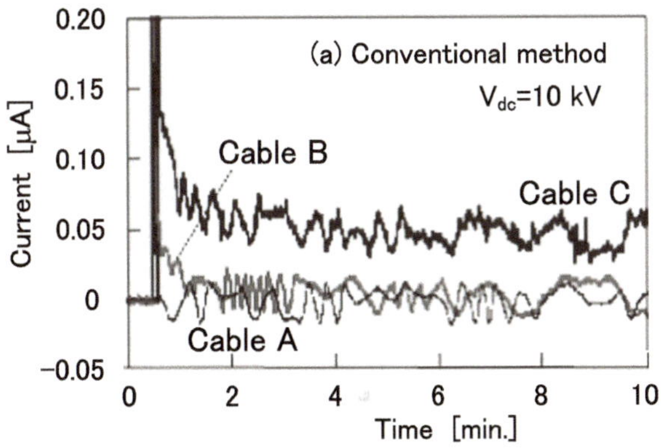

Fig. 6.18 The measured current results

The leakage current is obtained from the time derivative of I_m $(= I_{leak}) = \Delta Q(t)/\Delta t$. The results are also shown in Fig. 6.20.

The noise level of the Q(t) method is reduced as compared with that of the conventional method (Fig. 6.18). It can be seen that the noise level of Q(t) measurement on the high voltage side is low.

(3) How to pick up noise

Our surrounding on earth is filled with the electromagnetic noise. Electrical partial discharge of electric power distribution lines, communication radio waves, broadcast radio waves, and their likes become electromagnetic noise affecting the precise electric measurement equipment.

Measurement of external electromagnetic noise (Fig. 6.21): When a metal plate with an area of 1 m^2 is grounded with a resistance of $R = 1$ MΩ and the noise voltage is measured, the result shows RI noise. Its value is from 1 to 100 mV. The metal plate functions as an antenna.

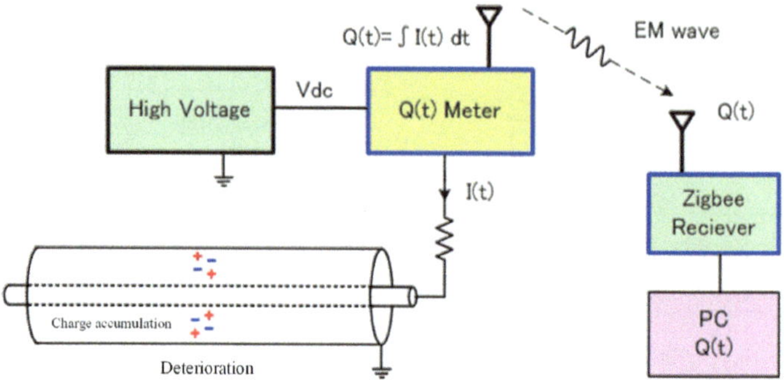

Fig. 6.19 The Q(t) measurement system

Fig. 6.20 The measured and calculated results

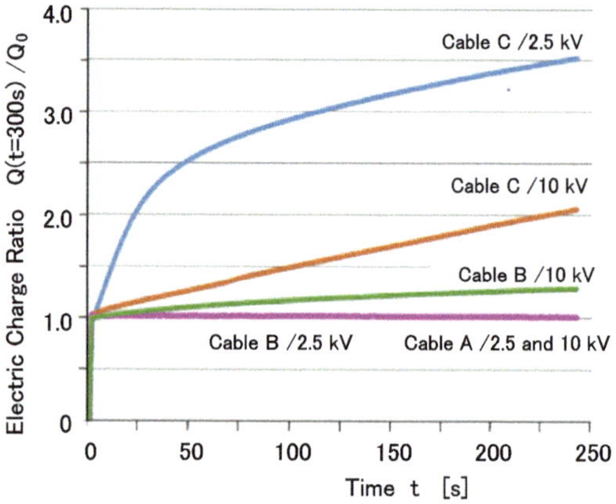

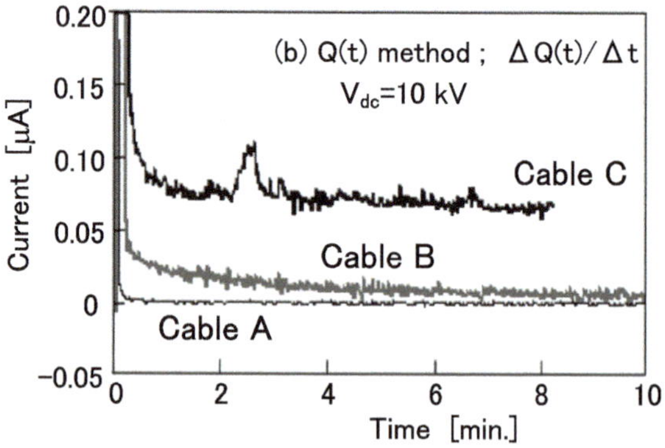

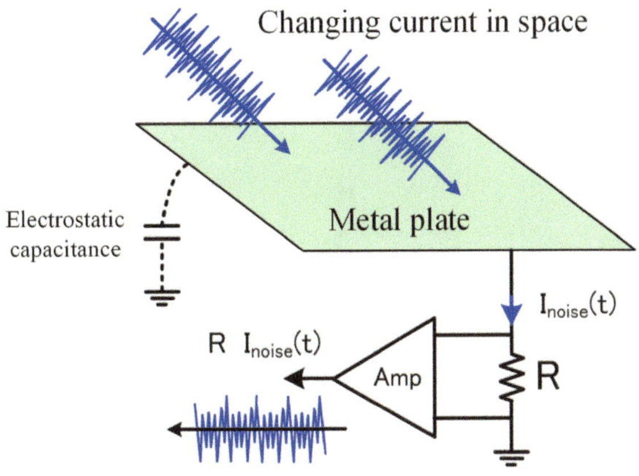

Fig. 6.21 The noise received by a metal plate

Conventional leakage current measurement (Fig. 6.22): An ammeter is connected to the cable's jacket conductor and ground earth. Since the jacket conductor plays a role of an antenna, a current ($I_{\mathrm{leak}} + I_{\mathrm{noise}}$) combining the leakage current (I_{leak}) with the noise current (I_{noise}) is measured. In Fig. 6.23, the measured current of undegraded Cable A is buried in the noise current I_{noise} due to $I_{\mathrm{leak}} << I_{\mathrm{noise}}$, and thus the leakage current (I_{leak}) cannot be evaluated.

Leakage current measurement on high voltage side (Fig. 6.24): Since the inner conductor of the cable is electrically shielded by the outer conductor, no noise current is induced in the inner conductor. Therefore, the ammeter does not measure the noise.

Just replace the ammeter in Fig. 6.22 with a Q(t) meter. The Q(t) data in Fig. 6.24 does not include noise. Since the charge ratio $Q(t)/Q_0 = 1$, it is determined that there is no charge accumulation.

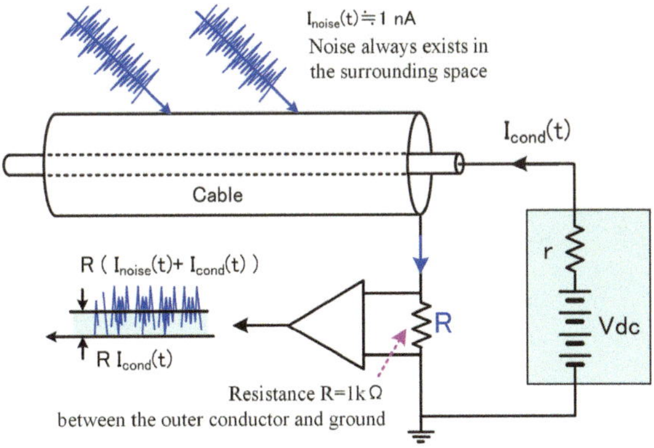

Fig. 6.22 The noise received by the cable

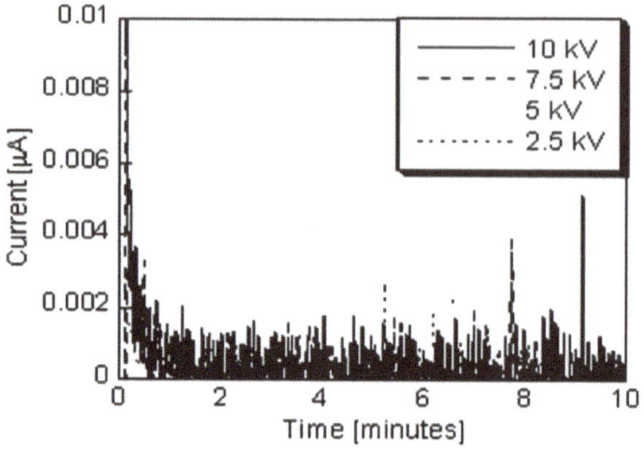

Fig. 6.23 The measured noise current

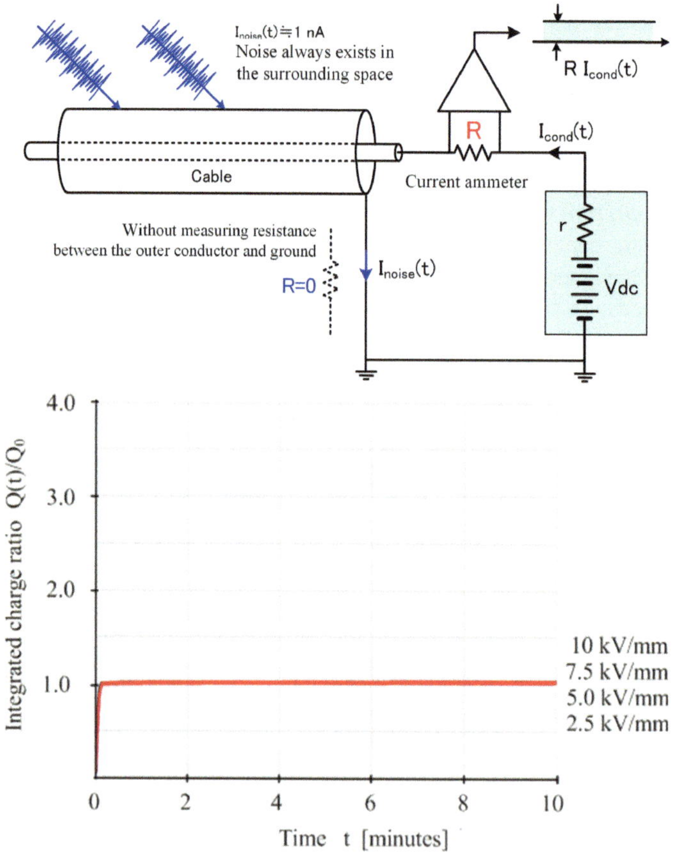

Fig. 6.24 The measurement system against noise

6.7 Double-Layer Dielectric Interface

(1) Double-layer dielectric interface in cable connection

Multilayer insulation is often used for the electric equipment and electronic devices. When a DC voltage is applied, charges are accumulated at the interface between the layers, and there is a concern that the charges may distort the designed electric field. There has been a desire to measure the charge accumulation at this interface. Since a double-layer dielectric interface exists at the connection point of the power cable (Fig. 6.25), the Q(t) measurement of charge accumulation at the LDPE/EPDM sample interface is performed [4].

(2) Q(t) measurement at the double-layer dielectric interface

The direct Q(t) measurement on the power cable connection has not yet been realized. Therefore, Q(t) measurement is performed on a double-layer dielectric sample in which two sheet samples are stacked. Figure 6.26 shows the Q(t) measurement system for the dielectric interface.

(3) Q(t) measurement results (Fig. 6.27)

(a) Q(t) data of a single-layer LDPE sample with a thickness of 140 μm. Up to 20 kV/mm, the Q(t) data does not increase. However, at 40 kV/mm, it sharply increases.
(b) Q(t) data of a single-layer EPDM sample with a thickness of 140 μm. The Q(t) data slightly increases for a while after the voltage application, but it does not increase rapidly even at the electric field of 60 kV/mm.
(c) Q(t) data of a double-layer EPDM/LDPE sample. It shows an intermediate Q(t) characteristic compared with that of two single-layer samples.
(d) The figure shows the electric field dependence of the charge ratio $Q(t)/Q_0$ of these samples. The EPDM/LDPE sample has a charge ratio close to that of the single-layer EPDM sample, which is in the region of $1 < Q(t)/Q_0 < 2$.

This indicates that the charge is accumulated in the EPDM/LDPE sample, but there is no significant migration thereafter.

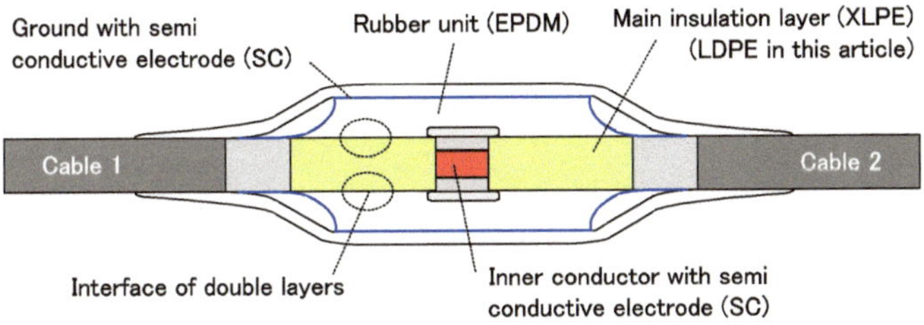

Fig. 6.25 The double-layer part in the cable

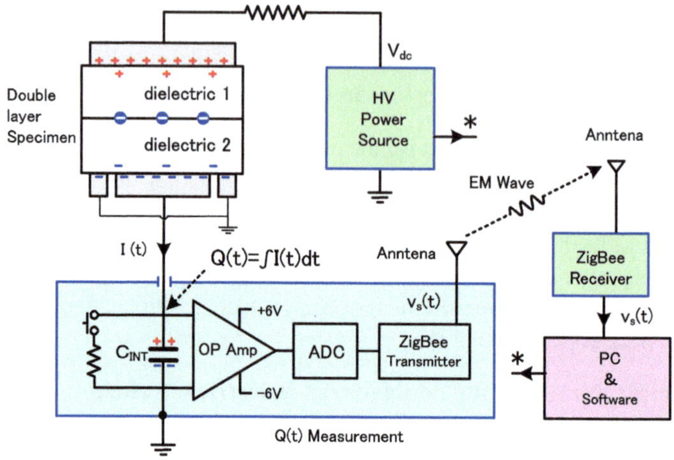

Fig. 6.26 The Q(t) system for the double-layer sample

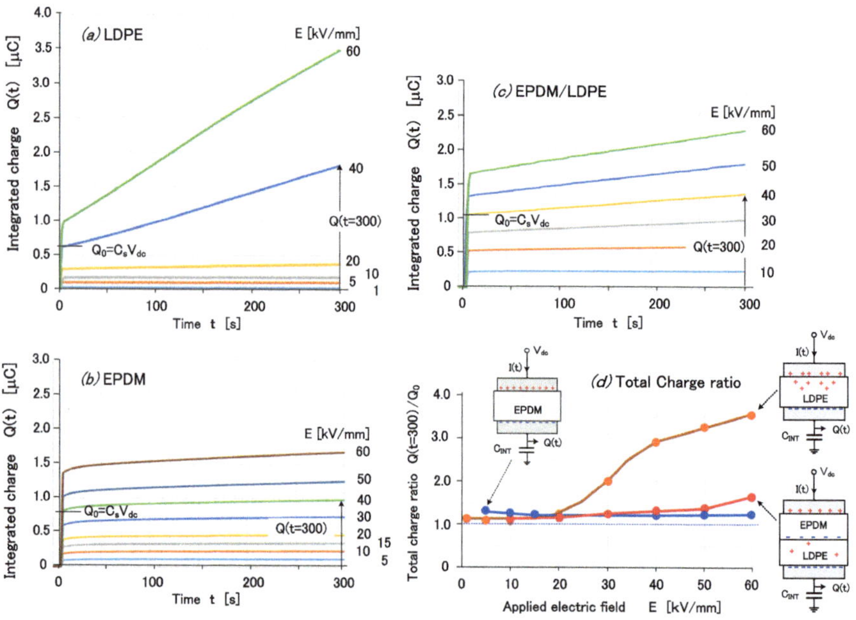

Fig. 6.27 The measured results of the samples

(4) Joint application of Q(t) measurement and PEA measurement

An accurate evaluation of charge accumulation is difficult to be achieved by the Q(t) characteristic alone. Therefore, we try to use the Q(t) and PEA measurements together.

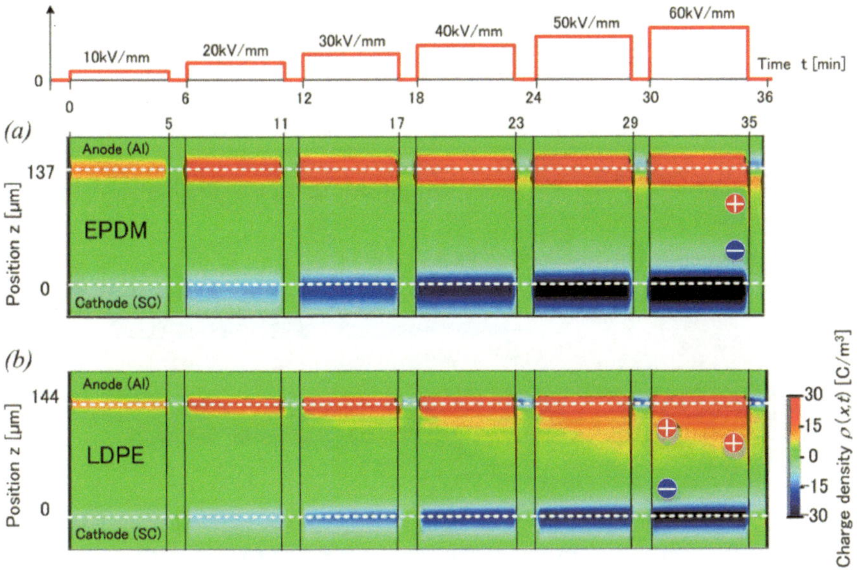

Fig. 6.28 The PEA results of the single-layer samples

First, Fig. 6.28 shows the PEA data of single-layer EPDM and LDPE samples. As with the Q(t) measurement, there is charge accumulation in LDPE, but no charge accumulation shows in EPDM. Both methods show the same results.

Figure 6.29 shows the PEA data of the double-layer samples. When a positive voltage is applied, negative charges are accumulated at the interface. Positive charges are accumulated at the interface when a negative voltage is applied. The results of Q(t) measurement also show charge accumulation.

Thus, by using the Q(t) measurement and the PEA measurement together, the behavior of charge accumulation can be comprehensively considered. A detailed study has been published [5].

6.8 Evaluation of Charge Accumulation in Power Devices

The application field of power devices is developing, such as GCT/GTO thyristor, HV-IGBT, HV-IPM, etc. Also, the operation control of thyristors and trains in the high-voltage power field, and IPM, AS-IPM, DIP-IPM and IGBT-M for operation control of electric vehicles and robots needs the power devices. IGBT modules are one kind of them, and they are required to be driven at high temperatures and high voltages.

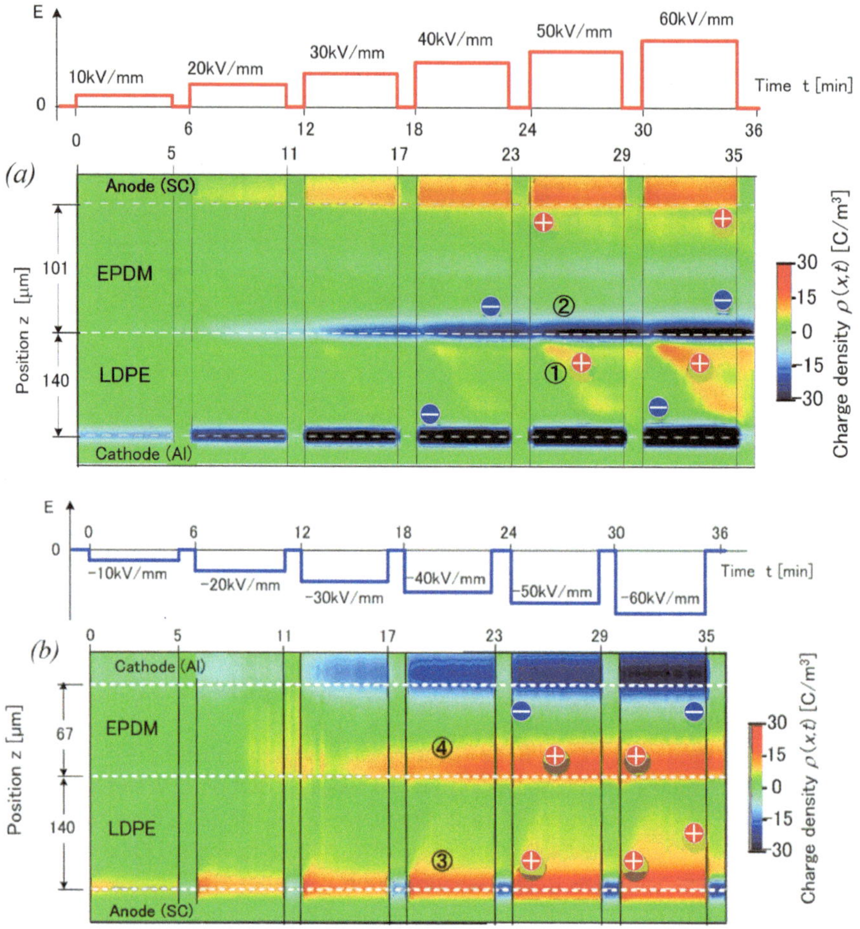

Fig. 6.29 The PEA results of the double-layer samples

(1) Evaluation items for power devices

Evaluation of heat conduction and charge storage: Since power devices generate a lot of heat, the heat dissipation is an important issue. Furthermore, since power devices are stacked, electrical insulation between the devices is also a major issue. That is, the electric insulating layer (insulated sheet) requires having a high thermal conductivity and good electric insulation.

Figure 6.30a shows the cross-sectional structure of an IGBT-M power device, which is used as an example for the Q(t) measurement [6]

The heat generated from the power tip is evenly distributed on the heat spreader, and it is radiated from the base plate to the insulated sheet.

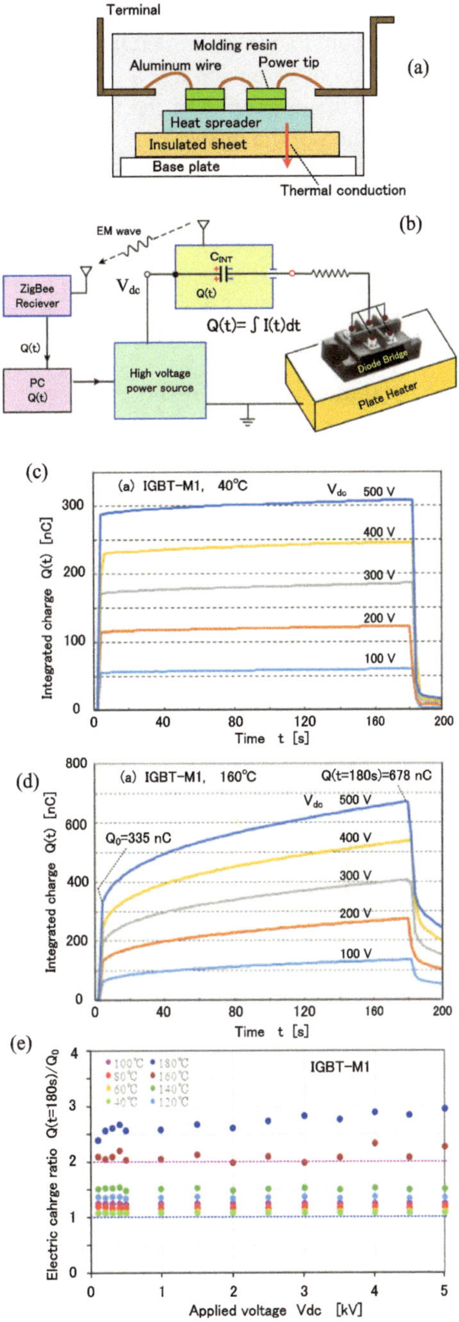

Fig. 6.30 The Q(t) results of the IGBT-M sample

Since a DC voltage is applied to this insulating layer, the characteristics of space charge accumulation are important for evaluating the electrical insulation. The insulation of a power device has a complicated shape. Therefore, the Q(t) method is applied to the evaluation of the voltage and temperature dependence of the charge accumulation of the insulation.

(2) Q(t) measurement system

Figure 6.30b shows the Q(t) measurement system for the IGBT-M whose temperature is increased by a plate heater to evaluate the charge accumulation and high temperature characteristics of the IGBT-M. The temperature of IGBT-M can be controlled from room temperature to 200 °C.

(3) Q(t) measurement results

The temperature is changed from 40 to 200 °C, and the Q(t) characteristic is measured in an applied voltage range of 100 V to 5 kV. Examples of the measurement results are shown in Fig. 6.30c (40 °C, 100 to 500 V) and Fig. 6.30d (100° C, 100 to 500 V). This charge accumulation is evaluated based on the charge ratio $Q(t = 180\ \text{s})/Q_0$ in Fig. 6.30e. At 40 °C, the charge ratio is 1, and it can be determined that there is no charge accumulation. When the temperature rises to 60 °C, the charge ratio exceeds 1. When the temperature reaches 100 °C and 160 °C, the charge ratio becomes larger than 1.5 and 2, respectively. A large amount of charge is accumulated. However, the dependence of the applied voltage is not very strong.

(4) Comparison of IGBT-M

The IGBT-M1 in Fig. 6.31b and IGBT-M2 in Fig. 6.31a with similar specifications are prepared from different manufacturers. Their Q(t) characteristics are measured. Figure 6.32 shows the temperature dependence of the charge ratio $Q(t = 180\ \text{s})/Q_0$.

The charge ratio of IGBT-M2 is 1 at 40 °C, and it increases to the range of 1.5–2.0 at 50 °C, and suddenly exceeds 4 at 80 °C. In contrast, the charge ratio of GBT-M1 is

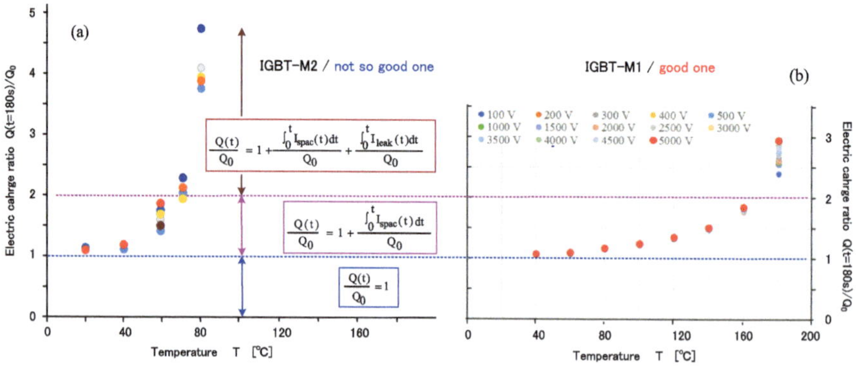

Fig. 6.31 The charge ratios of the two IGBT-M samples

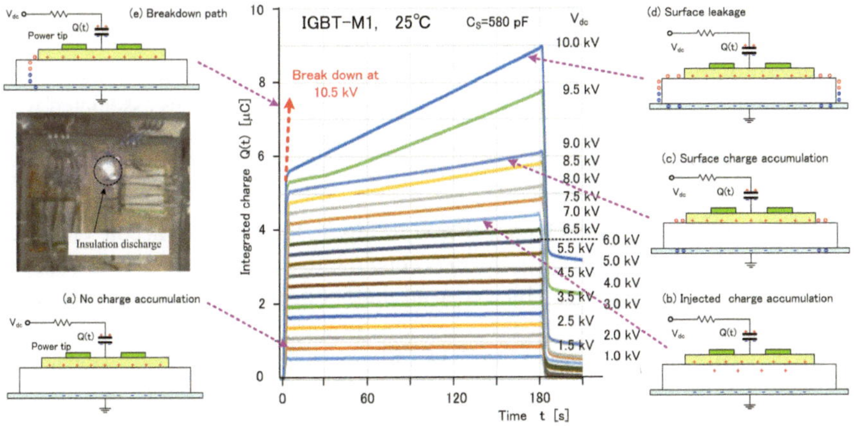

Fig. 6.32 The Q(t) results up to the breakdown of the sample

1.8 at 160 °C and above 2 at 160 °C. From the comparison between the two modules, IGBT-M1 is clearly superior.

Although the material of the insulating layer is unknown, it can be said that the Q(t) measurement is very useful for evaluating the charge accumulation of the insulating layer.

(5) Dielectric breakdown test of IGBT-M

The dielectric breakdown test of IGBT-M is performed using the Q(t) measurement system. The temperature is set to 25 °C, and the applied voltage is from 1.0 kV to the dielectric breakdown voltage of 10.5 kV at 180 s with of rise step of 0.5 kV. An increase in Q(t) data is observed at around 7.0 kV, and no breakdown occurs up to 10 kV. The next application of 10.5 kV leads to a burst of damage.

Figure 6.32 shows a photograph of the discharge light at the time of breakdown. As depicted in the sketch, it is broken at the electrode edge. Therefore, I think the charge behavior leads to the destruction. It shows that some charges leaked on the surface of the insulating layer due to the voltage rise.

6.9 Electrical Tree

The electrical tree in Fig. 6.36 is a kind of local dielectric breakdown. Electrons that gain energy accelerated by a local ultra-high electric field can break molecular bonds, generate small molecules (H_2, CH_4, etc.) and vaporize them. When the small molecules scatter, bubbles (voids) remain there. This is the electrical tree. This section introduces the results of the Q(t) method, in which the current in the process of generation and development of the electrical tree is integrated over time.

(1) Observation of electrical trees

Observation of electrical trees: Up to now, we have observed photographs of the generation and development of electrical trees and discharge pulses associated with them. Then, the generation and development conditions of the electrical tree have been considered. Here, the generation and progress of electrical trees are observed by the Q(t) method [7].

Sample with generated electrical tree in Fig. 6.33: A needle-needle electrode (gap 1 mm) is arranged, and epoxy resin is poured into a mold (7 × 75 × 120 mm) to make a sample. The sample is subjected to an alternating current to generate electrical trees.

Q(t) measurement system in Fig. 6.34: Q(t) meter is installed between the needle electrode and the ground, and DC voltages of $V_{DC} = 250$, 500, 1000 V are applied. The total current $I(t)$ flowing through the sample is integrated over time by the integration capacitor C_{Int}, and $Q(t) = \int I(t)\,dt$ is measured.

How to read the Q(t) data in Fig. 6.35: The vertical axis is the charge amount of integration of the total current $I(t)$. The horizontal axis is the time after the voltage application. Generally, the initial charge amount Q_0 is charged on the needle-needle electrode surface immediately after the application of the DC voltage, and thereafter the result increases as $Q(t) = Q_0 + \Delta Q(t)$. $\Delta Q(t)$ is the sum of the amount of charge accumulated in the sample and the amount of charge in the leakage current.

Capacitance of the sample in the needle-needle electrode system: Pay attention to the result at $V_{DC} = 250$ V before the electrical tree generation. Immediately after the application of the voltage, Q(t) data increases vertically. This is the initial charge amount $Q_0 = C \times V_{DC}$. C is the capacitance of the sample. From the data in Fig. 6.35a, $Q_0 = 0.6$ nC, and thus the capacitance of the sample in this needle-needle electrode system is about 2 pF.

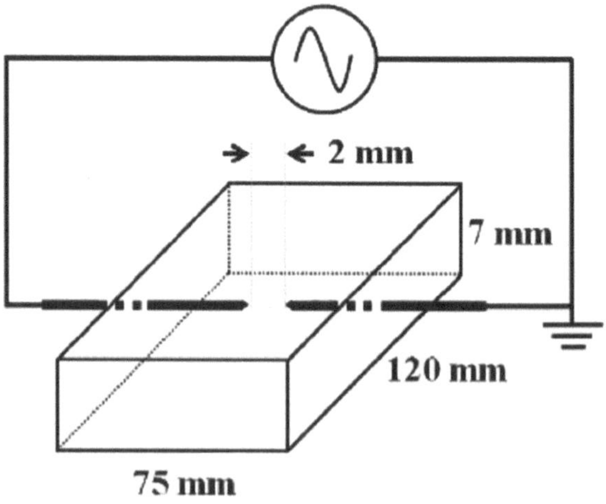

Fig. 6.33 The system for generating water tree

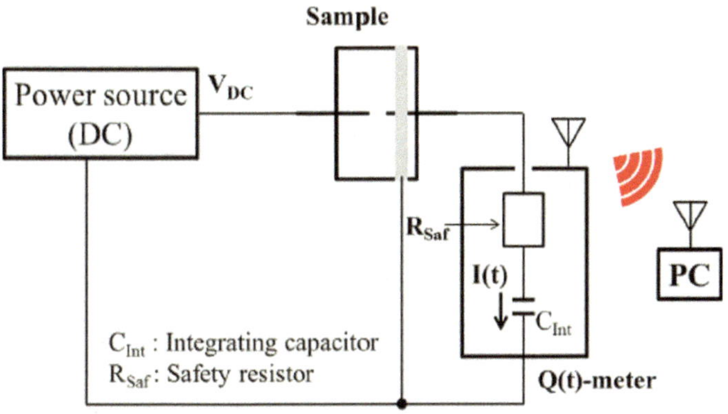

Fig. 6.34 The Q(t) system for the measurement

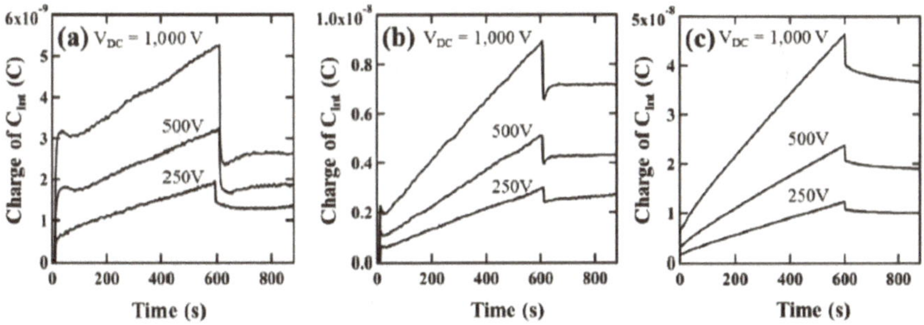

Fig. 6.35 The Q(t) results of the sample under different states

(a) **Before electrical tree generation in Fig. 6.35a:** Confirm that there is no electrical tree as shown in Fig. 6.36a. The initial charge at $V_{DC} = 250$ V clearly appears as $Q_0 = 0.6$ nC. However, at $V_{DC} = 500$ and 1000 V, the results shows a complicated phenomenon, i.e. the results firstly increase, then decrease, and finally increase again.

(b) **Early stage of electrical tree generation:** Fig. 6.36b confirms that electrical tree is generated. The initial charge amount Q_0 is not exactly the same as that in Fig. 6.36a before the generation of the electrical tree. The leakage current is also larger.

(c) **After electrical tree growth:** Fig. 6.36c confirms that the electrical tree has grown significantly. The initial charge amount Q_0 is not exactly the same value as that in Fig. 6.36a before the generation of the electrical tree. The leakage current is larger.

Although the Q(t) evaluation of electrical trees has begun, the Q(t) data contains a lot of information on the generation and progress of electrical trees. Future research is expected.

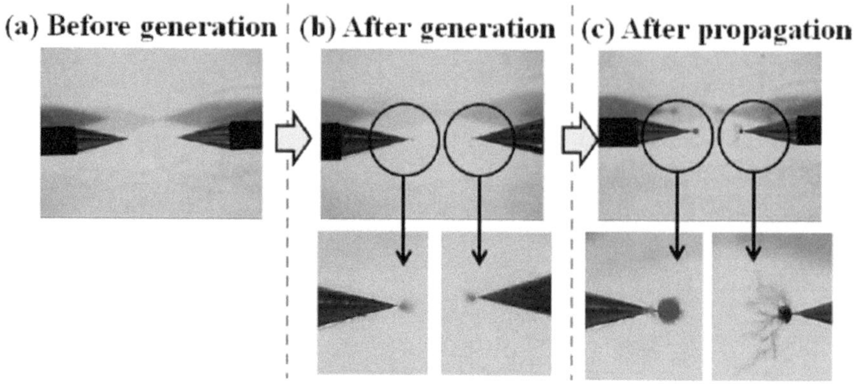

Fig. 6.36 The corresponding photos of water tree at different states

(2) **Model o f dharge transfer in generation and development of electrical tree**

In the previous section, we mentioned that the Q(t) data contains a lot of information on the generation and progress of electrical trees, but there are also issues that cannot be explained. Here, we present a model to analyze the issues of charge generation and transfer.

(a) **Electric Line of force and equipotential surface in Fig. 6.37:** A DC voltage V_{DC} is applied to the sample. Then, the electric lines of force exit from the positive charge at the anode toward the negative charge at the cathode. The equipotential surfaces show the same potential, and they are always orthogonal to the electric lines of force. Since the surface of the needle electrode is equipotential, the electric force lines are orthogonal to the electrode surface.
(b) **Electrode charge and dielectric polarization charge in Fig. 6.38:** Dielectric is polarized by the applied electric field. Negatively polarized charges are generated on the anode side and positively polarized charges are generated on the cathode side. At the stage before the electrical tree is generated, only polarized charges are present in the sample, as shown in Fig. 6.38. Then, the integrated charge becomes $Q(\mathrm{t}) = Q_0 = C \times V_{DC}$ as the initial charge.
(c) **Immediately after the generation of the electrical tree in the left of Fig. 6.39:** Only the upper half of the model is drawn, and the initial electrical tree from the anode is discussed here. The electrical tree is shown by the model with the conductive path. The conductive path is represented by the electric resistance R. Therefore, the integrated charge amount becomes $Q(t) = Q_0 + \Delta Q(t)$. $\Delta Q(t)$ is the amount of space charge supplied from the power supply and accumulated in the sample. Therefore, the Q(t) data measures the charge increased by $\Delta Q(t)$ from Q_0.
(d) **Electrical tree in progress shown by the right of Fig. 6.39:** The progress of electrical tree is indicated by the electric resistance R. Space charge $\Delta Q(t)$ increases through this resistance. Not all of the $\Delta Q(t)$ is the amount of space charge. $\Delta Q(t)$ also includes leakage current.

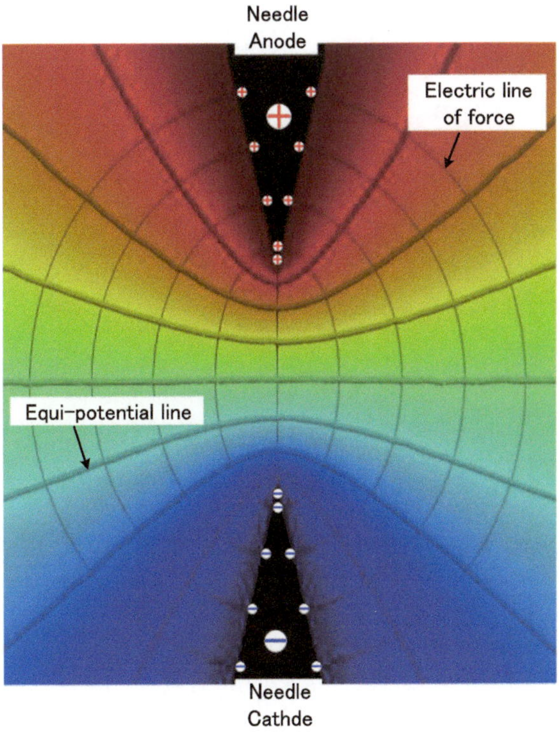

Fig. 6.37 The electric line of force and equipotential surface

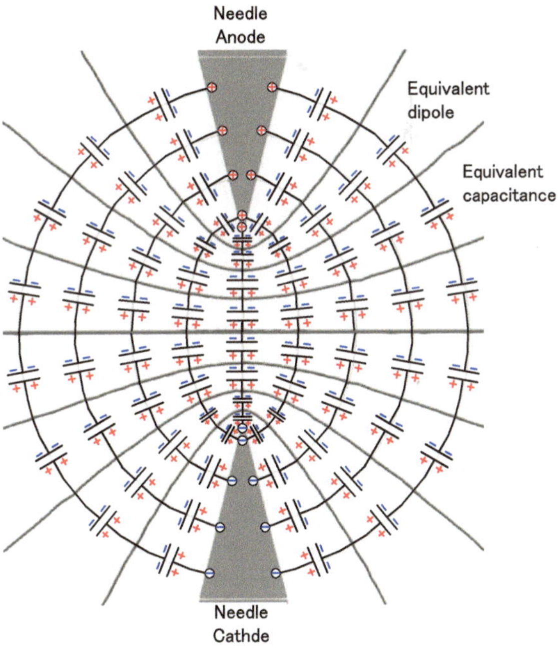

Fig. 6.38 The charge distribution

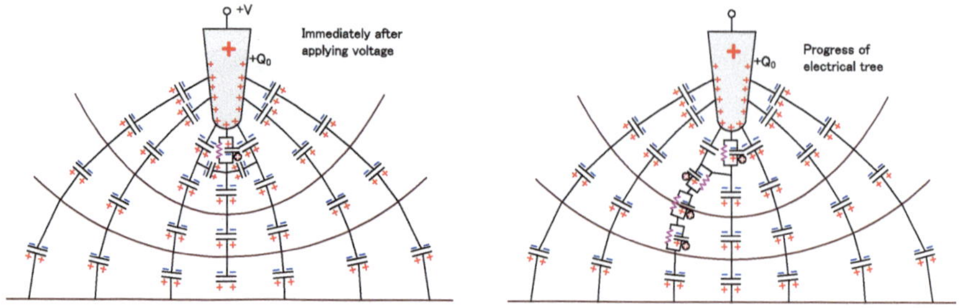

Fig. 6.39 The electrical tree in progress

Thus, a model of the shared capacitance and the electric resistance for electrical tree locus is presented.

6.10 Measurement of Two-Dimensional Current Distribution

For the measurement of the conductivity of the insulating material, a leakage current is measured by inserting a flat sample into parallel flat electrodes. In that case, it is assumed that the current is flowing uniformly in the sample. Does it really flow as a uniform current? In order to confirm this, the results by the two-dimensional Q(t) method are introduced [8].

(1) **Two-dimensional Q(t) measurement system**

As shown in Fig. 6.40, a flat sample with an array of 60 probe electrodes (14 × 14 mm) and high voltage electrodes is established. Each probe electrode (8 × 6 + 6 × 2 = 60) is connected to the current integration capacitor C_{int}, and the integration of the current $I(x, y, t)$ in Eq. (6.3) flows through the divided electrodes.

$$Q(x, y, t) = \int_0^t I(x, y, t)\, dt \tag{6.3}$$

$$\frac{Q(x, y, t)}{Q_0(x, y)} \tag{6.4}$$

$$I(x, y, t) = \frac{\Delta Q(x, y, t)}{\Delta t} \tag{6.5}$$

(2) **Dependence of $I(x, y, t)$ on electrode shapes**

(a) The air gap between needle electrode and sample is 10 mm, and LDPE sample is with the thickness of 30 μm.

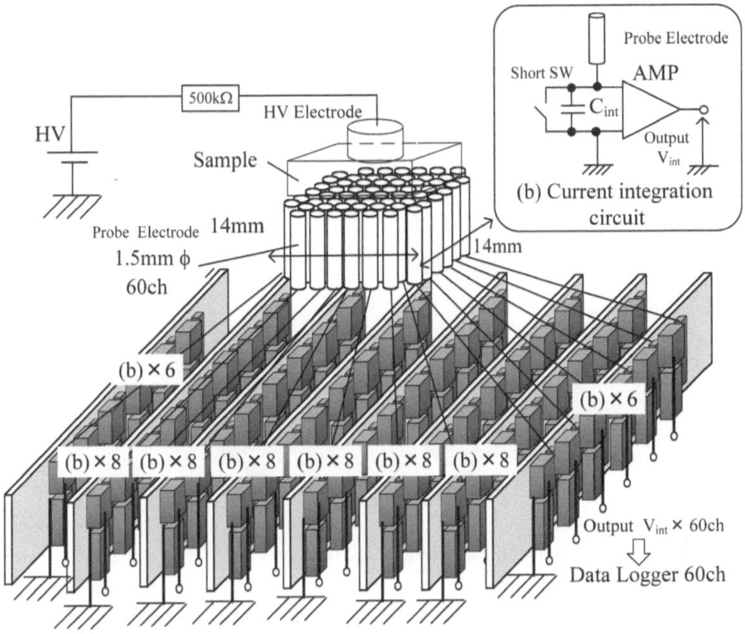

Fig. 6.40 The two-dimensional measurement system

(b) Plate electrode with the diameter of 10 mm: Leakage current does not flow uniformly on the sample surface. At the high voltages of 5 and 7 kV, the leakage current at the electrode ends is large. Edge effect appears.
(c) Spherical electrode with the diameter of 10 mm: Leakage current in the non-contact portion is larger than that flowing at the contact point between the sphere electrode and the sample (Fig. 6.41).

6.11 Q(t) Measurement of Enameled Wire Insulation

Power motors are used in all types of power sources including electric vehicles. Their demands are still increasing. Since this motor is driven by a unipolar pulse voltage, a unidirectional DC voltage is always applied to the insulation layer of the enamel winding. As a result, charges are accumulated in the insulating coating of the enameled wire. It is presumed that the accumulated charge causes insulation deterioration and dielectric breakdown. An example of applying the Q(t) method to the evaluation of the charge accumulation of the insulating coating is introduced.

(1) **Q(t) measurement circuit and enamel winding**

A test sample is prepared by twisting two enameled wires (each is 1.5 m long, but the thickness of insulating coating layer is unknown), as shown in Fig. 6.42. A high

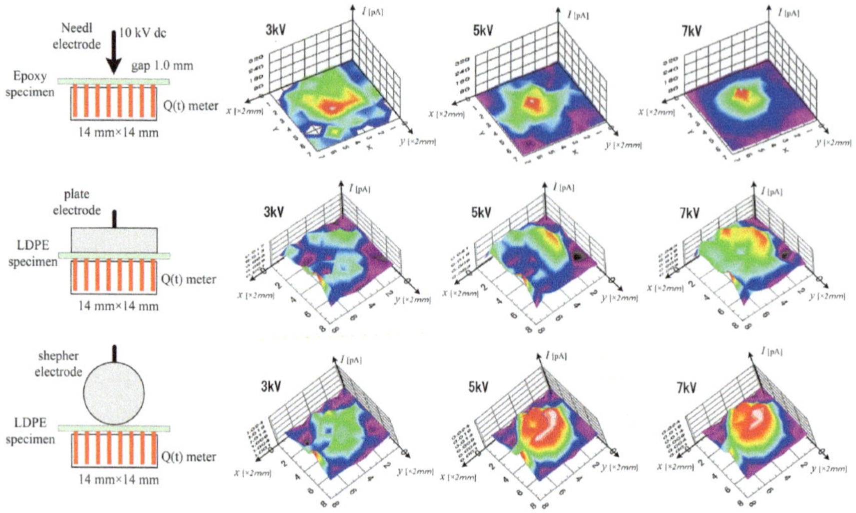

Fig. 6.41 The measured results under different electrodes

voltage is applied to one of them, and the other is grounded through the integration capacitor. The DC voltage is applied to the two insulating coating layers.

(2) **Q(t) measurement results**

Figure 6.43 shows the results of Q(t) data by applying a DC voltage (100 V to 2.0 kV) to the twisted enamel winding for 600 s. Figure 6.44 shows the voltage dependence of the charge ratio $Q(t = 600\ \text{s})/Q_0$. Even when the applied voltage is 100 V, the charge amount ratio $Q(t = 600\ \text{s})/Q_0 = 1.3 > 1.0$. This indicates that the charge accumulation occurs even at a low voltage. The thickness of the insulating coating layer is unknown. Just assume that the thickness of the double-layer coating is $a =$ 10 μm, the average electric field is $V_{dc} = 1000$ V. Thus, a high electric field of $E =$ 100 kV/mm can be applied.

As described above, it is shown that the charge accumulation in the coating layer of the enameled wire can be easily evaluated by the Q(t) method.

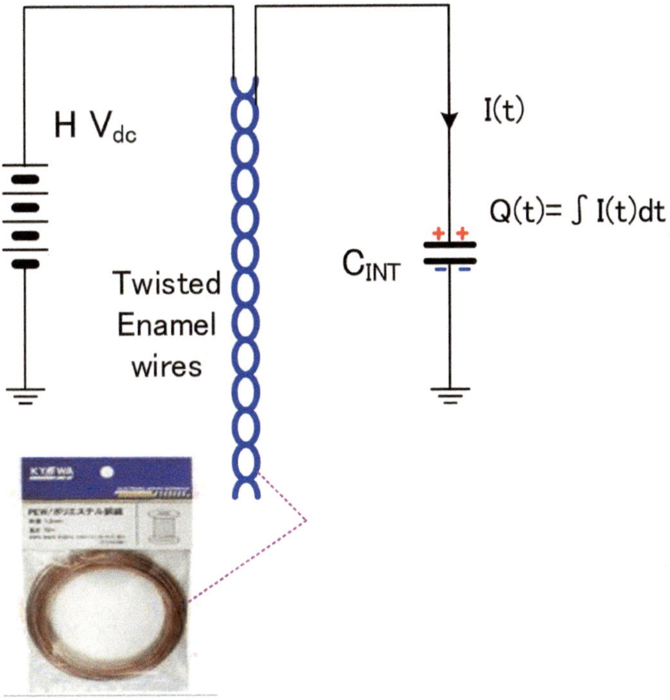

Fig. 6.42 The Q(t) system for the wire insulation

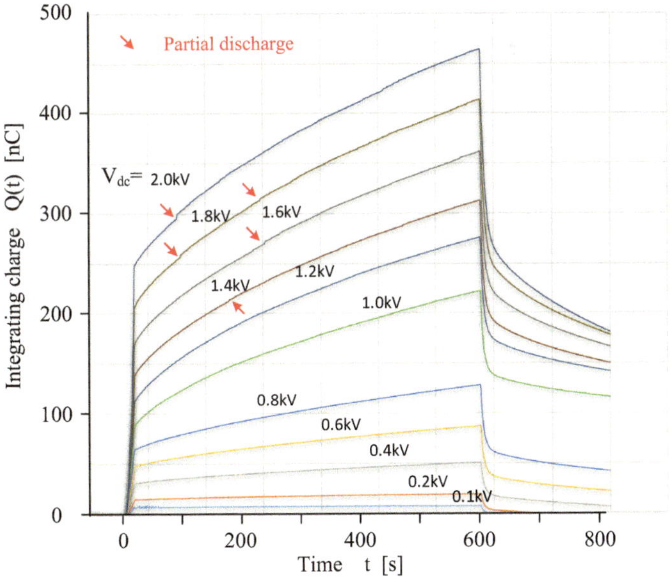

Fig. 6.43 The measured charge results

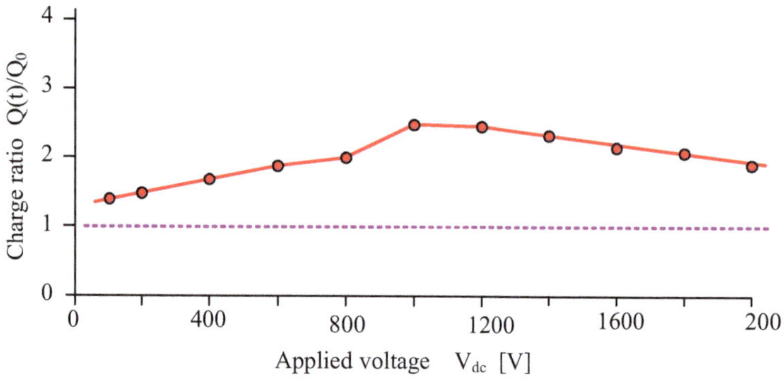

Fig. 6.44 The calculated charge ratio

References

1. T. Takada, T. Mori, T. Iwara, T. Fujitomi, Y. Ono, H. Miyake, Y. Tanaka, Proposal of insulation deterioration diagnosis by using Q(t) method, Paper presented at the papers of Technical Meeting of IEEJ, Tokyo, Japan, DEI-16-058 (2016).
2. D. Mori, M. Yag, T. Takada, Evaluation of Q(t) measurement for DC-XLPE cable, Paper presented at the annual meeting of IEEJ, Tokyo, Japan, 7–121 (2018).
3. T. Takada, Y. Fujitomi, T. Mori, T. Iwata, T. Ono, H. Miyake, Y. Tanaka, New diagnostic method of electrical insulation properties based on current integration, *IEEE Transactions on Dielectrics and Electrical Insulation*, 24(4), 2549–2558 (2017).
4. W. Wang, K. Sonoda, S. Yoshida, T. Takada, Y. Tanaka, Takashi Kurihara, Current integrated technique for insulation diagnosis of water-tree degraded cable, *IEEE Transactions on Dielectrics and Electrical Insulation*, 25(1), 94–101 (2018).
5. T. Takada, T. Tohmine, Y. Tanaka, J. Li, Space charge accumulation in double-layer dielectric systems-measurement methods and quantum chemical calculations, *IEEE Electrical Insulation Magazine*, 35(5), 36–46 (2019).
6. T. Takada, K. Hijikata, W. Wang, T. Inoue, Sensor for accumulated charge detection in packaged insulation layer of insulated gate bipolar transistor power devices, *Sensors and Materials*, 31, 2565–2578 (2019).
7. S. Iwata, R. Kitani, T. Takada, Diagnostic technique for electrical tree by current integration method, Paper presented at 37th IEEE Electrical Insulation Conference, Calgary, Canada, 317–320 (2019).
8. M. Fukuma, Y. Sekiguti, Current distribution measurement in insulating polymer cross section by current integration meter, Paper presented at International Symposium on Electrical Insulating Materials, Toyohashi, Japan, 73–76 (2017).

Part II
Fundamentals and Application of Pulsed Electro-Acoustic Method

Chapter 7
DC Insulation and Space Charge Accumulation

7.1 Measurement of Space Charge Distribution by PEA Method

(1) An example of measurement results

In order to understand the PEA method (Pulsed Electro-Acoustic method) for space charge distribution measurement, a representative result is firstly introduced. Figure 7.1 shows the PEA measurement results of charge behaviors when a positive DC electric field of 62.5 kV/mm is applied to a low-density polyethylene sample (LDPE with a thickness of 320 μm). < Results can also be seen in Takada Text, Vol. I, Part 2, Chapt. 5, Sect. 5.1 >

As shown in Fig. 7.1, only two seconds later after the electric field is applied, the injection of positive charge (red part) from the anode is observed, which is thought as the first stage of charge accumulation. The positive electric charge packet moves to the sample center after 30 s, at which time the maximum electric field strength increases to 80 kV/mm, which far exceeds the applied average field of 62.5 kV/mm. Meanwhile, at high electric fields, the injection of negative charges (blue part) from the cathode is observed. From 30 s to 2 min, the positive charge amount is still increasing. After 5 min, the neutralization of positive and negative charges is observed, and the distribution of them almost reaches an equilibrium state.

When the polarity of the applied electric field is reversed (-62.5 kV/mm), negative charges are injected from the cathode to neutralize the residual positive charges. After the same measurement time, the residual positive charges are almost neutralized by negative ones, but some charges still remain.

As described above, the PEA system is a powerful measurement instrument capable of observing the changes of the internal charge distribution $\rho(x,t)$ [C/m^3] and the internal electric field distribution $E(x,t)$ [V/m].

T. Takada et al., *Electric Charge Accumulation in Dielectrics: Measurement and Analysis*,
https://doi.org/10.1007/978-981-19-6156-4_7

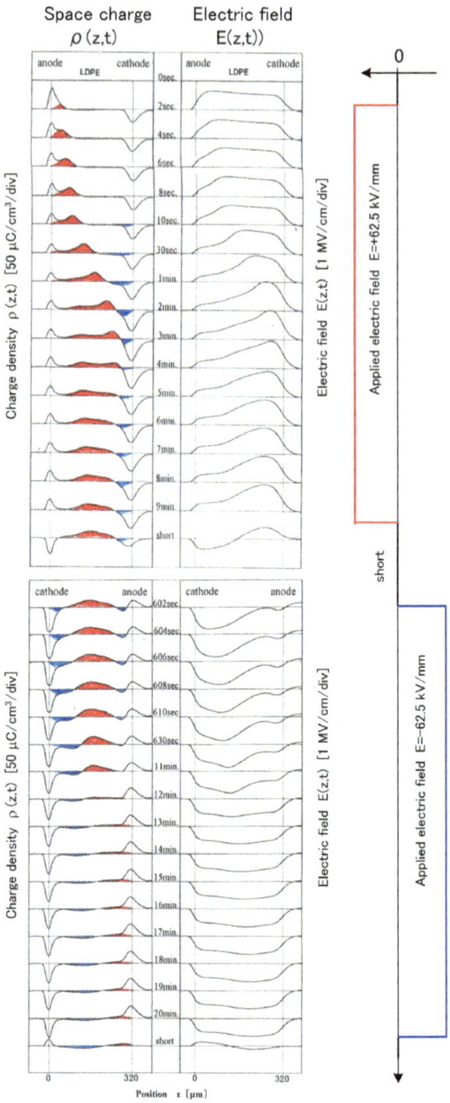

Fig. 7.1 PEA results in a LDPE sample

(2) **PEA equipment**

Figure 7.2 shows the basic measurement principle of the PEA method for a flat sample. In the actual measurement setup, a DC voltage V_{dc} and a pulse voltage $v_p(t)$ are superimposed and applied to a dielectric material. The pulse pressure wave $p(t)$ then generates including $p_1(t)$ from induced electrode charge $\sigma(0)$ at the ground anode, $p_2(t)$ from accumulated sheet positive charge $\rho(z, t)\Delta z$ inside the sample, and $p_3(t)$ from induced electrode charge $\sigma(a)$ at the high-voltage cathode. The $p(t)$ is finally converted into an electric signal $v_s(t)$ by the piezoelectric device attached to

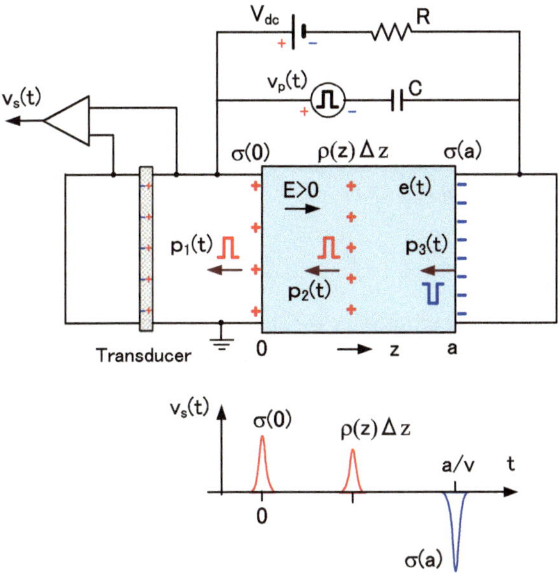

Fig. 7.2 PEA method for a flat sample

the back surface of the ground electrode. From the measured signal $v_s(t)$ with time, the electric charge profile of $\sigma(0)$, $\rho(z, t)\Delta z$ and $\sigma(a)$ can be presented.

7.2 DC Insulation and AC Insulation

(1) Why is it necessary to evaluate DC insulating materials?

Many people think that the required characteristics of the insulating materials are the same for AC (alternating current) and DC (direct current) environments. However, the actual characteristics between AC and DC insulations are very different. For example, there is usually not too much charge accumulation under AC voltages, due to the frequent neutralization of positive and negative charges. Therefore, there isn't much distortion of the internal electric field, which requires an easier insulation design. Conversely, a large amount of charge usually accumulates in insulating materials under DC voltages, and thus the internal electric field is distorted and locally exceeds the designed electric field, making insulation design difficult.

As a kind of energy-saving technology, the inverter is widely used in household appliances, electric vehicles, industrial equipment, power equipment, and railway equipment. The inverter can convert DC voltage to AC voltage by using pulse voltages with different frequencies and periods. Since the DC or unipolar pulse voltages are used, electric charges always accumulate in the insulation layer, causing an insulation degradation and breakdown. In order to obtain the information of the charge accumulation, a PEA instrument for measuring charge distribution has been developed in [1–5].

(2) **There is no charge accumulation inside AC insulation**

The AC characteristics can be evaluated by the dielectric constant ε and the dielectric loss tanδ in Eq. (7.1).

$$\varepsilon = \varepsilon' - j\varepsilon'' \quad \tan\delta = \varepsilon''/\varepsilon' \tag{7.1}$$

where the permittivity is $\varepsilon = \varepsilon_0\varepsilon_r$. ε_0 is the vacuum permittivity, and ε_r is the relative permittivity of the insulating material. Polymer insulating materials developed after 1950 have a small relative permittivity of $\varepsilon_r = 2$ to 3 and a small dielectric loss of tan$\delta = 10^{-3}$ to 10^{-4}, and these parameters have gotten enough studies.

In Fig. 7.3a, it is explained that space charge is not accumulated when AC voltages are applied. Since the voltage polarity is reversed in a half cycle (= 10 ms), the electrons and holes injected from the electrode can come back to it. Therefore, the charges cannot move into the depth of the sample, and there is no charge accumulation. Meanwhile, the internal electric field of the AC insulation basically remains the same without distortion.

(3) **There is charge accumulation inside DC insulation**

In the case of DC voltage (Fig. 7.3b), when the application time is sufficiently long, for example from 10 mins to 12 h, the injected charges including electrons and holes continue to move under the electric field and accumulate as space charges inside the sample. Further, these charges reach the counter electrode and become a leakage current in an equilibrium state.

Therefore, when the space charge density $\rho(x)$ is accumulated, the electric field $E(x)$ inside the insulating layer has a distortion. The PEA method for measuring space charge distribution has been developed that can evaluate these characteristics [1]. The next two chapters will explain the principle of the method.

7.3 Accumulated Charge and Internal Electric Field

The research field on charge storage and developing insulating materials always involve the relationship between charge density $\rho(z)$, electric field $E(z)$, and potential $V(z)$. The involved charge accumulation properties are also usually drawn and discussed. ***Solving Poisson's equation is a prerequisite condition for the charge research.***

(1) **Poisson's equation**

The relationship between $\rho(z)$, $E(z)$, and $V(z)$ is shown in Eqs. (7.3) to (7.5), which can be obtained by the one-dimensional Poisson's equation of Eq. (7.2).

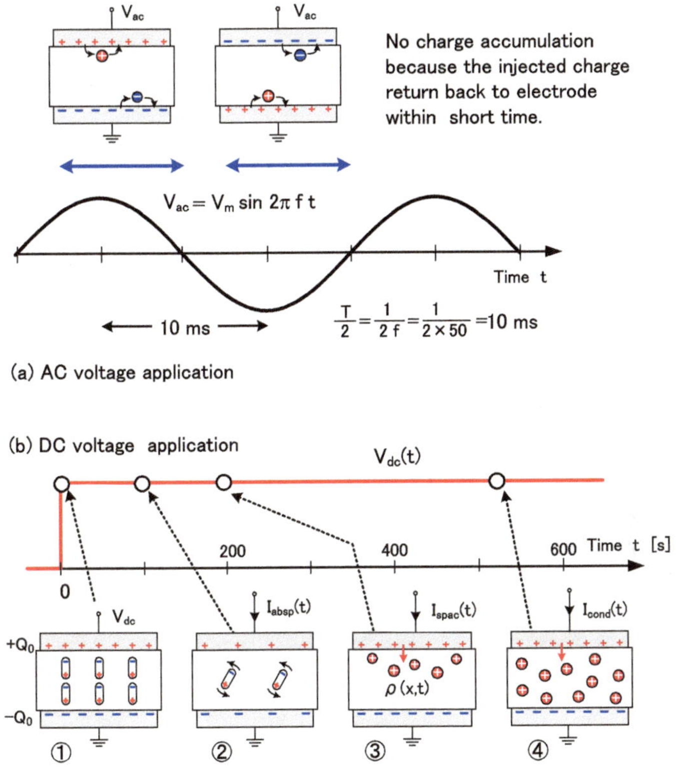

Fig. 7.3 PEA measurement under AC and DC voltages

Poisson's equation (7.2)

$$-\frac{\partial^2 V(z)}{\partial z^2} = \frac{\partial E(z)}{\partial z} = \frac{\rho(z)}{\varepsilon_0 \varepsilon_r} \tag{7.2}$$

$$E(z) = \frac{\rho}{\varepsilon_0 \varepsilon_r} z + C_1 \tag{7.3}$$

$$V(z) = -\frac{\rho}{2\varepsilon_0\varepsilon_r} z^2 - C_1 z + C_2$$

$$C_1 = \frac{V_{dc}}{a} - \frac{\rho a}{2\varepsilon_0\varepsilon_r} \tag{7.4}$$

$$E(z) = -\frac{V_{dc}}{a} + \frac{\rho}{\varepsilon_0 \varepsilon_r}\left(z - \frac{a}{2}\right) \tag{7.5}$$

(2) **Distributed charge density $\rho(z)$ and electric field distribution $E(z)$**

In order to solve the Poisson's equation conveniently as an example, the space charge density is firstly set to constant. The following analysis is given:

Boundary condition: The flat sample thickness is set to 100 μm. With the given boundary conditions of $V(0) = 0$ and $V(a) = V_{dc}$, the coefficient $C_2 = 0$ and the coefficient C_1 of Eq. (7.1.4) can be obtained. Also, the relationship between the charge density and the internal electric field in Eq. (7.1.3) can also be calculated. For the assumed situation of applied voltage $V_{dc} = +10$ kV, $V_{dc} = 0$, and $V_{dc} = -10$ kV with respective charge density $\rho = 0$, $\rho = +2.5$ C/m^3, and $\rho = +80$ C/m^3, the internal electric field distribution can be calculated as shown in Fig. 7.4.

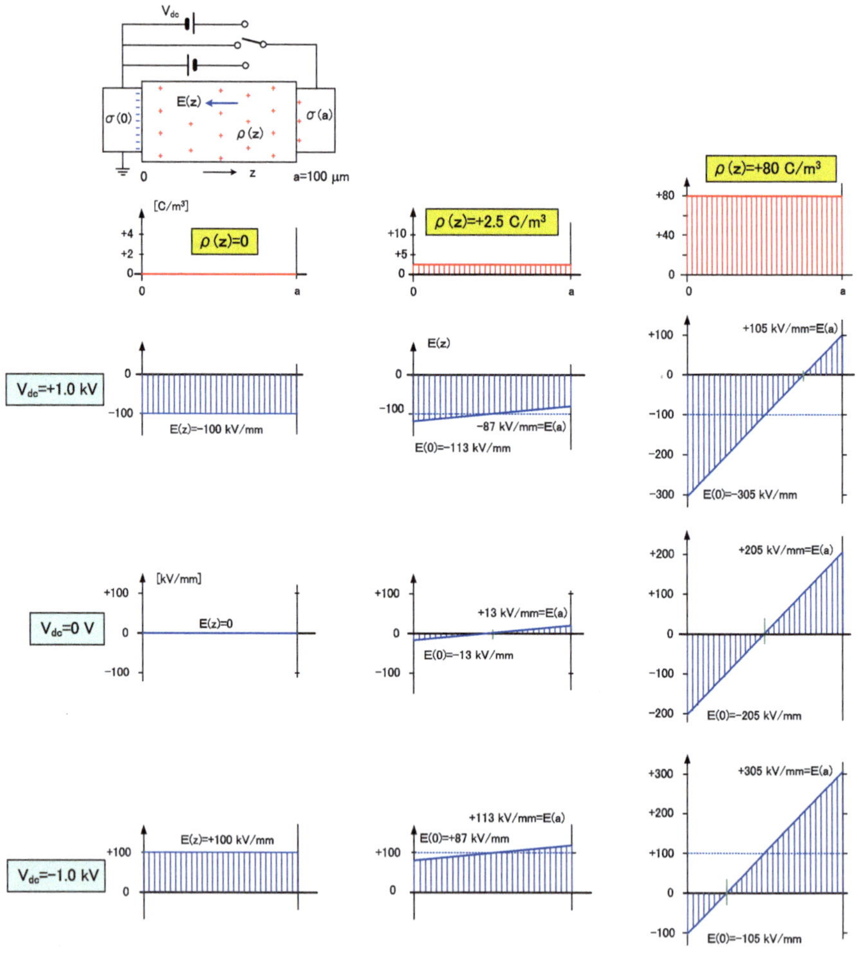

Fig. 7.4 Electric field distributions under different charge accumulations

Condition A, $\rho = 0$: The internal electric field is constant at the field strength of 100 kV/mm.

Condition B, $\rho = +2.5$ C/m^3: The change of the field strength is about 15% of the applied electric field.

Condition C, $\rho = +80$ C/m^3: The change of the field strength is three times the applied electric field. The accumulation of space charge distorts the electric field severely. The dielectric breakdown may occur if the internal electric field reaches the breakdown value.

7.4 Dielectric Deterioration and Breakdown Caused by Accumulated Electric Charge

(1) Assessment on DC insulation materials

Why is the research of charge accumulation in insulating materials important? Electrical insulating materials are usually used for many years under high electrical stresses and severe temperature changes from room temperature to over 100 °C. Inevitably, the insulating material deteriorates under these conditions, and eventually dielectric breakdown can occur. The breakdown accident usually can cause a wide area of blackout and a great deal of confusion in social life.

Recently, the development of renewable energy resources has led to the widespread adaption of DC equipment. Therefore, the electric charge can accumulate in the insulating materials under DC voltages with high temperature conditions, which can further distort the electric field, deteriorate insulation, and lead to dielectric breakdown. The explanation of this process needs deep research. Basically, there are three steps to solve this problem.

Firstly, observe the charge accumulation using an advanced measurement technology. Then, explain the charge accumulation mechanism using Quantum Chemical Calculation. Finally, develop the DC insulating materials applicable to high temperatures and high electric fields. The following content introduces the basic model for analyzing charge accumulation, insulation deterioration, and dielectric breakdown.

(2) Relationship between trapped charge and intermolecular distance expansion

Situation 1: As shown in Fig. 7.5a, if excess electrons are added to the molecule, will the bond-atom distance increase or decrease?

Observation result: The Quantum Chemical Calculation is used to calculate it. The linear PE molecule ($C_{24}H_{50}$) is negatively charged (one electron is added), and the electrostatic potential distribution and carbon–carbon distance are observed. As a result, the position of C_{15}-C_{16} part is the center of the negative potential distribution. The excess electron is trapped at this position. The C_{15}-C_{16} interval shows that the expansion is recognized from 1.5347 Å to 2.0880 Å. It is thus revealed that the existence of excess electrons widens the C–C interval distance.

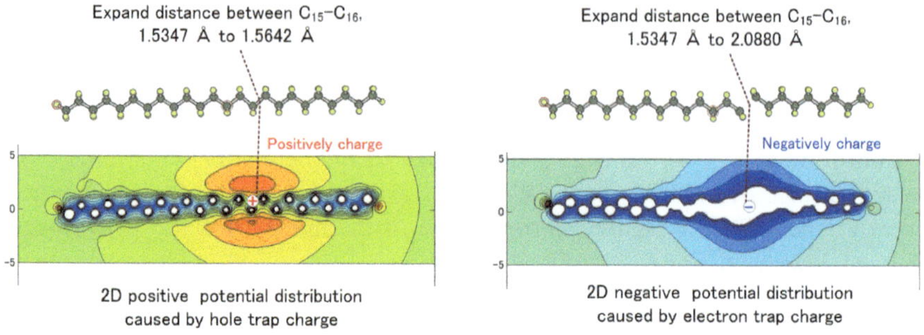

Fig. 7.5 Calculation results based on the Quantum Chemical Calculation

Situation 2: As shown in Fig. 7.5b, the lack of an electron causes the positive polarity situation. The position of the electron-lacking C_{15}-C_{16} is the center of the positive potential distribution. As a result of the Quantum Chemical Calculation, the change of the C_{15}-C_{16} interval is 0.03 Å, which is extremely small compared to the above negative situation.

Research issue on insulation deterioration

From the above observations by the Quantum Chemical Calculation, it can be said that the existence of excess electrons (or lack electrons) in a molecule means that energy locally centers, with electrons remaining in the vicinity. The energy distribution is significantly disturbed and the intermolecular bond distance greatly changes, leading to molecular breakage.

In 1986, as summarized in the Technical Report [6] from the Institute of Electrical Engineers of "Complex Factor Degradation", insulation degradation is caused by multiple factors (heat, mechanical stress, electric stress, space charge, oxygen, water, light, etc.). In 2020 after 34 years, it is expected to utilize the progress of measurement and calculation technologies to analyze the complex deterioration and develop new materials.

(3) Charge accumulation and dielectric breakdown

Many researchers think that dielectric breakdown occurs only caused by a high electric field. In fact, the electric field is not the only reason for it.

When electric charge exists in the insulating material with a high electric field in it, the electron is accelerated and charge multiplication (electron avalanche) is caused. Then, molecular braking and local current increase, and finally dielectric breakdown occurs.

Therefore, the fact of dielectric breakdown is the presence of electric charges in a high electric field, and space charges are the triggers for dielectric breakdown. From this point as well, it is important to carefully observe the conditions for charge accumulation by the PEA method to understand the mechanism of dielectric breakdown, and to work on the insulation design assisted by the Quantum Chemical Calculation.

References

1. H. Kushibe, T. Maeno, T. Takada, Measurement of accumulated charges inside dielectric by pulsed electric force techniques, *The Transactions of the Institute of Electrical Engineers of Japan.A*, 106(3), 118-124 (1986)
2. T. Takada, Space charge formation in dielectrics, *IEEE Transactions on Electrical Insulation*, EI-21(6), 873-879 (1986)
3. T. Takada, T. Maeno, H. Kushibe, An electric stress-pulse technique for the measurement of charge in a plastic plate irradiated by an electron beam, *IEEE Transactions on Electrical Insulation*, EI-22(4), 497-501 (1987)
4. T. Maeno, H. Kushibe, T. Futami, T. Takada, C.M. Cooke, Measurement of spatial charge distribution in thick dielectrics using the pulsed electroacoustic method, *IEEE Transactions on Electrical Insulation*, 23(3), 433-439 (1988)
5. T. Takada, Acoustic and optical methods for measuring electric charge distribution in dielectrics, *IEEE Transactions on Dielectrics and Electrical Insulation*, 6(5), 519-547 (1999)
6. T. Takada, Present state of research on multi-factor deterioration of insulation systems, Paper presented at the Technical Report of the IEEJ, Tokyo, Japan, 225 (1986)

Chapter 8
PEA Method: Pulsed Electro-Acoustic Method

8.1 Pulse Pressure Wave of PEA Method

(1) Pulse Pressure Wave Model

Figure 8.1 shows the principle of the PEA method to measure the accumulated charge distribution. When the DC voltage V_{dc} is applied to the dielectric with a thickness of a, the accumulation of space charge $\rho(z, t)$ takes place. As a result, $\sigma(0)$ and $\sigma(a)$ are induced at the cathode and the anode respectively, which can be calculated by Eqs. (8.1) and (8.2). Here, b is the position where the internal electric field is zero when only $\rho(z, t)$ exists and $V_{dc} = 0$ is assumed.

$$\sigma(0) = -\,\varepsilon\,\frac{V_{dc}}{a} - \frac{a-b}{a}\int_0^a \rho(z)\,dz \tag{8.1}$$

$$\sigma(a) = +\,\varepsilon\,\frac{V_{dc}}{a} - \frac{b}{a}\int_0^a \rho(z)\,dz \tag{8.2}$$

In order to measure these charge distributions, a nanosecond pulse electric field $e(t)$ is applied. As a result, each electric charge faces a pulsed electrostatic force. Then, its position changes and a strain amount $\xi(t)$ generates, which is described detailed by the basic piezoelectric equation in Chap. 10. A pulsed elastic wave $p(t)$ (= Zu) is generated which is proportional to the particle velocity $u = \Delta\xi(t)/\Delta t$, where Z is acoustic impedance.

The $p(t)$ propagates through the sample and the ground electrode, then it reaches the piezoelectric transducer attached to the back of ground electrode. Pressure force $F(t) = p(t)$ applies to the piezoelectric device and induces a charge signal $q(t)$. This is also shown by the pressure wave formula of the piezoelectric constant in Chap. 10. The charge signal $q(t)$ is converted into an electric signal waveform $v_s(t)$ by the OP amp. And this time-series waveform $v_s(t)$ shows the charge distributions including $\sigma(0)$, $\rho(z, t)$, and $\sigma(a)$.

T. Takada et al., *Electric Charge Accumulation in Dielectrics: Measurement and Analysis*,
https://doi.org/10.1007/978-981-19-6156-4_8

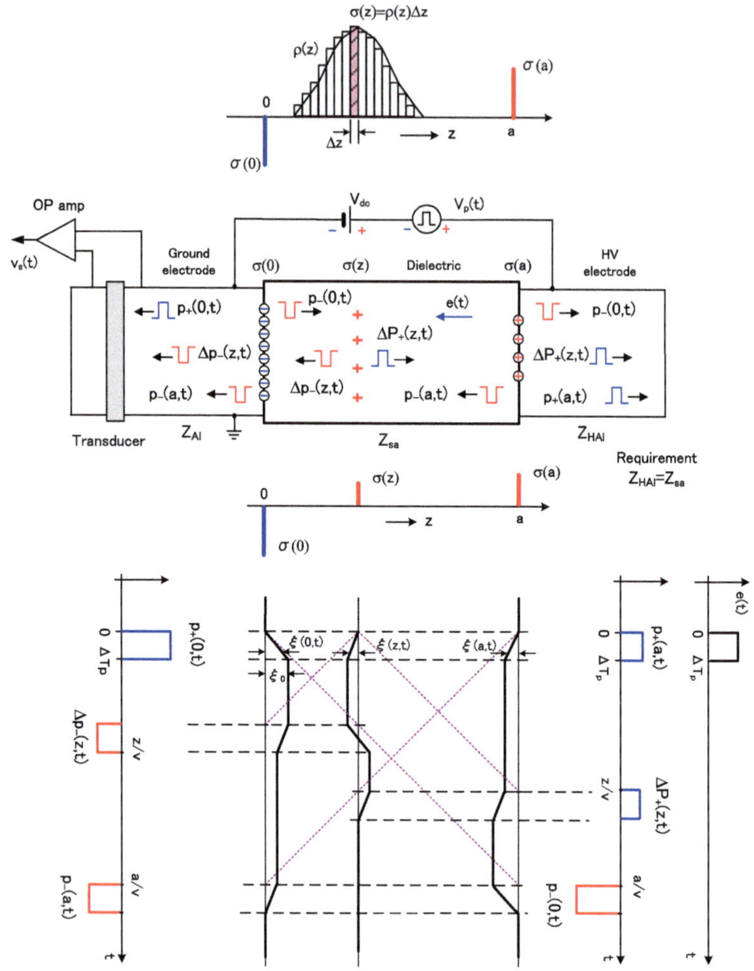

Fig. 8.1 Measurement principle of PEA method

(2) Generation and Propagation of Each Pressure Wave

Explanation of the generation and propagation of each pressure wave is drawn in Fig. 8.1. The sum of $p_+(0, t)$, $\Delta p_-(z, t)$, and $p_-(a, t)$ in Fig. 8.1 is given by the following Eq. (8.3). Here, the time t is the traveling time of pressure wave with the pulsed electric field applied at $t = 0$.

$$
\begin{aligned}
p(t) &= p_{+}(0,t) \quad + \quad \Sigma\, \Delta p_{-}(z,t) \quad + \quad p_{-}(a,t) \\
&= \frac{Z_{Al}}{Z_{sa}+Z_{Al}} \sigma(0)\, e(t) + \frac{2\,Z_{Al}}{Z_{sa}+Z_{Al}} \frac{1}{2} \int_{-\infty}^{+\infty} \rho(\tau)\, e(t-\tau) v_{sa}\, d\tau \\
&+ \frac{2\,Z_{Al}}{Z_{sa}+Z_{Al}} \frac{Z_{sa}}{Z_{sa}+Z_{HAl}} \sigma(a)\, e(t-a/v_{sa})
\end{aligned}
\tag{8.3}
$$

$p_{+}(0,t)$: The position of the cathode charge $\sigma(0)$ is disturbed by the strain $\xi(0, \mathrm{t})$ due to the application of the pulsed electric field. As a result, a wave (blue) generates on the cathode side. The coefficient $Z_{Al}/(Z_{sa}+Z_{Al})$ indicates the rate when the pressure wave generates at the interface of cathode and dielectric and then propagates to the ground electrode side. Z_{Al} and Z_{sa} are acoustic impedances of the aluminum electrode and the sample.

$\Delta p_{-}(z,t)$: The compressed pressure wave $p(z, t)$ generates from the distributed accumulated charge $\rho(z)$. The second term of Eq. (8.3) represents it. In fact, it is difficult to accurately express the whole accumulated charge by an equation due to its wide distribution, and thus it is assumed that the wave $\Delta p_{-}(z)$ generated from the sheet charge $\sigma(z) = \rho(z)\Delta z$ passes through the cathode surface after the time z/v.

When $e(t)$ is applied to the thin charge layer $\sigma(z)$, the acting force $f(z) = \rho(z)\Delta z e(t)$ generates. As a result, the strain $\xi(\mathrm{t})$ at that position increases with time. Elastic wave $\Delta p_{-}(z, t)$ generates by the change of strain amount $\Delta\xi(t)/\Delta t$ and moves, which is detailedly discussed in Chap. 9.

The coefficient 1/2 of the second term in Eq. (8.3) is because the pressure wave is generated inside the sample with the same acoustic impedance. Thus, it is divided into forward and backward pressure waves with the same magnitude, and the coefficient becomes 1/2. The coefficient $2Z_{Al}/(Z_{sa}+Z_{Al})$ is the ratio of $p(z,t)$ passing through the interface of cathode (Al) and dielectric.

$p_{-}(a,t)$: Compressed wave generates from the anode charge $\sigma(\mathrm{a})$. It passes through the cathode surface after the time a/v. The coefficient $Z_{sa}/(Z_{sa}+Z_{HAl})$ of the third term in Eq. (8.3) represents the ratio of the wave generating at the interface of anode (High voltage electrode, Al) and dielectric, which travels towards the dielectric side. The coefficient $2Z_{Al}/(Z_{sa}+Z_{Al})$ is the ratio of $p_{-}(a, t)$ passing through the interface of cathode (Al) and dielectric.

(3) Acoustic Impedance Matching

By matching the acoustic impedance $Z_{HAl} = Z_{sa}$ between the high voltage electrode and the dielectric sample, Eq. (8.3) can be simplified as Eq. (8.4). There is no coefficient in each component of the charge distribution ($\sigma(0)$, $\rho\ (z, t)$, $\sigma(a)$) in Eq. (8.4). This means that the charge distribution can be measured directly from the waveform of $p(t)$. Therefore, this acoustic matching becomes an important requirement for the PEA measurement.

$$p(t) = \frac{Z_{Al}}{Z_{sa} + Z_{Al}} \left[\sigma(0)\, e(t) + v_{sa} \int_{-\infty}^{+\infty} \rho(\tau)\, e(t-\tau)\, d\tau + \sigma(a)\, e\left(t - \frac{a}{v_{sa}}\right) \right] \tag{8.4}$$

$where\ dz = v_{sa} d\tau$

(4) Detection of Pressure Waves b y Piezoelectric Transducer

Next, the measurement of $p(t)$ in Eq. (8.4) will be described. The pressure wave $p(t)$ is converted into the charge signal $q(t)$ in Eq. (8.5) by the piezoelectric transducer with a thickness b and a piezoelectric constant d [m/V]. This is based on the piezoelectric effect described by the first term of Eq. (10.4) in Fig. 10.1. Equation (8.5) can be thought as the convolution calculation of the piezoelectric constant $d(\tau)$ and the pressure wave $p(t\text{-}\tau)$. Here, the coefficient $2Z_p/(Z_{Al} + Z_p)$ is the ratio of $p(t)$ passing through the interface of cathode (Al) and piezoelectric transducer.

The Eq. (8.4) is also based on when the acoustic impedances of the piezoelectric transducer and the backing material matches so that reflection does not occur at the interface. The charge signal $q(t)$ in Eq. (8.5) is converted into the voltage signal $v_s(t)$ in Eq. (8.6) by the OP amp, where C_p is the capacitance of the transducer, $A(f)$ is the amplification factor with the OP frequency characteristic, and $W(f)$ is the frequency characteristic of the electrical system for voltage signal detection including the transducer capacitance. ε_{rp} is the relative permittivity of the transducer.

$$\text{Electric charge signal} \quad q(t) = \frac{2\, Z_p}{Z_{Al} + Z_p} \int_{-\infty}^{+\infty} d(\tau)\, p(t-\tau)\, \frac{v_p\, d\tau}{b} \tag{8.5}$$

$$\text{Voltage signal} \quad v_s(t) = \frac{A(f)\, W(f)}{C_p}\, q(t) = \frac{A(f)\, W(f)}{C_p} \frac{2\, Z_p}{Z_{Al} + Z_p} \int_{-\infty}^{+\infty} d(\tau)\, p(t-\tau)\, \frac{v_p\, d\tau}{b} \tag{8.6}$$

where $C_p = \frac{\varepsilon_0\, \varepsilon_{rp}}{b}$

8.2 Signal Processing

(1) Effect of Signal Processing

The effect of signal processing is firstly introduced in Fig. 8.2. The PMMA with a thickness of 10 mm is irradiated with a high energy electron beam (1 meV) to make negative charge accumulation. The PEA signal is shown in Fig. 8.2a, where a negative peak v_{s2} is observed near the sample center, and positive signals v_{s1} and v_{s3} with the opposite polarity of the negative accumulated charge are observed near both ground electrode surfaces. A reflected waveform is observed behind the each

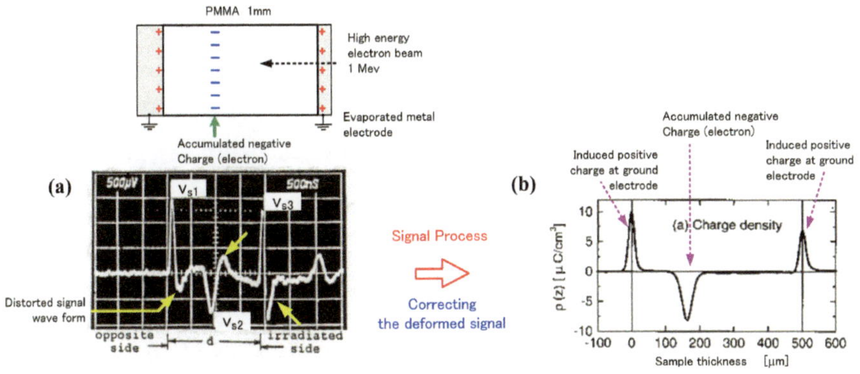

Fig. 8.2 The measured and corrected waveforms by PEA method

signal. This reflection is not the charge signal, it is due to the electric mismatch of the measurement system in fact. The result of calibrated waveform by signal processing is shown in the right side of Fig. 8.2. The mismatched waveforms are clearly removed, and the signals of "induced positive charge", "accumulated negative space charge", and "induced positive charge" are also shown in Fig. 8.2b.

(2) Signal Display in Time Domain and Frequency Domain

The signal processing algorithm is shown in Fig. 8.3. The measured signal $v(t)$ and reference signal $v_0(t)$ in the time domain are obtained by PEA system firstly.

Then, the measured signal $V(f)$ and reference signal $V_0(f)$ in the frequency domain are calculated by the software of Fourier transform (FT). In order to obtain the accumulated charge profile $U(\mathrm{f}) \times G(\mathrm{f})$ in (9), the signal processing calculation is finished in the frequency domain. After that, (9) is subjected to Inverse Fourier Transform (IFT) to obtain the "charge distribution waveform" in (10) in the time domain.

The computer software FFT (Fast Fourier Transform) performs the operation of transforming the measurement signal between the "time domain" and the "frequency domain". Therefore, $p(t)$ in Eq. (8.7), $q\,(t)$ in Eq. (8.9) and $v(t)$ in Eq. (8.11) in the time domain are transformed to $P(f)$ in Eq. (8.8), $Q(f)$ in Eq. (8.10) and $V(f)$ in Eq. (8.12), respectively. The transforming equations are shown below.

Pressure wave signal

In time domain
$$p\,(t) = \frac{Z_{Al}}{Z_{sa} + Z_{Al}} \left[\sigma(0)\,e\,(t) + \int_{-\infty}^{+\infty} \rho(\tau)\,e(t-\tau)\,v_{sa}\,d\tau + \sigma(a)\,e\left(t - \frac{a}{v_{sa}}\right)\right] \quad (8.7)$$

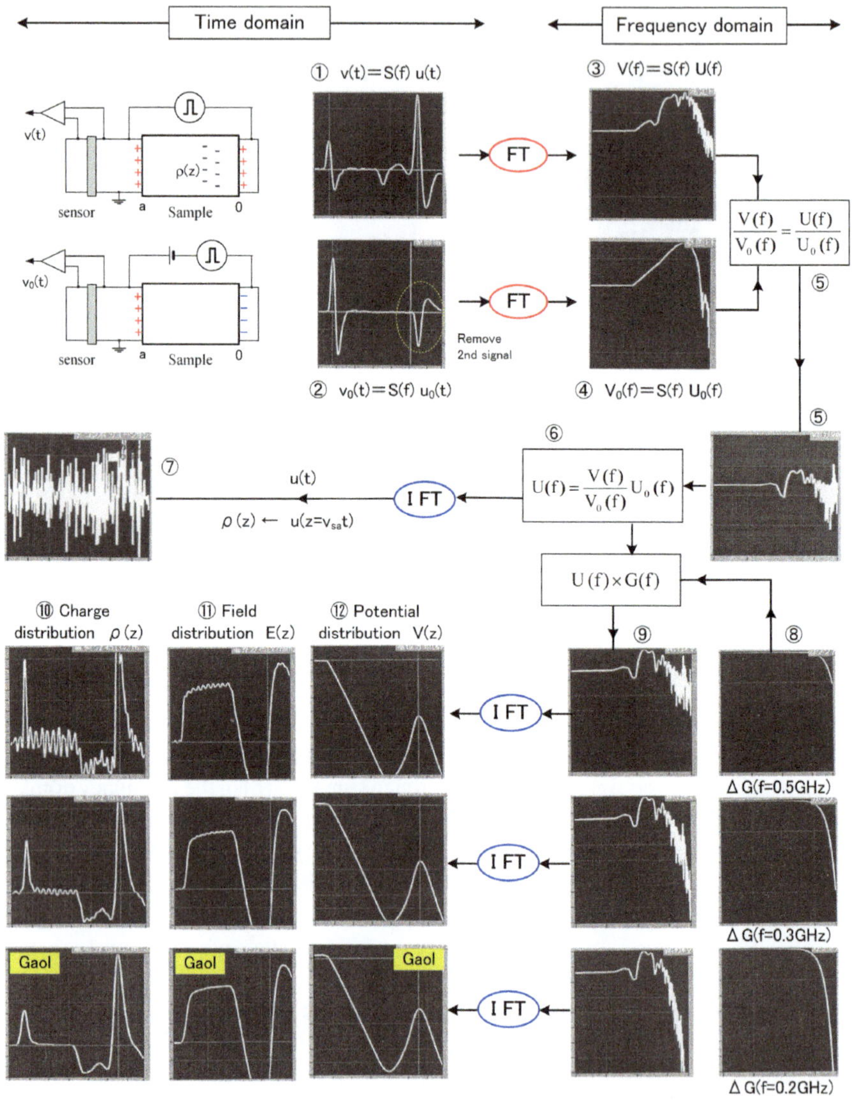

Fig. 8.3 The signal processing algorithm and process in PEA method

$$\text{In frequency domain } P(f) = \frac{Z_{Al}}{Z_{sa} + Z_{Al}} v_{sa} \Delta \tau \; E(f) \times \left[\frac{\sigma(0)}{v_{sa} \Delta \tau} + R(f) + \frac{\sigma(a)}{v_{sa} \Delta \tau} \exp\left(-j\, 2\pi\, f \frac{a}{v_{sa}} \right) \right] \tag{8.8}$$

Charge signal

$$\text{In time domain } q(t) = \frac{2\,Z_p}{Z_{Al} + Z_p}\,\frac{v_p}{b}\int_{-\infty}^{+\infty} d(\tau)\; p(t-\tau)\; d\,\tau \tag{8.9}$$

$$\text{In frequency domain } Q(f) = \frac{2\,Z_p}{Z_{Al} + Z_p}\,\frac{v_p\,\Delta\,\tau}{b}\, D(f)\; P(f) \tag{8.10}$$

Voltage signal

$$\text{In time domain } v_s(t) = \frac{A(f)\;W(f)}{C_p}\,\frac{2\,Z_p}{Z_{Al} + Z_p}\int_{-\infty}^{+\infty} d(\tau)\; p(t-\tau)\;\frac{v_p\,d\tau}{b} \tag{8.11}$$

$$\text{In frequency domain } V(f) = S(f) \times\left[\frac{\sigma(0)}{v_{sa}\,\Delta\,\tau} + R(f) + \frac{\sigma(a)}{v_{sa}\,\Delta\,\tau}\exp\left(-j\,2\pi\,f\frac{a}{v_{sa}}\right)\right] \tag{8.12}$$

System function

Finally, the measured voltage signals are transformed shown in Eq. (8.12) in the frequency domain. The part in the bracket [] of Eq. (8.12) indicates the charge distribution to be measured, and *S*(*f*) is the system function in Eq. (8.13) that is composed of the constant parameters and does not include the charge information.

$$\text{System function } S(f) = \frac{A(f)\;W(f)}{C_p}\,\frac{2\,Z_p}{Z_{Al} + Z_p}\,\frac{Z_{Al}}{Z_{sa} + Z_{Al}}\,\frac{v_p\,\Delta\tau}{b} \times v_{sa}\,\Delta\tau\; H(f)\,E(f) \tag{8.13}$$

Charge amount calibration

The unit of the voltage signal $v(t)$ in Eq. (8.11) and $V(f)$ in Eq. (8.12) is volt [V]. Therefore, it is necessary to calculate the voltage signal $v(t)$ to get the accumulated charge distribution $\rho(z, t)$. At the first step, the signal voltage $v_0(t)$ under a low DC applied voltage $V_{0\mathrm{dc}}$ is measured, under which no charge accumulates in the dielectric. $V_0(f)$ in the frequency domain of $v_0(t)$ and the charge density $\sigma_0(0)$ at the ground electrode is given by Eq. (8.14). The values of ε_0, ε_r, $V_{0\mathrm{dc}}$ and sample thickness a are known. Therefore, since $V_0(f)$ is measured and transformed, the calibrated system function $S(f)$ can be quantitatively obtained.

$$\text{Impulse response signal } V_0(f) = S(f)\,\frac{\boldsymbol{\sigma_0(0)}}{\boldsymbol{v_{sa}\,\Delta\,\tau}} \quad \text{where } \sigma_0(0) = \varepsilon_0\,\varepsilon_r\frac{V_{0dc}}{a} \tag{8.14}$$

Finally the calibration equation is summarized as Eq. (8.15).
Correction equation for calibration

$$\left[\frac{\sigma\,(0)}{v_{sa}\,\Delta\,\tau}+R\,(f)+\frac{\sigma\,(a)}{v_{sa}\,\Delta\,\tau}\,\exp\left(-j\,2\pi\,f\,\frac{a}{v_{sa}}\right)\right]$$
$$=\varepsilon_0\,\varepsilon_r\frac{V_{0dc}}{a}\frac{1}{v_{sa}\,\Delta\,\tau}\times\frac{V\,(f)}{V_0\,(f)} \tag{8.15}$$

Impulse response signal

The PEA measurement instrument has a unique system function $S(f)$ shown in Eq. (8.13), and the measured signal $v(t)$ and the reference signal $v_0(t)$ in Fig. 8.3 can be represented by Eqs. (8.16) and (8.17), where $u(t)$ is a voltage signal based on the actually accumulated charge distribution, and thus the measured voltage signal $v(t)$ can be calculated by the product of $u(t)$ and $S(f)$ in Eq. (8.16). Similarly, $v_0(t)$ can also be expressed by Eq. (8.17), where $u_0(t)$ is a voltage signal only composed of the surface charge $\sigma(0)$ on the ground-side electrode, which removes the surface charge signal on high-voltage side.

$$v(t)=S(f)\,u(t) \tag{8.16}$$

$$v_0(t)=S(f)\,u_0(t) \tag{8.17}$$

For Eq. (8.16), $v(t)$ is a measured value, but $u(t)$ and $S(f)$ are unknown. And in Eq. (8.17), $v_0(t)$ can be measured, and $u_0(t)$ is a known amount because the induced surface charge $\sigma(0)$ can be calculated. Therefore, the system function $S(f)$ can be calculated by Eq. (8.17). By substituting this $S(f)$ into Eq. (8.16), the accumulated charge distribution $u(t)$ can be further obtained. Therefore, in order to obtain the system function $S(f)$ as described above, it is necessary to obtain the voltage signal v_0(t) of the system impulse response. Since $\sigma(0)$ is the surface charge at the electrode, it is obviously possible to create an ideal impulse sound source without width.

So, we are grateful that

"The sound source of the impulse response is the gift that God has given to the PEA method."

(3) **Description of Signal Processing Process (Corresponding to Fig.** 8.3)

(1) $v\,(t)=S\,(f)\,u\,(t)$

This $v(t)$ is the measured signal of the negative charge (in Fig. 8.3) accumulated in the sample. The waveform is distorted and does not directly show the charge distribution. The main cause of this distortion is the oscillation of the pulse voltage excitation, mismatch of the measurement circuit and so on. Although $v(t)$ can be measured, $S(f)$ and $u(t)$ are unknown. A method of processing the waveform $v(t)$ to obtain the charge distribution $\rho(z)$ is described below.

(2) $v_0(t) = S(f)\, u_0(t)$
The reference signal $v_0(t)$ is measured in advance to obtain the unknown $S(f)$, and it has no accumulated charge information inside the sample. For this aim, the applied DC voltage V_{dc} is as low as possible. Since the surface charge at the ground electrode is a sheet charge without width, the corresponding pressure wave is an impulse signal. Although the pulsed pressure wave also generates from the electrode at high-voltage side, this signal is removed and only the signal from the ground electrode is adopted, and $v_0(t)$ is used to represent the corresponding impulse response.

(3) $V(f) = S(f)\, U(f)$
By using the Fourier Transform, the signal waveform (1) $v(t)$ is calculated to the data $V(f)$ in the frequency domain. At this stage, $S(f)$ and $U(f)$ are still unknown quantities.

(4) $V_0(f) = S(f)\, U_0(f)$
Also by the Fourier Transform, the reference signal waveform (2) $v_0(t)$ is transformed to $V_0(f)$ in the frequency domain. This $V_0(f)$ is a known amount because it is determined by the induced charge amount ($\sigma_0(0) = \varepsilon_0\varepsilon_r V_{0dc}/a$) and the sampling thickness ($v_{sa}\Delta\tau$) as shown in Eq. (8.14). Therefore, at this stage, only $S(f)$ is an unknown quantity. The ratio of $V(f)$ and $V_0(f)$ is calculated by computer software. The spectrum of this ratio ($V(f)\,/V_0(f)$) is shown in Fig. 8.3 (5).

(5) $\frac{V(f)}{V_0(f)} = \frac{U(f)}{U_0(f)}$
The ratio of $V(f)$ and $V_0(f)$ is calculated by the computer software. The spectrum of this ratio ($V(f)\,/V_0(f)$) is shown in Fig. 8.3 (5)

(6) $U(f) = \frac{V(f)}{V_0(f)} \times U_0(f) = \frac{V(f)}{S(f)}$
By multiplying the obtained ratio ($V(f)/V_0(f)$) by the impulse function $U_0(f)$, the function $U(f)$ of the charge distribution can be obtained.

In this case, $u_0(t)$ represents the impulse function of the electric charge on the electrode surface, and it is an impulse whose distribution area is "1" without width. Since such an impulse has a constant spectrum in the entire frequency range, its impulse function in the frequency domain is $U_0(f) = 1$. Using this result, $V_0(f)/U_0(f)$ of (4) becomes the system function $S(f)$. Therefore, $V_0(f) = S(f)$ and $V_0(f)$ is equal to the system function $S(f)$.

(7) $u(t)$
$u(t)$ is obtained by Inverse Fourier Transform of $U(f)$ in (6). $u(t)$ is a waveform showing the charge distribution, but the directly calculated data shows that it is buried in noise and cannot show the accurate charge distribution.

The reason is that in the process of division ($V(f)/V_0(f)$), the denominator $V_0(f)$ has extremely small values in the high-frequency region as seen in the spectrum of (4). As a result, a spectrum component with a relatively large value generates in the high-frequency region, which is like the spectrum of (5). To reduce this effect, high-frequency cutoff filter $G(f)$ described in (8) should be used.

(8) Gaussian filter $G(f)$.
Here, the $G(f)$ values in full width with a half maximum $\Delta G(f)$ are shown for comparison at 0.5, 0.3, and 0.2 GHz. The detailed information is explained in Chap. 10.

(9) $U(f) = G(f) \times \frac{V(f)}{V_0(f)} \times U_0(f)$
To reduce the zero-division effect described in (7), multiply the $U(f)$ in (6) by the high-frequency cutoff filter $G(f)$. The results are shown in (9). $\Delta G(f = 0.5\ \text{GHz})$ has a small high-frequency cutoff effect, while $\Delta G(f = 0.2\ \text{GHz})$ has a large high-frequency cutoff effect. A similar effect can be seen in the frequency characteristic of $U(f)$ multiplied by $G(f)$.

(10) $u(z = v\,t) \rightarrow \rho(z)$ Charge distribution
The $U(f)$ multiplied by $G(f)$ is then inverse-Fourier transformed to $u(t)$ in the time domain. Furthermore, when $u(t)$ in the time domain is transformed within the position $z = v_{sa}t$, the charge distribution $\rho(z)$ is obtained. This result is shown in the waveform at the lower left figure of (10). As illustrated in this waveform, the charge distribution of the electrode charge $\sigma(0)$, the space charge distribution $\rho(z)$, and the electrode charge $\sigma(0)$ can be obtained. The best result is based on $\Delta G(f) = 0.2$ GHz.

(11) E(z) electric field distribution.
The electric charge distribution ρ(z) is integrated to obtain the electric field distribution $E(z) = \int (\rho(z)/\varepsilon)\,dz$.

(12) V(z) potential distribution.
The electric field distribution E(z) of (11) is further integrated to obtain the electric potential distribution

$$V(z) = -\int E(z)\,dz$$

Chapter 9
Generation of Pulse Pressure Wave

9.1 Generation of Pulse Pressure Wave from Electrode Induced Charge

(1) Strain amount at the electrode interface

When a DC voltage V_{dc} is applied to the dielectric without charge accumulation as shown in Fig. 9.1, $\sigma(0)$ and $\sigma(a)$ are induced on the electrode surfaces. When a pulsed electric field $e(t)$ is applied to this system, the pulsed electrostatic stress $\sigma(0)e(t)$ acts on the charge $\sigma(0)$. As a result, the strain amount $\xi(t)$ generates at the interface as shown in the figure. The particle velocity $u = \Delta\ \xi(t)/\Delta t$ is derived from Eq. (9.1) based on the electrical equivalent circuit, which is detailedly shown in Chap. 10. If the Eq. (9.1) is integrated over time, the strain amount $\xi(t)$ in Eq. (9.2) can be obtained. In addition, for the relationship of $Z = 1/sv$, refer to Chap. 10.

$$\frac{\Delta\,\xi(t)}{\Delta\,t} = \frac{\sigma(0)\,e(t)}{Z_{sa} + Z_{Al}} \tag{9.1}$$

$$\begin{aligned}\xi(t) &= \frac{Z_{Al}}{Z_{sa} + Z_{Al}} \int_{-\infty}^{+\infty} s_{Al}(\tau)\,\sigma(0)\,e(t)\,v_{Al} d\tau \quad \text{where} \quad Z_{Al} = \frac{1}{s_{Al} v_{Al}} \\ &= \frac{Z_{sa}}{Z_{sa} + Z_{Al}} \int_{-\infty}^{+\infty} s_{sa}(\tau)\,\sigma(0)\,e(t)\,v_{sa} d\tau \quad \text{where} \quad Z_{sa} = \frac{1}{s_{sa} v_{sa}}\end{aligned} \tag{9.2}$$

(2) Generation of pulse pressure wave

Pressure waves $p_1(t)$ and $p_2(t)$ from the cathode electrode charge $\sigma(a)$

A pressure wave p_1(t) generates on the dielectric side and a pressure wave $p_2(t)$ generates on the electrode side, and they propagate on the opposite directions. These

T. Takada et al., *Electric Charge Accumulation in Dielectrics: Measurement and Analysis*,
https://doi.org/10.1007/978-981-19-6156-4_9

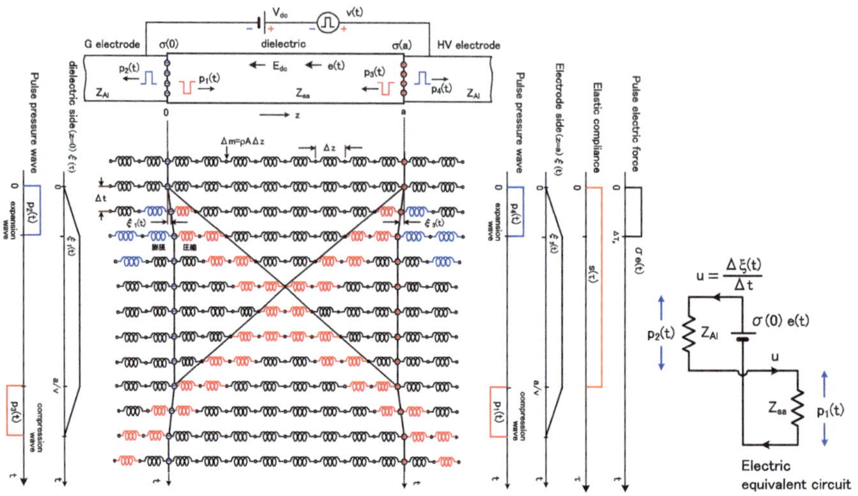

Fig. 9.1 Equivalent measurement system under a DC voltage

$p_1(t)$ and $p_2(t)$ are derived from the electrical equivalent circuit in the figure and shown in Eqs. (9.3) and (9.4).

Compressive wave in a dielectric ($z = 0$)

$$p_1(t) = Z_{sa}\frac{\partial \xi_1}{\partial t} = \frac{Z_{sa}}{Z_{sa} + Z_{Al}}\sigma(0)\,e(t) \tag{9.3}$$

Expansion wave in the electrode ($z = 0$)

$$p_2(t) = Z_{Al}\frac{\partial \xi_1}{\partial t} = \frac{Z_{Al}}{Z_{sa} + Z_{Al}}\sigma(0)\,e(t) \tag{9.4}$$

Pressure waves $p_3(t)$ and $p_4(t)$ from the anode electrode charge $\sigma(a)$

Similarly, pressure waves p_3(t) and $p_4(t)$ from the anode electrode charge $\sigma(a)$ are also drived.

Compressive wave in a dielectric ($z = a$)

$$p_3(t) = Z_{sa}\frac{\partial \xi_3}{\partial t} = \frac{Z_{sa}}{Z_{sa} + Z_{Al}}\sigma(a)\,e(t) \tag{9.5}$$

Expansion wave in the electrode ($z = a$)

$$p_4(t) = Z_{Al}\frac{\partial \xi_3}{\partial t} = \frac{Z_{Al}}{Z_{sa} + Z_{Al}}\sigma(a)\,e(t) \tag{9.6}$$

The magnitudes of pressure waves $p_1(t)$ and $p_2(t)$, $p_3(t)$ and $p_4(t)$ generated from the interfaces are determined by the acoustic impedances of both materials and opposite propagating directions.

9.2 Generation of Pressure Wave from Space Charge

(1) Strain amount due to sheet charge

To simplify the explanation, the volume charge density $\rho(z)$ [C/m^3] is decomposed into the sheet charge $\rho(z)\Delta z$ [C /m^2] as shown in the figure, and the pressure wave $\Delta p(t)$ from the sheet charge is considered. Based on this, $\Delta p(t)$ is integrated to derive the pressure wave $p(t)$ from all $\rho(z)$.

The relationship between the acting force $\rho(z)\Delta z e(t)$, the elastic compliance s [m^2/N] and the strain $\Delta \xi(t)/\Delta z$ obeys the local Hooke's law in Eq. (9.7). By spatially integrating Eq. (9.7), the strain amount $\xi(t)$ in Eq. (9.8) is obtained. It is a convolution calculation of the elastic compliance s of the dielectric and the acting force $F(t) = \rho(z)\Delta z e(t)$. Please also refer to the first term of Eq. (9.5) in Chap. 10 (Fig. 9.2).

$$\frac{\Delta \xi(t)}{\Delta z} = s\ \rho\ (z)\Delta z\ e(t)\quad \text{where}\quad \Delta z = v\ \Delta \tau \tag{9.7}$$

$$\xi\ (t) = \int_{-\infty}^{+\infty} s\ (\tau)\ \rho\ (z)\Delta z\ e\ (t - \tau)\ v\ d\ \tau \tag{9.8}$$

(2) Pressure waves from space charge

The sectional pressure wave $\Delta p_1(t)$ generated from the sheet charge $\rho(z)\Delta z$ is given by Eq. (9.9). It can be easily derived from the electrical equivalent circuit in the figure. The coefficient 1/2 is because the wave generates inside the sample with the same acoustic impedances. Integrating this $\Delta p_1(t)$ gives the calculation method of the pressure wave from the space charge $\rho(z)$.

In the figure, $\Delta p_1(t)$ in Eq. (9.10) is a compressive wave, and $\Delta p_2(t)$ in Eq. (9.11) is an expansion wave. Of course, the compressive wave and the expansion wave may be inverted depending on the polarity of the sheet charge and the direction of the pulsed electric field.

Pressure wave from sheet charge $\rho(z)\Delta z$

$$\Delta p_1(t) = \ \Delta p_2(t) = \frac{1}{2}\rho\ (\tau)\Delta z\ e(t)\quad \text{where}\quad \Delta z = v\ \Delta \tau \tag{9.9}$$

Compressive wave in a dielectric from the space charge $\rho(z)$

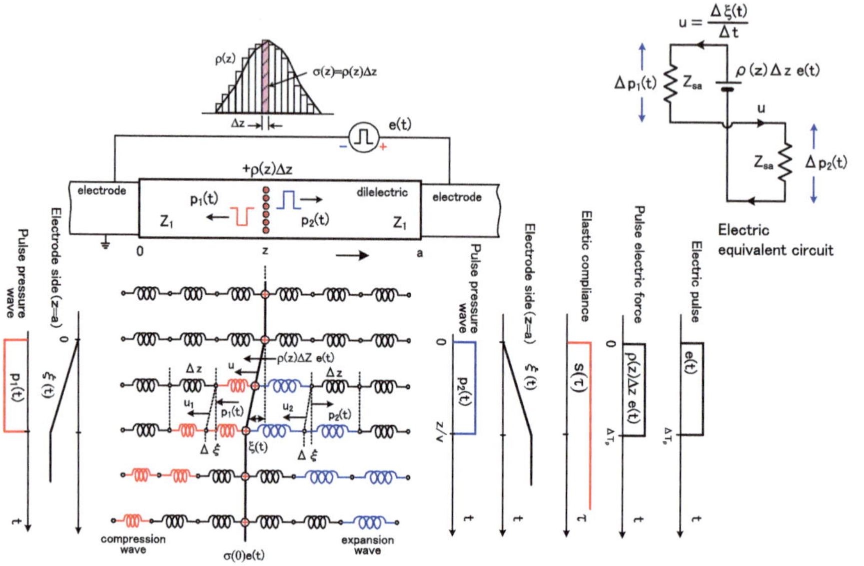

Fig. 9.2 The generation process of a pressure wave under the action of electric pulses

$$p_1(t) = \Sigma\, \Delta p_1(t) = \frac{1}{2}\int_{-\infty}^{+\infty} \rho(\tau)\, e(t-\tau)\, v\, d\tau \tag{9.10}$$

Expansion wave in a dielectric from the space charge $\rho(z)$

$$p_2(t) = \Sigma\, \Delta p_2(t) = \frac{1}{2}\int_{-\infty}^{+\infty} \rho(\tau)\, e(t-\tau)\, v\, d\tau \tag{9.11}$$

(3) Question: Which of e (t) and e (t-τ) is correct in the convolution integral?

The pressure wave generated from the space charge $\rho(z)$ is given by Eq. (9.12). In this case, the space charge distribution $\rho(\tau)$ is a convolutional expression in which the pulsed electric field $e(t\text{-}\tau)$ changes. Why does the pulsed electric field $e(t)$ changes with a time delay τ like $e(t\text{-}\tau)$ even though it is uniformly applied to the dielectric at $t = 0$?

The reason is illustrated in Fig. 9.3, where t is the real time of measurement, $\tau = z/v$ is the time representing the location of space charge. For the τ and t plane, when a pulsed electric field $e(t)$ is applied to the charge distribution $\rho(\tau)$, the force $\xi(\tau, t) = \rho(\tau)\Delta z \times e(t)$ (⑥ in Fig. 9.3) acts simultaneously on the (τ-t) plane. At the same time, pulsed pressure waves generate and propagate from each distributed charge.

The pressure wave $p(t)$ that propagates to the ground electrode surface $z = 0$ is depicted in ⑦ (in Fig. 9.3). The wave $p(t)$ generated at each position $z = v\tau$ is delayed

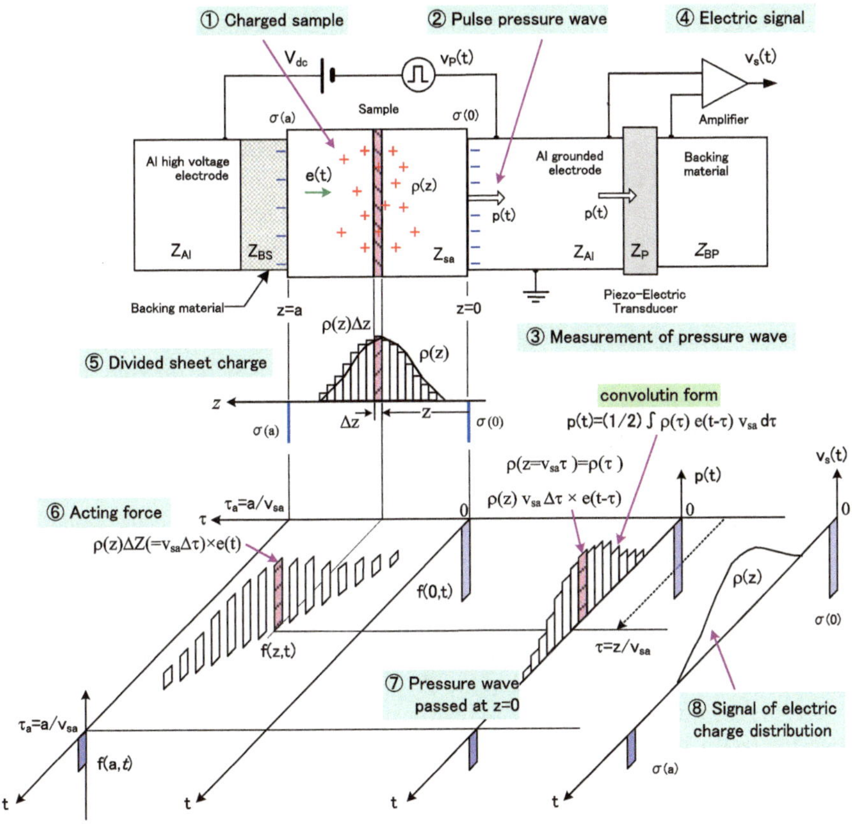

Fig. 9.3 The transform process from space charge to measured signals

by time $\tau = z/v$. The time delay τ is thus incorporated in the applied electric field $e(t\text{-}\tau)$. The charge density $\rho\ (\tau = z/v)$ is because the action of applied electric field $e(t\text{-}\tau)$ has no time delay.

$$p\,(t) = \frac{1}{2}\int_{-\infty}^{+\infty} \rho\,(\tau)\; e\,(t-\tau)\; v\, d\,\tau \tag{9.12}$$

9.3 Generation of Pressure Wave from Polarized Charge

(1) Pressure wave generation model

When a DC voltage is applied to the dielectric sample, the orientational polarization σ_p [C/m^2] occurs. For example, there is also a case of an electret-film sample in which a DC voltage is applied at a high temperature and the orientation polarization σ_p remains. As shown in Fig. 9.4, the orientation polarization of the dielectric surface at $z = 0$ side is $+\sigma_p(0)$, and the orientation polarization of the dielectric surface at $z = a$ side is $-\sigma_p(a)$.

When the pulsed electric field $e(t)$ is superposed with DC voltage to the sample, the strain amount $\xi(t)$ of the dielectric thickness expands, and the pressure wave $p(t)$ generates. The generation of the pressure wave $p(t)$ can be described by drawing the $\xi(t)$ and $p(t)$ in Fig. 9.4.

(2) Strain amount at electrode interface ξ(t)

Focus on the interface between the dielectric and the electrode at $z = 0$ side. The interface can be represented by the electrical equivalent circuit in Fig. 9.4. The force is equal to $\sigma_p(0)e(t)$, and the circuit impedance contains the electrode Z_{Al} and the dielectric Z_{sa}. The circuit current in electrical equivalent circuit is the particle velocity u [m/s]. Moreover, the particle velocity $u = \Delta\ \xi(t)/\Delta t$ can be derived from the electrical equivalent circuit as in Eq. (9.13).

The strain amount $\xi(t)$ in Eq. (9.14) can be obtained by time-integrating Eq. (9.13). Focusing on Eq. (9.14), the strain amount is displayed by the convolution calculation of the piezoelectric constant $d(\tau)$ and the pulsed electric field $e\ (t\text{-}\tau)$. This is the second term of the Eq. (9.5) in the dynamic Hooke's law in Chap. 10.

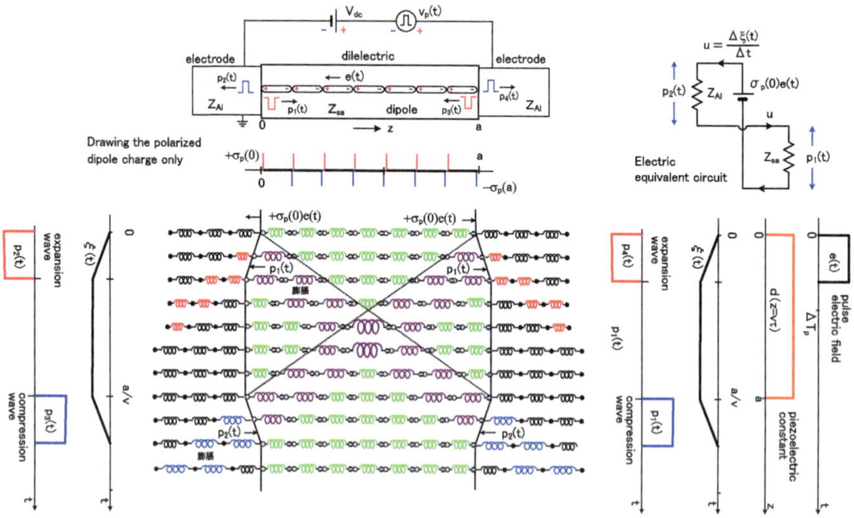

Fig. 9.4 The generation of the pressure wave

$$\frac{\Delta\xi(t)}{\Delta t} = \frac{\sigma_p(0)\,e(t)}{Z_{sa} + Z_{Al}} \tag{9.13}$$

$$\begin{aligned}\xi(t) &= \frac{1}{Z_{sa} + Z_{Al}} \int_{-\infty}^{+\infty} \sigma_p(0)\,e(t)\,dt = \frac{1}{Z_{sa} + Z_{Al}} \frac{1}{s_{sa}} \int_{-\infty}^{+\infty} d(\tau)\,e(t-\tau)\,d\tau \\ &= \frac{1}{Z_{sa} + Z_{Al}} \frac{v_{sa}}{s_{sa} v_{sa}} \int_{-\infty}^{+\infty} d(\tau)\,e(t-\tau)\,d\tau = \frac{Z_{sa}}{Z_{sa} + Z_{Al}} \int_{-\infty}^{+\infty} d(\tau)\,e(t-\tau)\,v_{sa} d\tau\end{aligned} \tag{9.14}$$

where $d(\tau) = \sigma_p(0)\,s_{sa}(\tau) \qquad Z_{sa} s_{sa} v_{sa} = 1$

(3) **Pressure wave from polarized charge**

The pressure wave $p(t)$ is defined by Eq. (9.12) in Fig. 9.3 and $p_1(t)$ to $p_4(t)$ in Fig. 9.4 are also derived in Eqs. (9.15)–(9.18). In conclusion, the pressure wave $p(t)$ is the product of the piezoelectric constant $d(\tau)$ and the pulsed electric field $e(t)$. The amount $p(t)$ depends on the acoustic impedance ratio.

Expansion wave on the dielectric side ($z = 0$)

$$p_1\,(0,t) = Z_{sa} \frac{\partial\xi}{\partial t} = \frac{Z_{sa}}{Z_{sa} + Z_{Al}} \frac{1}{s_{sa}} d\,(\tau)\,e\,(t) \tag{9.15}$$

Compressive wave on the electrode side ($z = 0$)

$$p_2\,(0,t) = Z_{Al} \frac{\partial\xi}{\partial t} = \frac{Z_{Al}}{Z_{sa} + Z_{Al}} \frac{1}{s_{sa}} d(\tau)\,e(t) \tag{9.16}$$

Expansion wave on the dielectric side ($z = a$)

$$p_3\,(0,t) = Z_{sa} \frac{\partial\xi}{\partial t} = \frac{Z_{sa}}{Z_{sa} + Z_{Al}} \frac{1}{s_{sa}} d\,(\tau)\,e\,(t) \tag{9.17}$$

Compressive wave on the electrode side ($z = a$)

$$p_4\,(0,t) = Z_{Al} \frac{\partial\xi}{\partial t} = \frac{Z_{Al}}{Z_{sa} + Z_{Al}} \frac{1}{s_{sa}} d(\tau)\,e(t) \tag{9.18}$$

Chapter 10
Basics of Electrodynamics and Elastic Mechanics

10.1 Basic Piezoelectric Formula

The measurement results of the space charge distribution by the PEA method have been introduced in the above chapters. The measurement principle is based on the basic piezoelectric Eqs. (10.1)–(10.5), the pressure wave generation of Eq. (10.6), and the pressure wave measurement of Eq. (10.7) in Fig. 10.1. The physical models and the equations of these viscoelasticities are also shown in the figure. <See Takada Text Vol. I, Part 1, Chap. 2>.

(1) **Basic piezoelectric formulas** [1]

$\boldsymbol{S} = \boldsymbol{s}\,\boldsymbol{F}$ The first tearm in Eq. (10.1): Strain S [0] occurs when mechanical stress F [N/m^2] is applied to the elastic dielectric material. The characteristics of S and F are in a proportional relationship and are drawn below the figure. The slope is the elastic compliance s [m^2/N].

$\boldsymbol{S} = \boldsymbol{d}\,\boldsymbol{E}$ The second tearm in Eq. (10.1): Similarly, when an electric field E [V/m] is applied, strain S [0] also generates. The characteristics of S and E are proportional to each other and are drawn below. The slope is the piezoelectric constant d [m/V].

$\boldsymbol{D} = \boldsymbol{d}\,\boldsymbol{F}$ The first tearm in Eq. (10.2): When mechanical stress F [N/m^2] is applied to an elastic dielectric material, a charge density D [C/m^2] generates on the dielectric surface. The characteristics of D and F are proportional to each other and are drawn below. The slope is the piezoelectric constant d [m/V].

$\boldsymbol{D} = \boldsymbol{\varepsilon}\,\boldsymbol{E}$ The second tearm in Eq. (10.2): Similarly, when an electric field E [V/m] is applied, a charge density D [C/m^2] also generates. The characteristics of D and E are proportional to each other and are drawn below. The slope is the dielectric constant (permittivity) ε [F/m].

All the above expressions are under the equilibrium state after the application time a/v of the mechanical stress F and the electric field E, where v is the sound velocity. Next, the dynamic relational expression under the transient state is dealt with before the strain $S(t)$ and the charge density $D(t)$ reach the equilibrium state.

T. Takada et al., *Electric Charge Accumulation in Dielectrics: Measurement and Analysis*,
https://doi.org/10.1007/978-981-19-6156-4_10

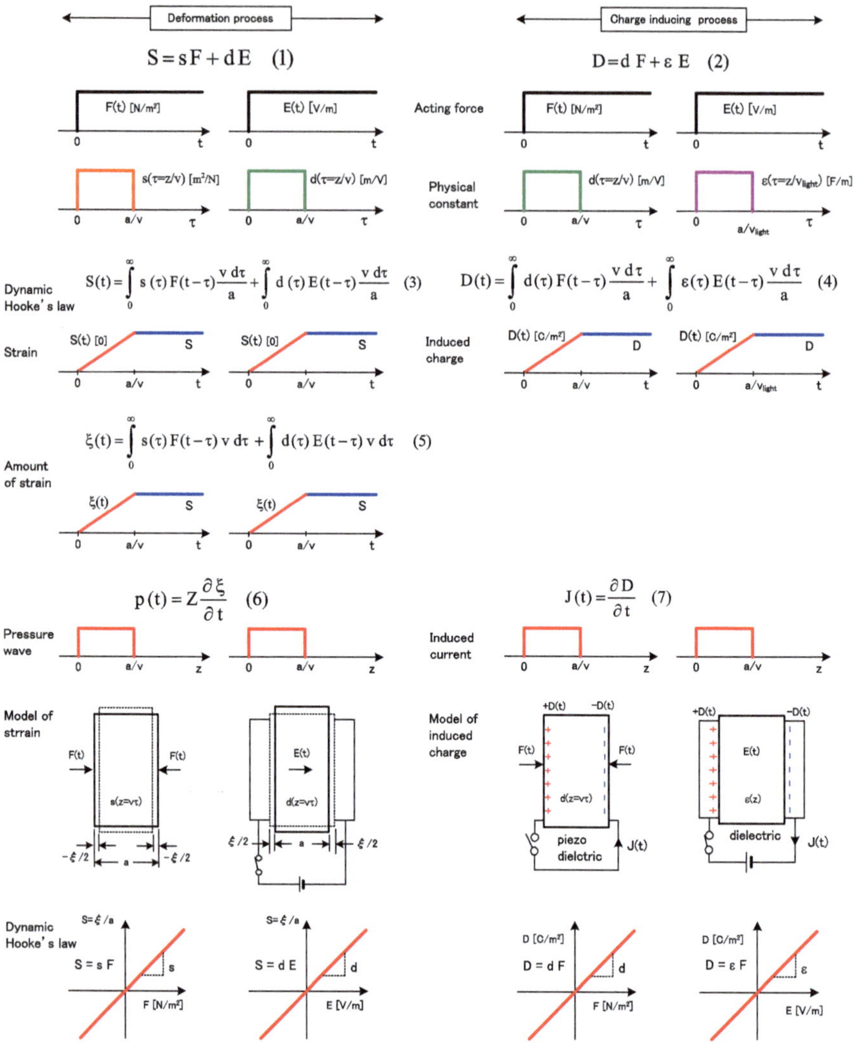

$$S(t)=\int_0^\infty s(\tau)F(t-\tau)\frac{v\,d\tau}{a}+\int_0^\infty d(\tau)E(t-\tau)\frac{v\,d\tau}{a} \quad (3)$$

$$D(t)=\int_0^\infty d(\tau)F(t-\tau)\frac{v\,d\tau}{a}+\int_0^\infty \varepsilon(\tau)E(t-\tau)\frac{v\,d\tau}{a} \quad (4)$$

$$\xi(t)=\int_0^\infty s(\tau)F(t-\tau)\,v\,d\tau+\int_0^\infty d(\tau)E(t-\tau)\,v\,d\tau \quad (5)$$

$$p(t)=Z\frac{\partial\xi}{\partial t} \quad (6)$$

$$J(t)=\frac{\partial D}{\partial t} \quad (7)$$

Fig. 10.1 The basic piezoelectric formulas and diagrammatic sketches

(2) Dynamic Hooke's law

The PEA method is a measurement instrument based on a phenomenon according to the dynamic Hooke's law in which the strain $S(t)$ changes with time before reaching equilibrium.

1st term in Eq. (10.3): The dynamic strain $S(t)$ before reaching equilibrium can be expressed by the convolution integral of $s(\tau)$ and the mechanical action stress $F(t-\tau)$. Since the elastic compliance $s(\tau = z/v)$ is uniform in the elastic body, it becomes a rectangular function. Therefore, $S(t)$ rises linearly with time.

2nd term in Eq. (10.3): This is also an equation of the convolution integral of $d(\tau = z/v)$ and the electric field $E(t-\tau)$ before reaching the equilibrium. Here, since the piezoelectric constant $d(\tau = z/v)$ is uniform in the dielectric body, it becomes a rectangular function. Therefore, the dynamic strain $S(t)$ becomes a linearly rising function.

1st term in Eq. (10.4): The charge density $D(t)$ before reaching equilibrium can be expressed by the convolution integral of $d(\tau)$ and the acting mechanical stress $F(t-\tau)$. Here, since the piezoelectric constant $d(\tau = z/v)$ is uniform, it becomes a rectangular function. Therefore, $D(t)$ rises linearly with time.

2nd term in Eq. (10.4): Further, the charge density $D(t)$ before reaching equilibrium is an expression of the convolution integral of the permittivity $\varepsilon(\tau = z/v_{\text{light}})$ and the electric field $E(t-\tau)$. Here, since the permittivity $\varepsilon(\tau = z/v_{\text{light}})$ is uniform in the dielectric body, it becomes a rectangular function. Therefore, the dynamic strain $D(t)$ becomes a linearly rising function. Even so, since the light has a velocity of 3 $\times$ 10^8 m/s, the $\tau = 10^{-12}$ s is very short when $z = 300$ μm. Therefore, the rise time can be ignored.

Detection of pressure wave: PEA measurement is based on the relationship between pulsed pressure wave $p(t)$ (= acting force $F(t-\tau)$) and its convolution integral with piezoelectric element $d(\tau = z/v)$. Then, the charge distribution is measured.

(3) **Dynamic strain amount $\xi(t)$**

Here, the strain $S(t)$ and the strain amount $\xi(t)$ have a relation of $S(t) = \xi(t)/a$. Therefore, Eq. (10.5) is obtained by multiplying Eq. (10.3) by the thickness a.

1st term in Eq. (10.5): Dynamic strain amount $\xi(t)$ can be calculated by the convolution integral of elastic compliance $s(\tau)$ and acting mechanical force $F(t-\tau)$. $\xi(t)$ also rises linearly with time.

Detection of space charge: This term is used to generate $\xi(t)$ due to electrostatic force acting on electrode charges $\sigma(0)$ and $\sigma(a)$ and space charge $\rho(z)$ in PEA measurement.

2nd term in Eq. (10.5): Further, the dynamic strain amount $\xi(t)$ can be expressed by the convolution integral of the piezoelectric constant $d(\tau)$ and the electric field $E(t-\tau)$. Therefore, $\xi(t)$ also rises linearly with time.

Polarization charge detection: This term is used to generate $\xi(t)$ due to the electrostatic force acting on the polarization charge σ_p [C/m^2] i n PEA measurement.

(4) **Pressure wave $p(t)$ from the strain**

Equation (10.6): The pressure wave $p(t)$ is generated by the product of the time change $\Delta\xi(t)/\Delta t$ of $\xi(t)$ and the acoustic impedance Z.

I would like to write something special. "Electrostatic force $\sigma(0)e(t)$ is not the same as pressure wave $p(t)$".

When an electric field is applied to electric charges, electrostatic forces also act on them. The dielectric deforms and the strain amount $\xi(t)$ increases or decreases. The pressure wave $p(t)$ generates from the time change ($\Delta\xi(t)/\Delta t$) of the dynamic strain amount $\xi(t)$.

In general, many researchers think that the electrostatic force and the pressure wave $p(t)$ are the same. Without considering the elastic mechanics in the above way, the correct charge behavior model cannot be gotten from the measured signal. The combination of electrostatic and viscoelastic mechanics create the basic principle of the PEA method.

Equation (10.7): The current $J(t)$ flows during the time change of the dynamic charge density $D(t)$.

10.2 Local Hooke's Law and Dynamic Hooke's Law

(1) Mass point and spring model

If a tensile force or a compressive force is applied to the elastic body, the body will expand and shrink. Even when a sound wave propagates, the elastic body can still expand and shrink, which further causes the propagation of the sound wave. This physical model is drawn in Fig. 10.2. The elastic body is divided into small parts of $A\Delta z$, and the mass point is $\Delta m = \rho_m A \Delta z$. Here, ρ_m [kg/m^3] is the density, A [m^2] is the cross-sectional area, and Δz [m] is the infinitesimal section. When an acting force is applied to it, the infinitesimal section Δz expands and shrinks by $\Delta\xi$. This expansion and shrink is displayed by a "spring". As shown in the figure, the elastic body is expressed by the "mass point Δm" and "mass point spring model" to creat the motion equation.

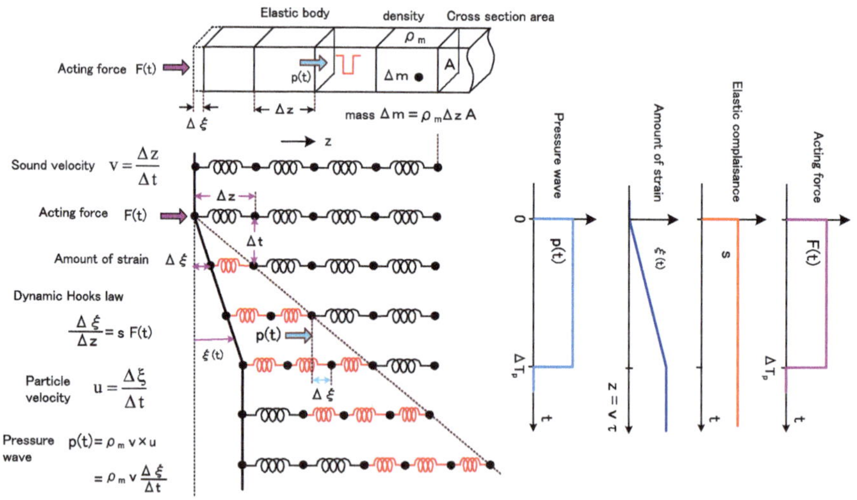

Fig. 10.2 The equivalent spring model of the measured elastic body

(2) **Local Hooke's law**

As shown in Fig. 10.2, when a rectangular acting force $F(t)$ [N/m^2] is applied to the elastic body from the border, the infinitesimal section Δz shrinks by $\Delta\xi$ after an infinitesimal time Δt. By applying Hooke's law to this infinitesimal section, the local Hooke's law of Eq. (10.1) is given. Here, the proportionality coefficient is the elastic compliance s [m^2/N].

$$\frac{\Delta \xi}{\Delta z} = sF(t) \tag{10.1}$$

$$\frac{\Delta \xi}{\Delta z = v\,\Delta \tau} = s\,(\tau)\,F(t) \tag{10.2}$$

$$\xi\,(t) = \int_{-\infty}^{+\infty} s\,(\tau)\,F\,(t-\tau)\,v\,d\,\tau \tag{10.3}$$

(3) **Dynamic Hooke's law**

Further, if the infinitesimal segment is set as $\Delta z = v\Delta\tau$, Eq. (10.2) is obtained, and by integrating Eq. (10.2) with $v\Delta\tau$, $s(\tau)$ and $F(t)$ in Eq. (10.3) are obtained. It is the convolution integral, and thus Eq. (10.3) becomes the dynamic Hooke's law. On the right side of the figure, the linearly increasing strain amount $\xi(t)$ is drawn based on the convolution integral of $s(\tau)$ and $F(t)$ (also see Fig. 10.1). When the acting force $F(t)$ stops, the increase of strain amount $\xi(t)$ also stops. Then, the energy of the mass point Δm is conserved, and the "spring in a shrinking state" continues to move due to inertia.

(4) **Particle velocity and sound velocity**

The difference between the sound speed and the particle speed is summarized here. Equation (10.4) is the particle velocity u [m/s], and it is because the infinitesimal section Δz shrinks by $\Delta\xi$ after Δt. Equation (10.5) is the sound velocity v [m/s], which is the speed that advances by the infinitesimal section Δz after Δt. For example, the particle velocity of aluminum is $u = 0.059$ m/s (in case of P = 100 kg/cm^2) and the sound velocity is $v = 6420$ m/s. The particle velocity is extremely small.

$$\text{particle velocity} \quad u = \frac{\Delta\,\xi(t)}{\Delta\,t} \tag{10.4}$$

$$\text{sound velocity(propagation velosity)} \quad v = \frac{\Delta\,z}{\Delta\,t} \tag{10.5}$$

(5) **Characterization of pressure wave**

The elastic pressure wave $p(t)$ is shown in Eq. (10.6). As a result, $p(t)$ is the product of the acoustic impedance Z (=$\rho_m\,v$) and the particle velocity u (=$\Delta\xi\,(t)/\Delta t$). This $p(t)$ can be derived from Eqs. (10.7)–(10.9).

$$\text{elastic pressure wave} \quad p\,(t) = \rho_m\, v\;\, u = \rho_m\, v\, \frac{\partial\, \xi(t)}{\partial\, t} = Z\, \frac{\partial\, \xi(t)}{\partial\, t} \tag{10.6}$$

$$\text{Momentum of mass point} \quad \Delta m\; u \tag{10.7}$$

$$\text{Momentum and impulse } \Delta m\; u = p\,(t) A\;\Delta t \tag{10.8}$$

$$\text{acoustic impedance } Z = \rho_m\;\, v \tag{10.9}$$

(6) **Important relational expressions**

The relationship between density ρ_{m} [kg/m^3], sound velocity v [m/s], and elastic compliance s [m^2/N] is given by Eq. (10.10). It is a basic relational expression in viscoelastic mechanics, and there is a relation of "force F" = "mass m" × "acceleration α", as shown in Eq. (10.11). Equation (10.14) can be derived from the relationship between Eqs. (10.12) and (10.13). As a result, the acoustic impedance in Eq. (10.9) can be derived to the relationship in Eq. (10.14).

$$\rho_m\; v^2 s = 1 \tag{10.10}$$

$$\frac{1}{s}\left[\frac{N}{m^2}\right] = \rho_m \left[\frac{kg}{m^3}\right] \times v^2 \left[\frac{m^2}{s^2}\right] \qquad F\;[N] = m\;[kg] \times \alpha \left[\frac{m}{s^2}\right] \tag{10.11}$$

$$\frac{\Delta\xi(t)}{v\;\Delta t} = sp(t) \tag{10.12}$$

$$p\,(t) = \frac{1}{s\;v}\,\frac{\partial\,\xi(t)}{\partial\, t} = Z\,\frac{\partial\,\xi(t)}{\partial\, t} \tag{10.13}$$

$$Z = \rho_m\;\, v = \frac{1}{v\; s} \tag{10.14}$$

(7) **Electric equivalent circuit and pressure wave distribution**

When the acting force ($F(t) = \sigma(0)e(t)$) acts on the interface of different elastic bodies (dielectric and electrode), the interface is displaced by the strain amount $\xi(t)$ with the particle velocity $u = \Delta\xi(\mathrm{t})/\Delta t$. Then, the pressure wave $p_1(t)$ generates on the dielectric side (Z_1) and the pressure wave $p_2(t)$ generates on the electrode side (Z_2). The relationship between the acting force $F(t)$, the particle velocity u, and the pressure wave $p(t)$ can be displayed by an electric equivalent circuit.

Electric equivalent circuit

The acting force $F(t) = \sigma(0)e(t)$ can be represented by the power source. The acoustic impedances Z_1 and Z_2 correspond to the electric circuit impedance, and the particle

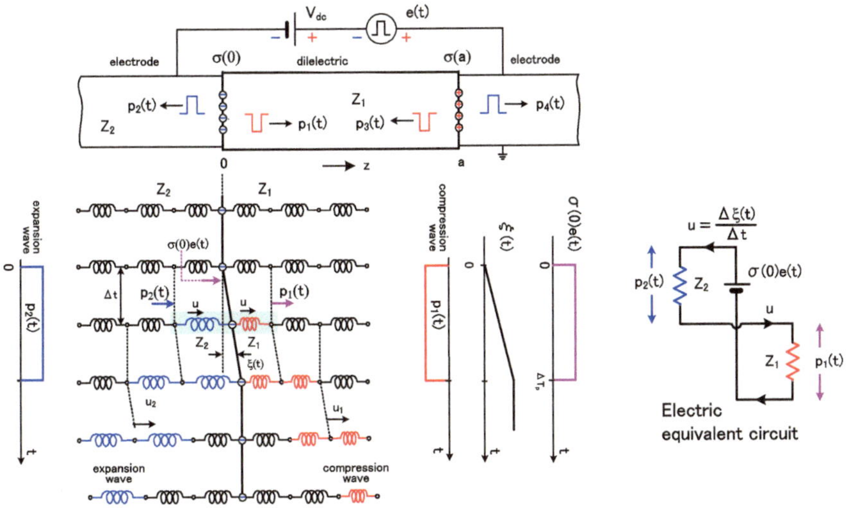

Fig. 10.3 The electric equivalent circuit under an acting force

velocity $u = \Delta\xi(t)/\Delta t$ can correspond to the current. The blue part on the left of the figure is applied to the electric equivalent circuit (right in the figure).

Equation (10.15) can be gotten when applying Ohm's law to this equivalent circuit. Here, the particle velocity is $u = \Delta\xi(t)/\Delta t$. Each pressure wave $p(t)$ is given by the product of the acoustic impedance Z and the particle velocity. Therefore, the calculations of pressure waves $p_1(t)$ and $p_2(t)$ become Eqs. (10.16) and (10.17), respectively. That is, the acting force $F(\mathrm{t}) = \sigma(0)e(t)$ is distributed to the pressure waves $p_1(t)$ and $p_2(t)$ by the acoustic impedance ratio (Fig. 10.3).

$$
\begin{aligned}
F(t) = \sigma(0)\, e(t) &= (Z_1 + Z_2)\, u = (Z_1 + Z_2)\frac{\partial\, \xi(t)}{\partial\, t} \\
&= Z_1 \frac{\partial\, \xi(t)}{\partial\, t} + Z_2 \frac{\partial\, \xi(t)}{\partial\, t} = p_1(t) + p_2(t)
\end{aligned}
\tag{10.15}
$$

Secondary action stress on the dielectric side (z = 0)

$$
p_1(t) = Z_1 \frac{\partial\, \xi(t)}{\partial\, t} = \frac{Z_1}{Z_1 + Z_2}\, \sigma(0)\, e(t)
\tag{10.16}
$$

Secondary action stress on the electrode side (z = 0)

$$
p_2(t) = Z_2 \frac{\partial\, \xi(t)}{\partial\, t} = \frac{Z_2}{Z_1 + Z_2}\, \sigma(0)\, e(t)
\tag{10.17}
$$

10.3 Explanation of Convolution Integral

The convolution integral of Eq. (10.18) will be explained using a physical model. When a step-like acting pressure $F(t)$ is applied to an elastic body with an elastic compliance $s(\tau)$, the strain amount $\xi(t)$ increases with time. This is the basic piezoelectric equation for the first term of the Eq. (10.5) in Fig. 10.1. The mathematical model of this convolution integral will be described.

$$\xi(t) = \int_{-\infty}^{+\infty} s(\tau)\, F(t-\tau)\, v d\tau \tag{10.18}$$

Mathematical model

(1) **Elastic body (thickness *a*)**: The elastic compliance $s(z = \tau/v)$ is uniform. Therefore, the figure of $s(z = \tau/v)$ shows a rectangle. Then, the thickness a is converted into the time axis $\tau = a/v$ and drawn.
(2) **Acting force**: Draw the acting force $F(\tau)$ based on that the space z is a function of time $\tau(=z/v)$.
(3) **Folding**: $F(\tau)$ is folded at the origin of the time axis and $F(-\tau)$ is drawn.
(4) **Displacement**: When $F(-\tau)$ changes on the time axis τ, it becomes $F(t-\tau)$ after the elapsed time t.
(5) **Convolution**: The pressure wave $F(t-\tau)$ propagates in the elastic body with the elastic compliance $s(\tau)$.
(6) **Integral**: The part where $F(t-\tau)$ and $s(\tau)$ overlap is the solution $\xi(t)$ of the convolution integral (Fig. 10.4).

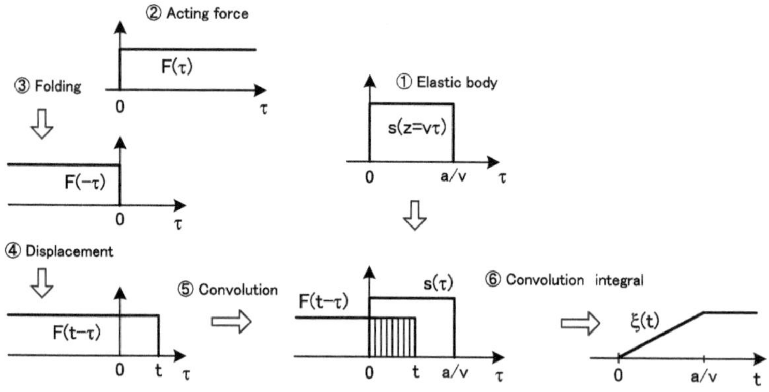

Fig. 10.4 Convolution integral process of the two parts

10.4 Expansion of Dielectric Material Under Applied Electric Fields—Force Acting on Polarized Charge

Virtual experiment and question

A voltage is applied to the air between parallel plate electrodes (Fig. 10.5a).

Q1: Are the electrodes separated? · Do they approach? Explain needs models and formulas.

A dielectric is sandwiched between electrodes and a voltage is applied (Fig. 10.5b).

Q2: Are the electrodes separated? · Do they approach? Explain needs models and formulas.

Q3: Does the dielectric expand? · Should it shrink? Explain needs models and formulas.

A dielectric is placed between the charged electrodes under an applied voltage (Fig. 10.5c).

Q4: Does the dielectric expand? · Should it shrink? Explain needs models and formulas.

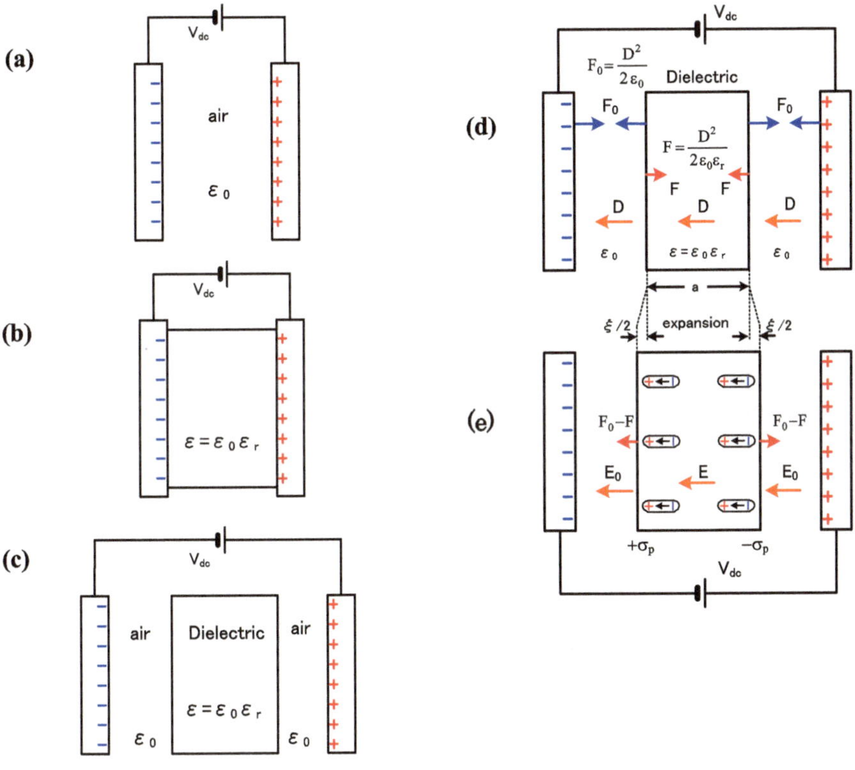

Fig. 10.5 Models for different mediums

<Consideration>

Consider the Q1 model (Fig. 10.5a)

The force F withstood by the electrodes is the product of the induced charge $\sigma(0)$ and the electric field $E_0/2$ due to the opposite $\sigma(a)$ as in the following equation. Please think about why the electric field is $E_0/2$.

$$F = \sigma(0) \times \frac{E_0}{2} = \sigma(a) \times \frac{E_0}{2} = \frac{\varepsilon_0 E_0^2}{2} \quad \text{where} \quad \sigma(0) = \sigma(a) = \varepsilon_0 E_0 \tag{10.19}$$

Consider the Q4 model (Fig. 10.5d)

Applying a DC voltage V_{dc} to the dielectric increases the thickness of the dielectric. Figure 10.5d shows the model. The Maxwell stress, which shrinks and is represented by both Eqs. (10.20) and (10.21), acts on the air layer (ε_0) and the dielectric ($\varepsilon = \varepsilon_0\varepsilon_r$). Since the relative permittivity is $\varepsilon_r > 1$ with $F_0 > F$, the force of $F_0 - F$ in Eq. (10.22), which is the difference between the two forces, acts on the dielectric surface and thus it expands.

$$F_0 = \frac{D^2}{2\varepsilon_0} \tag{10.20}$$

$$F = \frac{D^2}{2\varepsilon_0 \varepsilon_r} \tag{10.21}$$

$$F_0 - F = \frac{D^2}{2\varepsilon_0} - \frac{D^2}{2\varepsilon_0 \varepsilon_r} = \sigma_p \times \frac{E_0}{2} \tag{10.22}$$

When this $F_0 - F$ is transformed, it can be represented by the product of the polarization charge σ_p [C/m^2] in the final term of Eq. (10.23) and the electric field $E/2$. The polarization charge σ_p is expressed in Eq. (10.24).

$$\begin{aligned} F_0 - F &= \frac{D^2}{2\varepsilon_0} - \frac{D^2}{2\varepsilon_0 \varepsilon_r} = \frac{D^2(\varepsilon_r - 1)}{2\varepsilon_0 \varepsilon_r} = \frac{\varepsilon_0 \varepsilon_r E \times \varepsilon_0 E_0 \times \chi}{2\varepsilon_0 \varepsilon_r} \\ &= \varepsilon_0 \chi E \times \frac{E_0}{2} = \sigma_p \times \frac{E_0}{2} \end{aligned} \tag{10.23}$$

$$\begin{aligned} \text{where} \quad & D = \varepsilon_0 E + \sigma_p = \varepsilon_0 E + \varepsilon_0 \chi E, = \varepsilon_0 (1 + \chi) E = \varepsilon_0 \varepsilon_r E \\ & \sigma_p = \varepsilon_0 \chi E \qquad \varepsilon_r = (1 + \chi) \qquad \varepsilon_r - 1 = \chi \end{aligned} \tag{10.24}$$

Similar answers can be obtained from Eqs. (10.25) and (10.26).

$$F_0 - F = \frac{D^2}{2\,\varepsilon_0} - \frac{D^2}{2\,\varepsilon_0\,\varepsilon_r} = \frac{\varepsilon_0^2\,E_0^2}{2\,\varepsilon_0} - \frac{\varepsilon_0^2\,E_0^2}{2\,\varepsilon_0\,\varepsilon_r}$$
$$= \left(\varepsilon_0 E_0 - \frac{\varepsilon_0 E_0}{\varepsilon_r}\right) \times \frac{E_0}{2} = (D - \varepsilon_0 E) \times \frac{E_0}{2} = \sigma_p \times \frac{E_0}{2} \quad (10.25)$$

$$\text{where} \quad D = \varepsilon_0\,E_0 = \varepsilon_0\,\varepsilon_r\,E \quad D = \varepsilon_0\,E + \sigma_p \quad \sigma_p = D - \varepsilon_0\,E \quad (10.26)$$

Summary

The anode and cathode pull each other. Conversely, the dielectric expands due to the extension of the orientation polarization.

10.5 Fourier Transform

(1) **Fourier Transform**

The charge signal $q(t)$ based on the convolution integral of the pulse pressure wave $p(t-\tau)$ and the piezoelectric element $h(z = u\tau)$ is shown in Eq. (10.27). When the convolution integral in Eq. (10.27) is transformed by Fourier Transform, it becomes $Q(f)$ in Eq. (10.29).

This $Q(f)$ is in the form of the product of $H(f)$ and $P(f)$ which are the functions in the frequency domain. Similarly, $H(f)$ is the piezoelectric function $h(\tau)$, and $P(f)$ is the Fourier transform of the pressure wave $p(t)$.

Convolution integral in the time domain

$$q(t) = \frac{u}{b} \int_{-\infty}^{+\infty} h(\tau)\; p(t - \tau)\; d\tau \quad (10.27)$$

Product in frequency domain

$$\Leftrightarrow \quad Q(f) = \frac{u}{b}\; H(f)\; P(f) \quad (10.28)$$

Fourier transform

$$Q\,(f) = \int_{-\infty}^{+\infty} q\,(t)\; \exp\,(-j2\pi f\; t)\; dt \quad (10.29)$$

$$H(f) = \int_{-\infty}^{+\infty} h(t)\, \exp(-j2\pi f\; t)\; dt \quad (10.30)$$

$$P(f) = \int_{-\infty}^{+\infty} p(t) \exp(-j\,2\pi\; f\; t)\, dt \tag{10.31}$$

In Eq. (10.29) in the Fourier transform, the time function $q(t)$ is multiplied by the exponential function $\exp(-j2\pi ft)$ and then they are integrated in time from $-\infty$ to $+\infty$. As a result, $Q(f)$ becomes a frequency function.

(2) Fourier Transform model (Fig. 10.6)

$q(t)$ is the time function: the waveform drawn in the Time domain (time function) in the figure.

$Q(f)$ is the frequency function: the spectrum drawn in the Frequency domain in the figure.

The spectrum represents the amplitude value of each frequency component (DC component, fundamental wave, second harmonic, third harmonic, …). Therefore, $Q(f)$ indicates the amplitude value of each frequency. Conversely, if the amplitude value of each frequency is calculated, the time function $q(t)$ can be obtained.

(3) Fourier transform of convolution

The convolution integral in Eq. (10.27) can be expressed in the form of product of the frequency functions, as shown in the following equation.

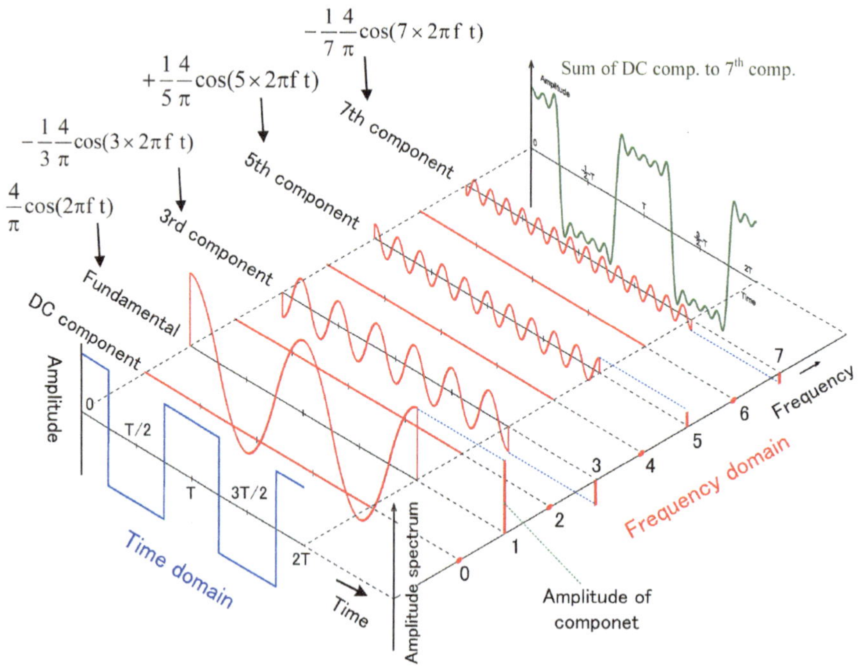

Fig. 10.6 Signal transformation from time domain to frequency domain

$$Q(f) = \int_{-\infty}^{+\infty} q(t)\exp(-\mathrm{j}2\pi f\,t)\,\mathrm{d}t = \int_{-\infty}^{+\infty}\left[\frac{u}{b}\int_{-\infty}^{+\infty} h(\tau)p(t-\tau)\,\mathrm{d}\tau\right]\exp(-\mathrm{j}2\pi f\,t)\,\mathrm{d}t$$

$$= \int_{-\infty}^{+\infty}\frac{u}{b}h(\tau)\left[\int_{-\infty}^{+\infty} p(t-\tau)\exp(-\mathrm{j}2\pi f\,t)\,\mathrm{d}t\right]\mathrm{d}\tau$$

$$\int_{-\infty}^{+\infty} p(t-\tau)\exp(-\mathrm{j}2\pi f\,t)\,\mathrm{d}t = \int_{-\infty}^{+\infty} p(w)\exp(-\mathrm{j}2\pi f\,(w+\tau))\,\mathrm{d}w$$

$$= \exp(-\mathrm{j}2\pi f\,\tau)\int_{-\infty}^{+\infty} p(w)\exp(-\mathrm{j}2\pi f\,w)\,\mathrm{d}w$$

$$= \exp(-\mathrm{j}2\pi f\,\tau)\,P(f)$$

$$Q(f) = \int_{-\infty}^{+\infty}\frac{u}{b}h(\tau)\exp(-\mathrm{j}2\pi f\,\tau)\,P(f)\,\mathrm{d}\tau = \frac{u}{b}P(f)\int_{-\infty}^{+\infty} h(\tau)\exp(-\mathrm{j}2\pi f\,\tau)\,\mathrm{d}\tau = \frac{u}{b}H(f)P(f) \quad (10.32)$$

(4) **Application of Fourier Transform**

Figure 10.6 shows an example of signal processing in which measurement data $v(t)$ as a time function (in Time domain) is converted into a frequency function (in Frequency domain) $V(f)$ by Fourier Transform. In this way, the Fourier Transform is used for signal processing that corrects the distortion of the measured waveform (Photos 10.1 and 10.2).

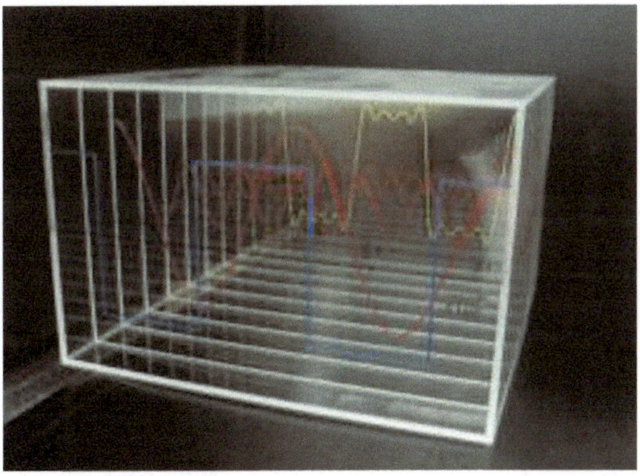

Photo 10.1 Fourier square wave (Front)

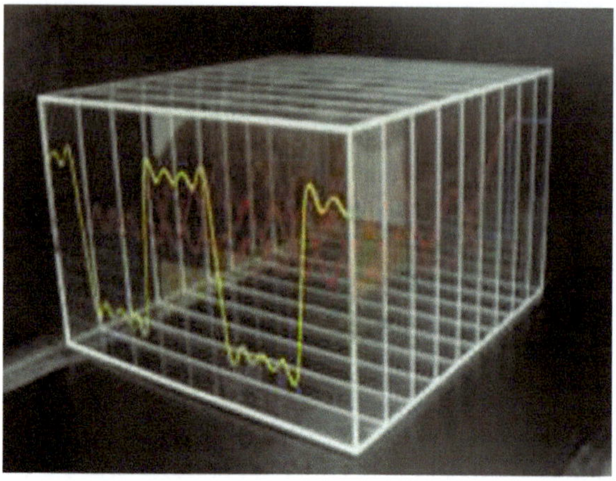

Photo 10.2 Fourier square wave (Back)

10.6 Gaussian Filter

(1) Display of Gaussian distribution in time domain and frequency domain

The Gaussian distribution function $g(t)$ is given by the left of Eq. (10.33) and has a symmetrical shape with respect to the function t on the horizontal axis, as shown in the Fig. 10.7. The coefficient $\surd(a/\pi)$ is standardized so that the integrated value becomes "1" when $g(t)$ is integrated from $-\infty$ to $+\infty$.

$$g\,(t) = \sqrt{\frac{a}{\pi}}\, e^{-a\,t^2} \quad \begin{array}{c}\Rightarrow\\ \overline{}\\ \Leftarrow\end{array} \quad \frac{F\ T}{I\ F\ T} \quad \begin{array}{c}\Rightarrow\\ \overline{}\\ \Leftarrow\end{array} \quad G(f) = e^{-\frac{(2\,\pi\,f)^2}{4\,a}} \tag{10.33}$$

FT is Fourier Transform, and IF is Inverse Fourier Transform.

See Takada Text Vol.1, Part 2, Chap. 4, Appendix 4 for derivation and explanation of these equations.

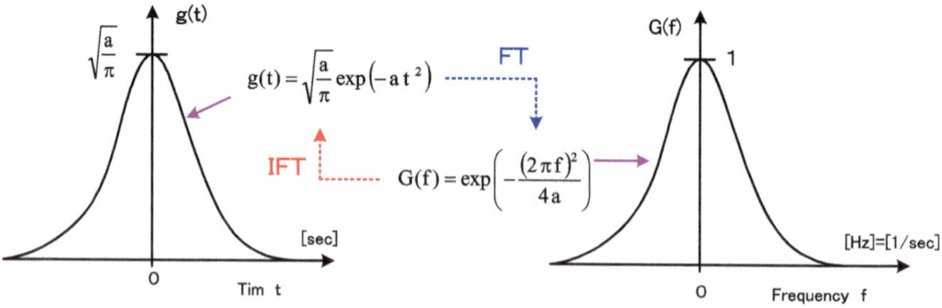

Fig. 10.7 The Gaussian functions in the time and frequency domains

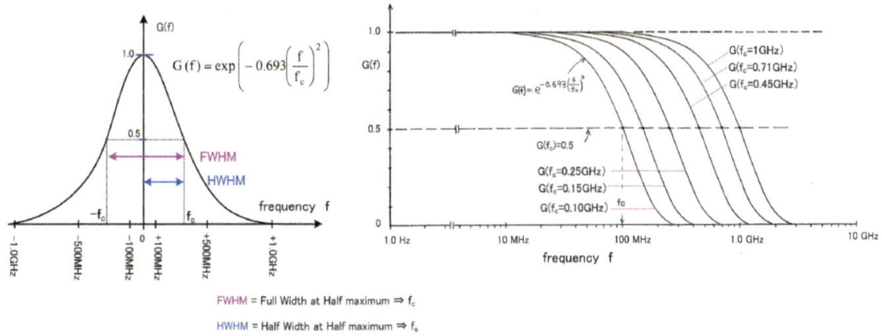

Fig. 10.8 The frequency characteristic of the Gaussian Filter

(2) Gaussian Filter and half-width frequency

The characteristic of the Gaussian distribution filter $G(f)$ in Eq. (10.33) is drawn in Fig. 10.8. The maximum value attenuates symmetrically around the direct current component ($f = 0$) with $G(f = 0) = 1$. There are positive frequencies on the horizontal axis without negative frequencies. Therefore, $G(f)$ covers only the positive frequency range.

The frequency width f_c corresponding to half of the maximum value is used as a parameter indicating the band characteristic of the filter. The frequency is called the half-width frequency f_c, at which the value of the Gaussian function is equal to $G(f = f_c) = 0.5$, and its value is given by Eq. (10.35). By substituting Eq. (10.35) into Eq. (10.34), the Gaussian Filter formula is obtained as in Eq. (10.36).

$$G(f = f_c) = 0.5 = \exp\left(- \frac{(2\,\pi \ f_c)^2}{4\,a} \right) \tag{10.34}$$

$$f_c = \sqrt{\frac{a \ \ln(0.5)}{\pi^2}} \quad \text{where } \ln(0.5) = 0.693 \tag{10.35}$$

$$G\,(f) = \exp\left(-\,0.693 \left(\frac{f}{f_c} \right)^2 \right) \tag{10.36}$$

(3) Semi-logarithmic graph display of Gaussian Filter

Generally, the frequency characteristic of an amplifier and the frequency characteristic of a filter are represented by a common logarithmic scale of frequency on the horizontal axis. The logarithmic scale is used because it is easy to clearly indicate the frequency value from high to low frequencies, and it is easy to read the frequency value.

Therefore, when only the positive frequency range of the Gaussian distribution function $G(f)$ in Eq. (10.34) is targeted, the characteristic is shown in Fig. 10.8 with a semi-logarithmic graph. $G(f) \approx 1$ in the frequency range from $f = 1.0$ Hz to 10 GHz, but the *G*(f) characteristic changes in the three-digit range from f = 10 MHz to 10 GHz.

The intersection between the value of *G*(f_c) = 0.5 on the vertical axis and the characteristic curve represents the half-width frequency f_c in Eq. (10.35).

(4) **Application of Gaussian filter**

The explanation of signal processing process in Fig. 2.3 of Chap. 2 introduces an example using this Gaussian Filter *G*(f). Thus, the Gaussian Filter *G*(f) can be used in the signal processing process.

10.7 Charged Elastic Body

Question

Does an action force act on an electron (electric charge) with a small mass to generate a sound wave?

It has been described that the strain amount $\xi(t)$ generates when an electric field is applied to the charged particles.

In this case, the sound wave cannot propagate unless the electric charge is kind of particle with the same mass as the surrounding medium (dielectric). If the charge is an electron or a hole, the mass is overwhelmingly small and cannot be a measured target.

Therefore, we consider the model of the charged charge as follows.

Way of thinking

Trapped charges (electrons and holes) are part of the surrounding medium.

We consider a model in which the electrons and holes are trapped in the trap sites of the dielectric material.

When the center of the polyethylene molecule in the figure is oxidized, the electron trap depth is ϕ_{te} = 2.62 eV and the hole trap depth is ϕ_{th} = 1.28 eV.

Even if a pulsed electric field is applied to the trapped electrons therein, the electrons are not released from the trap and remain trapped without move.

As a result, trapped electrons (negative charges) become integral with the surrounding medium and can be understood as "particles with equivalent mass".

Therefore, it is understood that a pressure wave can generate by making the dielectric substance get a strain amount $\Delta\xi(t)$ by the electrostatic force acting on the trapped electrons and holes (Fig. 10.9).

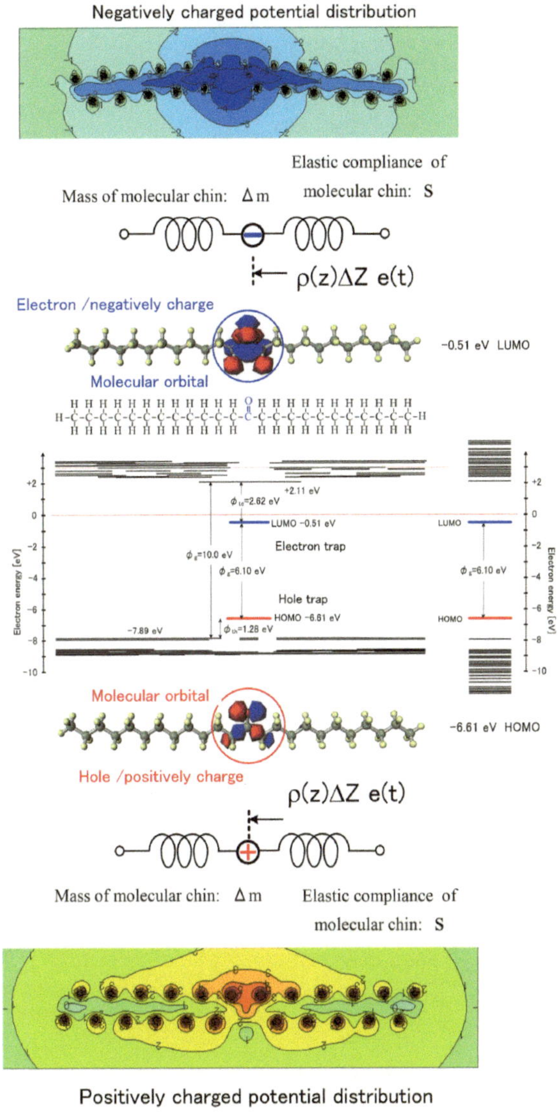

Fig. 10.9 Analysis of the trapped electrons and holes

Reference

1. T. Ogawa, *Basics of Crystal Engineering*, (Shoukabo, 1998), Section 5.1.

Chapter 11
Examples of PEA Measurement Results

11.1 Accumulation of Homo and Hetero Charges

(1) **Positive charge injection in Polyethylene (PE)**

Figure 11.1a: Positive DC electric field of +300 kV/mm is applied to a PE sample (with a thickness of 150 μm). Positive charges are injected from the anode on high-voltage side (semiconductive layer electrode) to form space charges.

Figure 11.1b: Negative DC electric field of −300 kV/mm is applied. Positive charges are injected from the anode on ground side (aluminum electrode) to form space charges. The polyethylene sample always has a positive charge injection from the anode. <See Takada Text Vol. II, Part 4, Chap. 7, Sect. 7.6>

(2) **Observed data of homo and hetero charges inside Polyimide (PI) Kapton-H (Toray/DuPont)**

Figure 11.1c: At a temperature of 80 °C, a homo electric charge ditribution is formed by applying a low electric field, and a hetero electric charge ditribution is formed by applying a high electric field.

- When a low voltage is applied ($E = 30$ kV/mm to 90 kV/mm), positive charge is injected from the anode and negative charge is injected from the cathode to form homo charges.
- When a high electric field is applied ($E = 120$ kV/mm), hetero electric charge clearly accumulates. This hetero charge is the generation of bulk charge. That is, electrons at the HOMO level are excited to the LUMO level, and holes remain at the HOMO level, so-called electron–hole pairs generate [1, 2].

T. Takada et al., *Electric Charge Accumulation in Dielectrics: Measurement and Analysis*,
https://doi.org/10.1007/978-981-19-6156-4_11

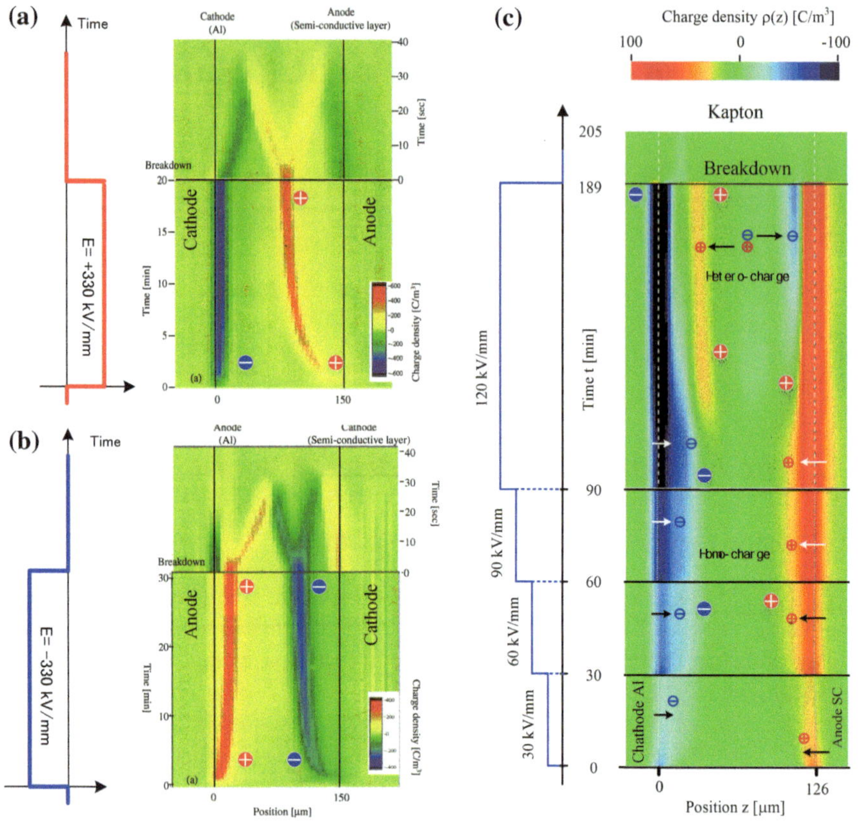

Fig. 11.1 Accumulated charge results inside different materials

11.2 Frequency Dependence of Space Charge Accumulation

Nearly no charge builds up in the insulating material layer of the AC high voltage power cable, but it does build up in the DC high voltage power cable. Then, we experimentally confirmed how slowly the accumulation of space charge appears under the AC voltages with very low frequencies (or periods).

<See Takada Text Vol. I, Part 2, Chap. 5, Sect. 5.2>

(1) **PEA measurement system for variable frequency condition**

Figure 11.2a shows an automatic PEA measurement system with variable frequencies.

(1) **PEA system**: The sample is between the high-voltage electrode (upper electrode) and the ground electrode (lower electrode). A system including a series

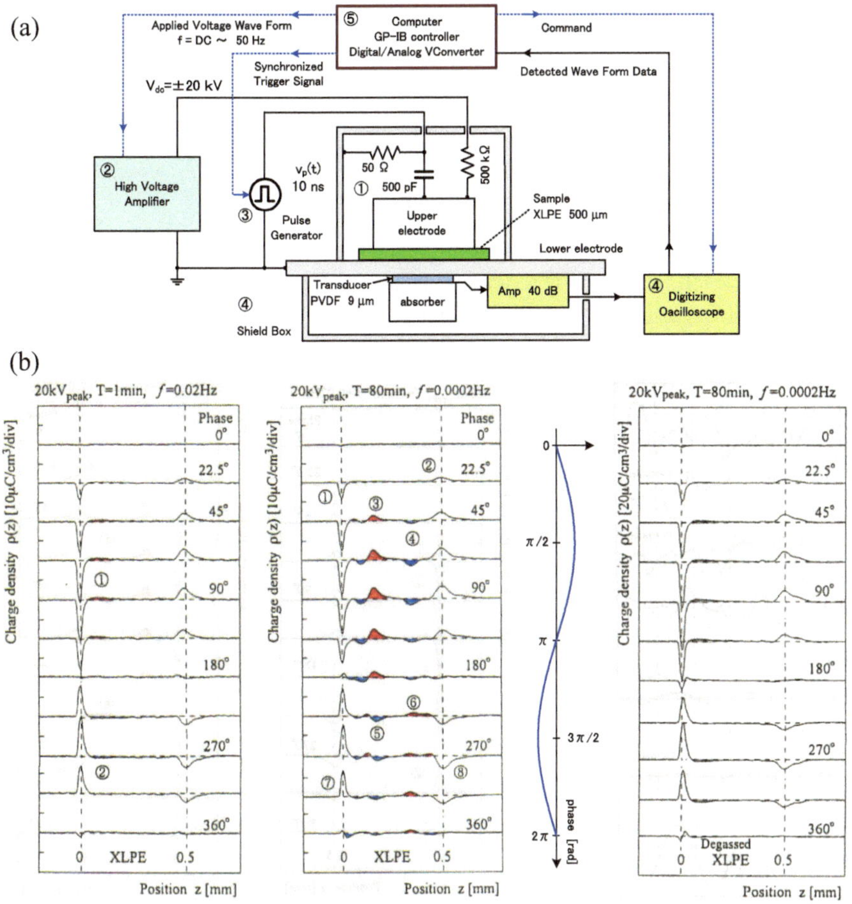

Fig. 11.2 Measured charge results under AC voltages

resistor, matching resistor, coupling capacitor, Transducer and Amplifier is established.

(2) **High voltage amplifier**: The applied voltage (sine wave, etc.) from the computer control is up to ± 20 kV.

(3) Nanosecond pulse voltage generator (Pulse Generator): it can generate pulse voltages (2 kV, 10 ns width).

(4) **Digital Oscilloscope**: ReCroy Wave Runner 6100, 1 GHz.

(5) **Computer control**: Command the AC voltage with vourious frequency (f = 0.0002 Hz (T = 80 min) to f = 50 Hz (T = 20 ms)) to the high voltage amplifier. Com mand synchronized trigger signal to ③ Pulse Generator. The detected waveform data is also commanded to transfer from ④ Digitizing Oscilloscope.

(2) **Sample: XLPE and degassed XLPE**

XLPE: It contains decomposition residue of crosslinking agent (acetophenone, etc.).

Degassed XLPE: A sample whose decomposition residual cross-linking byproduct has been removed by vacuum degassing (48 h at 80 °C).

(3) **Observed results of charge accumulation**

Figure 11.2b shows the accumulation results of the space charge distribution measured at each frequency (Frequency; f) and each phase (Phase; θ).

Figure 11.3 summarizes the dependence of the accumulated charge $Q(f)$ on the frequency (Frequency; f) and period (Period; T).

XLPE (left in Fig. 11.2b): Charge distribution $\rho(z, \theta)$ at frequency $f = 50$ Hz, period $T = 0.02$ seconds. Only the electrode induced charges of ① and ② are observed, and no charge accumulates inside the sample.

XLPE (middle in Fig. 11.2b): Ultra low frequency $f = 0.0002$ Hz, period $T = 80$ minutes, charge distribution $\rho(z,\theta)$ shows that hetero charge accumulates at positive voltage phases ③ and ④, and negative voltage phases ⑤ and ⑥.

Degassed XLPE (right in Fig. 11.2b): Charge distribution $\rho(z, \theta)$ at ultra low frequency $f = 0.0002$ Hz, period $T = 80$ minutes. Only the electrode induced charges of ① and ② are observed, and no charge accumulates inside the sample. It becomes clear that the accumulation of space charges is observed when an AC voltage is applied for longer than the application time $T = 1$ min.

(4) Effect of cross-linking decomposition residue

XLPE containing decomposition residue of cross-linking agent: The accumulated charge amount $Q(f)$ increases rapidly after the time reaches $T = 1$ min or longer.

Degassed XLPE: The accumulated charge amount $Q(f)$ is observed when the period is longer than $T = 1$ min, but it is smaller than that of XLPE (Fig. 11.3).

11.3 Charge Accumulation in a Double-Layer Dielectric

The high-voltage equipment is usually based on a multi-layer insulation in order to improve its withstand voltage.

For example, the joint of the high-voltage power cable is a double-layer insulation of XLPE as main insulation and EPDM as rubber insulation. In this section, we give the case of DC high voltage insulation and introduce the results of observing the charge distribution accumulated at the interface between the double-layer insulation by the PEA method [5–8]. < See Takada Text Vol. II, Part 4, Chap. 8>

(1) **PEA measurement of double-layer insulation**

Figure 11.4: DC voltage V_{dc} and pulse voltage $v_p(t)$ are applied to the two insulating layers (dielectrics 1 and 2) in a superimposed mode.

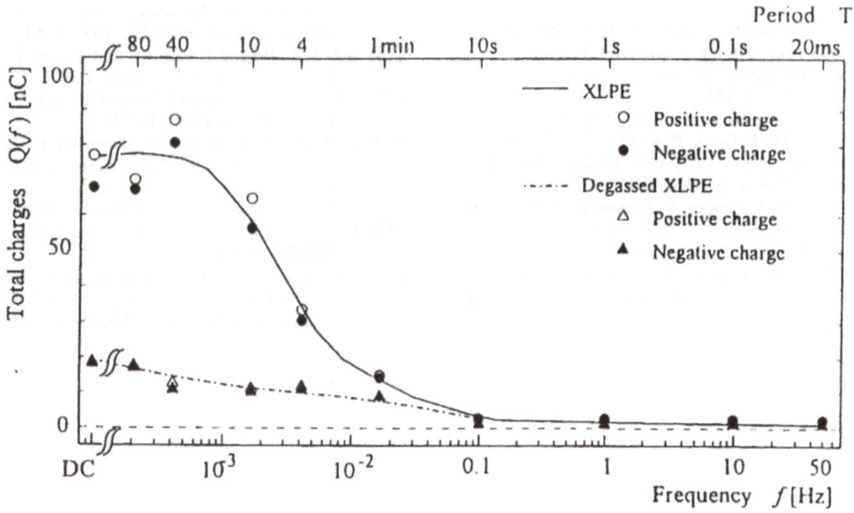

Fig. 11.3 The changed charge amounts with AC frequencies

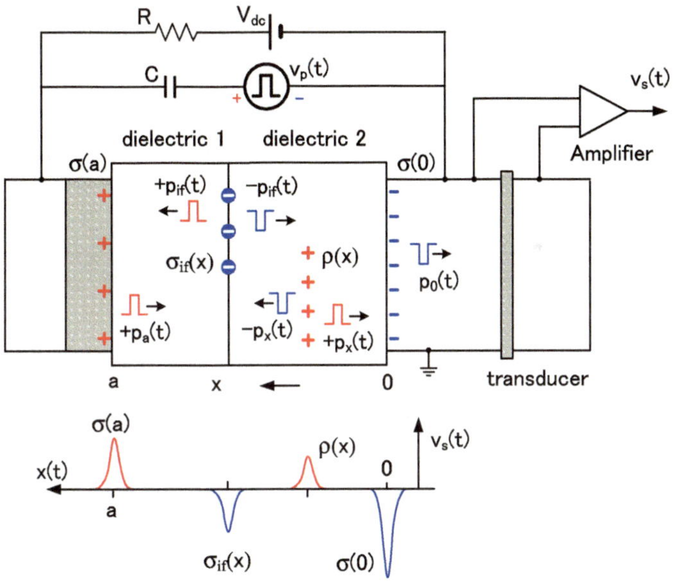

Fig. 11.4 The PEA system for double-layer insulation

The signals composed of volume space charge $\rho(x)$, interface charge $\sigma_{\text{if}}(x)$, electrode charge $\sigma(a)$ and $\sigma(0)$ are measured by a piezoelectric device attached to the back surface of ground electrode.

(2) **Positive voltage application:** Fig. 11.5a

The measured results of storage charge distribution in double-layer insulation are shown.

Double-layer insulation:

LDPE (Low density polyethylene),
EPDM (Ethylene-propylene terpolymer).

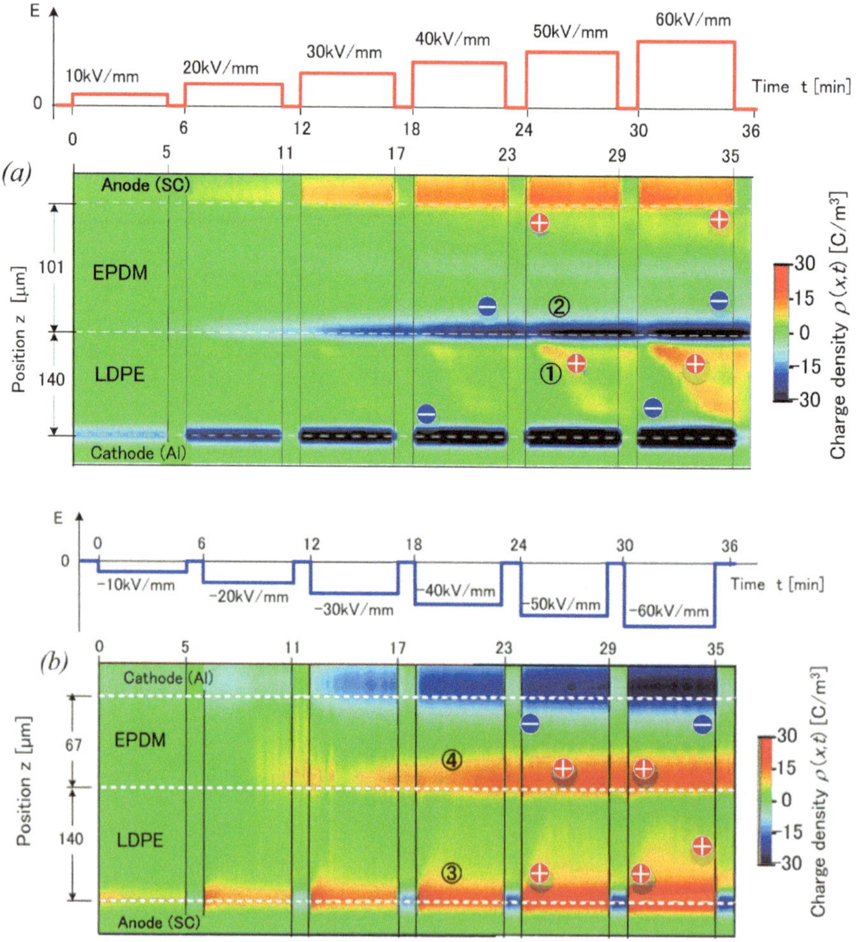

Fig. 11.5 The PEA results in double-layer samples

When the applied electric field reaches 20 kV/mm, ① Positive charge generates at the interface and moves toward the cathode. As a result, ② Negative charge accumulates at the interface.

When a higher electric field of 30 to 60 kV/mm is applied, ① More positive charge generates from the interface and clearly moves toward the cathode. ② Negative charge clearly accumulates at the interface.

(3) **Negative voltage application:** Fig. 11.5b

When the applied electric field reaches 20 kV/mm, ③ positive charge generates from the anode and moves toward the cathode side. At the same time, ④ positive charge generates from the interface and moves to the cathode. The negative charge remaining at the interface is neutralized by ③ positive charge.

(4) Summary

When a high DC voltage is applied to the multi-layer insulation after tenminutes, charges accumulate at the interface and the voltage distribution inside the sample changes, and a local high electric field generates. It shows a very complicated charge behavior. Refer to the following references for more analysis.

11.4 Effect of MgO Addition for Blocking Space Charge

Cross-linked polyethylene (XLPE), which is developed to improve heat resistance (90 °C), has been adopted as the insulating material for AC cables. Meanwhile, XLPE is also adopted as the insulating material of the DC cable. When a polarity reversal test is conducted on a DC cable, it is found that the breakdown voltage of the cable remarkably reduces. We think that the cause is the space charge accumulated in XLPE, and have succeeded in developing a DC-XLPE insulating material containing MgO that can block the charge accumulation [8, 9]. <see Takada Text Vol. II, Part 4, Chap. 10>

(1) **Measured samples**

Sample: Low density polyethylene (LDPE); Thickness is 100 μm.

MgO-added low density polyethylene (LDPE/MgO); Thickness is 100 μm.

(2) **PEA measurement results**

A high electric field of 200 kV/mm is applied, and the accumulation characteristics of space charge in PEA measurement are shown in the figure.

No MgO (non-added sample): Positive charge (red) is injected from the anode and moves toward the cathode.

Space charge density distribution: ① The stored charge distribution is $\rho(x,t) = 0$ at the initial time of voltage application ($t = 0$), but ② the maximum charge density $\rho(x,t) = 600$ C/m^3 after a certain time ($t = 360$ s) is found.

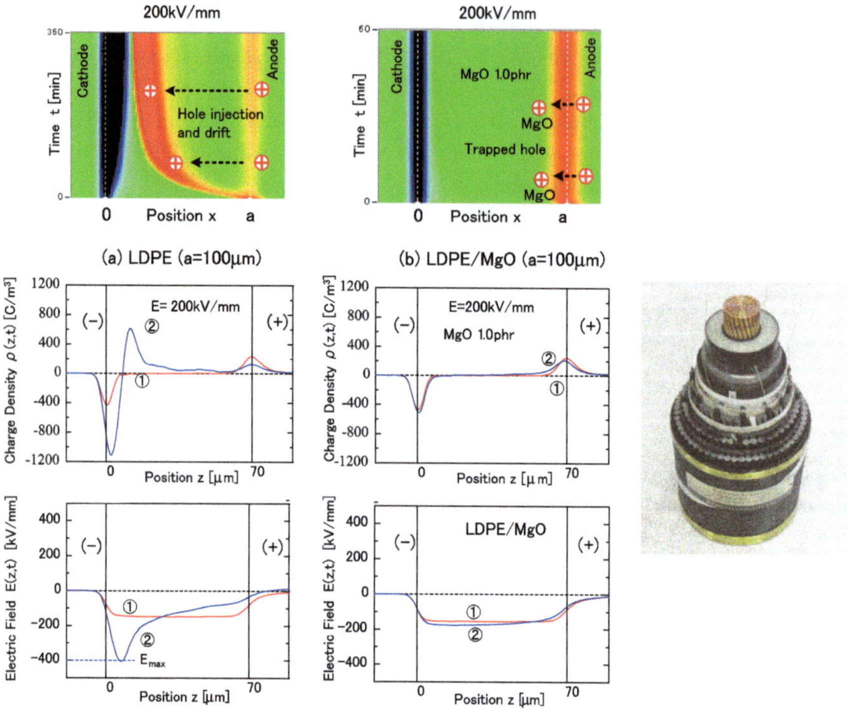

Fig. 11.6 The PEA results in LDPE and MgO-added LDPE samples

Local electric field distribution: ① The electric field distribution immediately after voltage application ($t = 0$) is constant $E(x) = V_{dc}/a$, but ② the maximum field strength is above $E_{max}(x) = -400$ kV/mm after a certain time ($t = 360$ s), and it reaches the breakdown electric field of 400 kV/mm (Fig. 11.6).

MgO 1.0 phr (added sample): No charge accumulation is observed in PEA measurement. (phr: per hundred resin).

Space charge density distribution: ① $\rho(x,t) = 0$ immediately after voltage application ($t = 0$) and ② after a certain time ($t = 360$ s).

Local electric field distribution: $E(x) = V_{dc}/a$ is constant at ① $t = 0$ and ② $t = 360$ s.

Therefore, the MgO-added LDPE sample (1.0 phr) is able to block the space charge effect in the PEA measurement. The mechanism of MgO with no charge accumulation effect is described in Part C, Chap. 3, Sect. 3.5.

11.5 Charge Accumulation Inside Coaxial Cables

The measured results of the charge distribution inside the coaxial cable insulation by the PEA method are introduced. < see Takada Text Vol. I, Part 2, Chap. 6, Sect. 6.5 >

(1) **Measurement system**

Figure 11.7a shows a measurement system that can measure the accumulation characteristics of space charge in a coaxial cable under a DC voltage [10].

Coaxial cable sample: Measured part of a linear coaxial cable (the inner radius is 7 mm, the outer radius is 9 mm, and the length is 2 m).

- It is possible to measure space charge distribution by applying DC voltages ($\pm$100 kV) and AC voltages (peak value: 17 kV).
- Superimpose a high pulse voltage (peak value: 4 kV, width: 50 nsec) on the DC voltage.
- The spherical electrodes on both ends of the tested cable are shields for corona discharge.

PVDF transducer: Piezoelectric PVDF film is used for detection of pulse sound waves. This PVDF film is sandwiched by a fan-shaped backing material and a fan-shaped electrode that touches the outer jacket of the coaxial cable.

Temperature rise: The temperature rise of the cable sample is achieved by the Joule heating current in the current transformer. The cable forms a closed circuit of one-turn coil with the auxiliary cable, and the closed circuit passes through the current transformer (500 turns). When 100 V/1 A is applied to the current transformer, 0.2 V/500 A is supplied to the closed circuit in the cable side. Then, 500 A current flows to the core wire of cable sample, and the temperature of the core wire can be raised to 360 Celsius degree.

(2) **Observation of heterocharges**

A DC voltage of −40 kV is applied to the inner electrode of XLPE coaxial cable for 480 mins (8 h), and the measured results of charge distribution are shown in Fig. 11.7b. An increase in the accumulation of hetero space charge is observed with the development of application time.

Figure 11.8 shows the results of the behavior of the hetero space charge when the polarity of the applied voltage is reversed. A voltage of −40 kV → +40 kV → −40 kV → +40 kV is applied to the core wire at intervals of 8 h. During the initial stage from $t = 0$ to $t = 8$ h, ② positive heterocharge accumulates near the core as the cathode, and ① negative heterocharge similarly accumulates near the outer conductor as the anode. After that, when the polarity of the applied voltage is reversed, the polarities of ③ and ④ heterocharges accumulated during the time from $t = 8$ h to $t = 16$ h are also reversed.

Each time the polarity of the applied voltage is reversed, the space charge moving into the XLPE cable forms the heterocharge. Thus, we can imagine that the actually measured data on site is also the accumulation of hetero space charges.

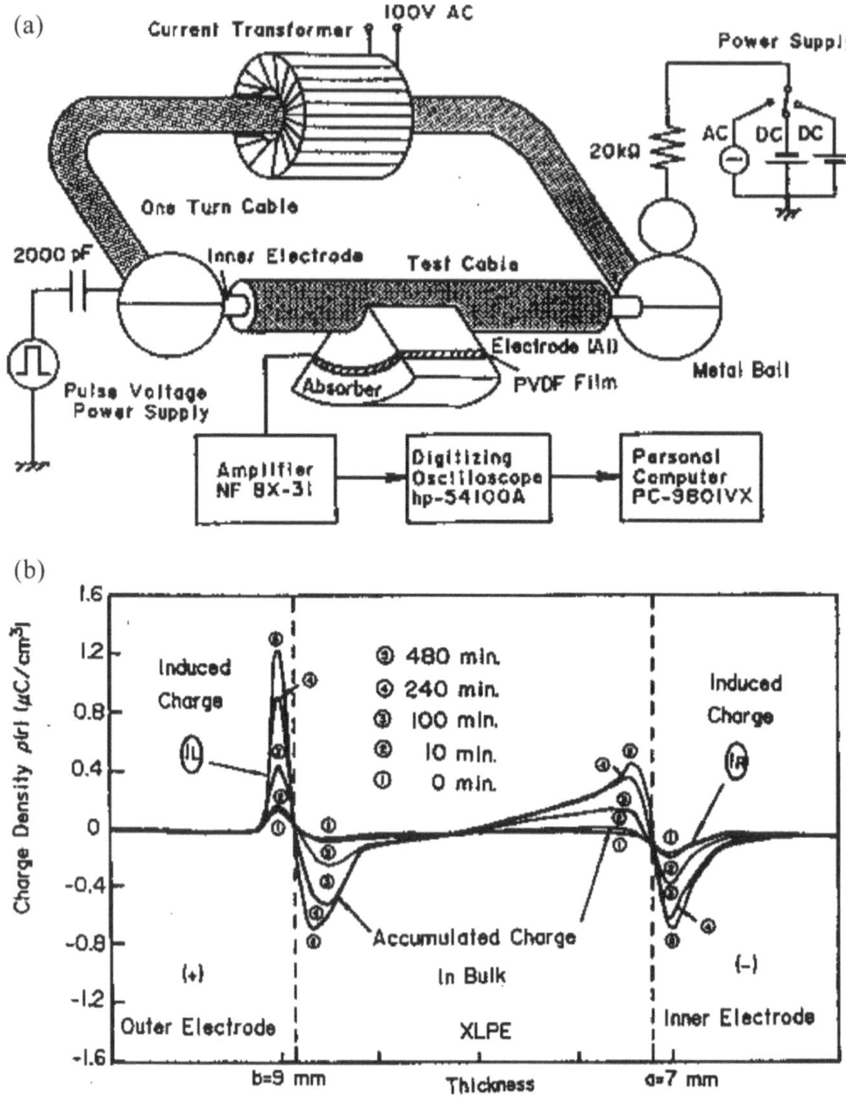

Fig. 11.7 Measurement system and results for a coaxial cable

(3) Observation of heterocharges during temperature rise

The formed hetero space charge remains even if the temperature is raised. Due to the change of the thickness of the insulating layer and the change of the acoustic velocity caused by the temperature rise, the distance between the induced charges of the anode and the cathode also changes.

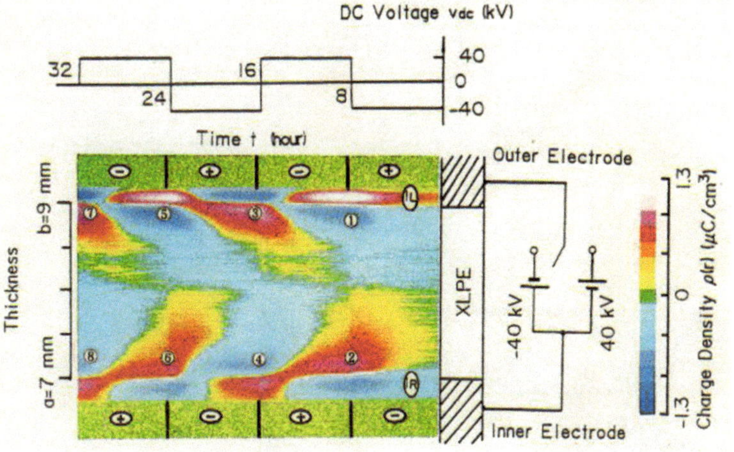

Fig. 11.8 The behavior of the hetero space charges in XLPE cable

11.6 Distribution of Accumulated Charge in the Cross Section of the Coaxial Cable

The measurement results of the charge distribution $\rho(r,\theta)$ in the cross section of the insulation layer of the coaxial cable by the PEA method are introduced. <see Takada Text Vol. I, Part 2, Chap. 6, Sect. 6.7>

(1) **PEA measurement of the cross section of a coaxial cable in a flat electrode type**

Figure 11.9: Explanatory diagram of measured results of space charge distribution $\rho(r,\theta)$ inside cross section (r,θ) of cable. The coaxial cable is put on the flat plate electrode. By rotating the coaxial cable with a small angle in the θ direction, the charge distribution is then measured.

Plate electrode type: The circumference length of the outer peripheral surface of the cable in contact with the plate electrode is $\Delta B = b\Delta\theta$. b is the outer radius of the cable (20 mm), and $\Delta\theta$ is the fan angle. When a coaxial cable is attached to a flat plate electrode by applying an external mechanical force, ΔB is about 3 mm, and thus the fan angle is $\Delta\theta = 0.15$ rad (=8.6 degrees).

Detection of pressure wave: The pressure wave propagating through the contact surface is measured by the piezoelectric device (PVDF film) set on the back surface of the flat plate electrode.

(2) **Space charge distribution $\rho(r,\theta)$ in the insulating layer cross section (r,θ)**

Sample: A XLPE cable with water tree deterioration.

Figure 11.9: Measurement results of space charge distribution in cross section (r, θ). Positive induced charge under positive voltage application is observed in the inner conductor (semiconductive layer), and negative induced charge is observed in the

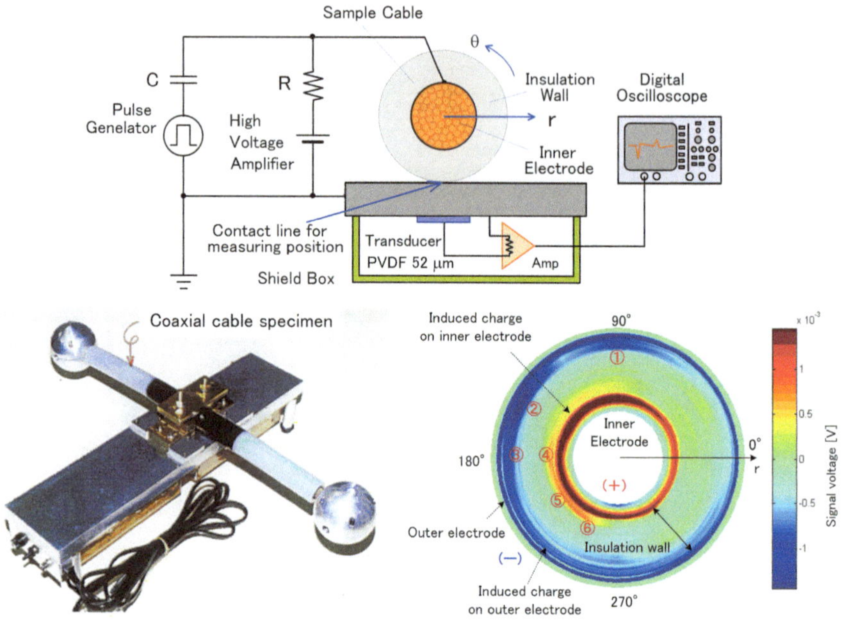

Fig. 11.9 The measurement system and results for the cross section of the coaxial cable

outer conductor (semiconductive layer). In particular, homocharge accumulation is observed in the numbered locations of the XLPE insulation layer. The traces of water trees are also observed in this part, and it is confirmed that homocharges accumulate at the places of traces of water trees [11].

11.7 Space Charge Measurement of Long Cables

The PEA method has been widely applied to a flat plate sample, and the accumulated charge distribution $\rho(x,t)$ of space charge can be observed. Central Research Institute of Electric Power Industry (Hozumi) has developed an system for the accumulated charge distribution $\rho(r,t)$ in the insulating layer of a long actual cable. The long cable has a XLPE insulation layer with a thickness o f 3 m m and the cable length is 15 m, as shown in the figure. When performing PEA measurement, how to apply the pulse voltage $v_p(t)$ and measure the pressure wave signal $v_s(t)$ becomes a problem. < see Takada Text Vol. I, Part 2, Chap. 6, Sect. 6.6>.

(1) **Applied method of pulse voltage $v_p(t)$**

(a) Remove the outer layer of metal conductor over a length three times the part to be measured (5 cm width), and make the semiconductive layer surface exposed.

(b) The central position on the surface of this semiconductive layer is processed to closely contact with detection electrode. Two guard ring electrodes are set on both sides of the detection electrode, and they are grounded.
(c) The pulse voltage is applied between the ground ring electrode and the left with right outer conductors.
(d) The applied pulse voltage is shared by the capacitance C_x of XLPE in the measurement section and the characteristic impedance Z_0 of the cable. Z_0 of the long cable in the non-measurement section serves as a coupling capacitor for pulse voltage application. The figure shows the experimental results of space charge measurement inside a long high-voltage cable. Also see the references for more details.

(2) **Measurement results**

Typical measurement results are shown in Fig. 11.10 (a) 70 kV and (b) 350 kV, respectively.

(a) When the average electric field is $E = 23$ kV/mm at 70 kV, hetero space charge accumulates.
(b) In the case of $E = 120$ kV/mm at 350 kV, space charge accumulates in XLPE with the mode of repeated generation, migration, and disappearance. Accumulation and migration of a large amount of packet charge are observed in

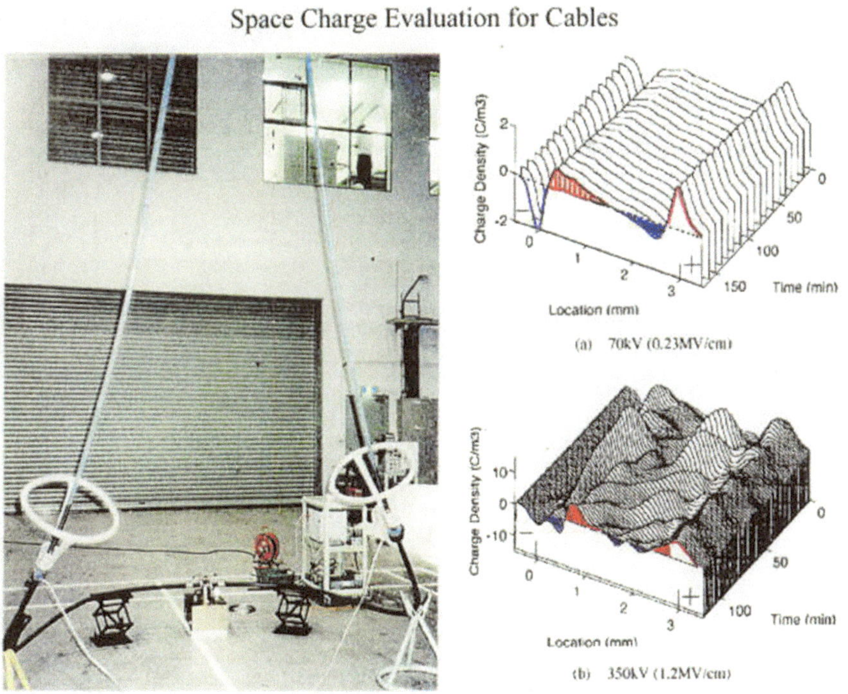

Fig. 11.10 The measurement system and results of long cables

the XLPE insulating layer, and thus it is hard to say the layer is an insulating material [12, 13].

11.8 Light Irradiation and Space Charge Formation

Sample: Kapton-H film is light brown and translucent. The wave with long wavelength (Red) can be transmitted inside the material, but the short one (Violet) cannot make it. The opacity is defined as the energy loss of light in absorbing light and generating electron–hole pairs in Kapton-H [14].

(1) **Improved PEA equipment for light irradiation**

Visible light (Red: ~2.0 eV to Voilet: ~3.1 eV) is irradiated to a Kapton-H film (band gap: $\phi_g = 2.45$ eV) with an optical fiber (Fig. 11.11b). Under the application of a DC electric field (30 kV/mm), the space charge formation process is observed by PEA method for 10 mins (Fig. 11.11a). In this case, the ground electrode uses Al material and the high-voltage electrode uses ITO film.

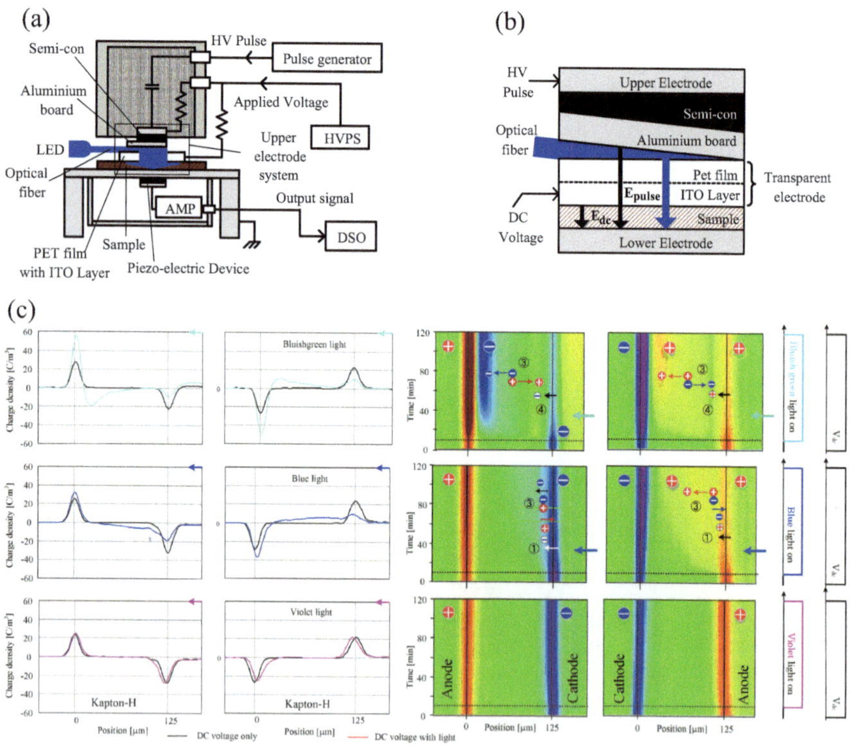

Fig. 11.11 The measurement system and results under light irradiation conditions

(2) **Charge accumulation data of Kapton-H during the visible light irradiation**

Irradiation with the Bulish light wavelength ($h\nu = 2.46$ eV) (Fig. 11.11c top): Electron–hole pairs uniformly generate in the bulk. Negative charge (electron) on the anode side and positive charge (hole) movement are observed near the cathode side. However, it is easy to inject charge from the ITO electrode, which easily neutralized with the migrated charge from the opposite side. Since the energy of the Bulish light ($h\nu = 2.46$ eV) is about the same as the band gap of Kapton-H ($\phi_g = 2.28$ eV), the generation of electron–hole pairs occurs.

Bule light wavelength ($h\nu = 2.64$ eV) (Fig. 11.11c): Short wavelength light is significantly absorbed, and this light absorption occurs near the sample surface. Electron–hole pairs generate near the surface. The generation of electron–hole pairs occurs on the surface of the irradiation side, which is different from the situation under the irradiation of Bulish light.

Irradiation with Violet light wavelength ($h\nu = 3.10$ eV) (bottom of Fig. 11.11c): Shorter light is more significantly absorbed, and this light cannot reach the sample inside. The generation of electron–hole pairs also occurs near the ITO film electrode, but no charge accumulation is observed.

<see Takada Text Vol. III, Part 4, Chap. 21>

11.9 Electron Beam Irradiation and Space Charge Formation

(1) **Electron beam irradiation device**

Electrons are emitted from the heating filament in the vacuum vessel, and the high DC voltage (40, 60, 80 kV) is applied to accelerate the electrons. The samples (PI and PTFE) with the PWP measurement set are irradiated (Fig. 11.12a). The accumulated charge distribution changes with time. <see Takada Text Vol. III, Part 4, Chap. 22>

(2) **PWP measurement of charge accumulation during electron beam irradiation**

Figure 11.12b shows the PWP measurement set that can measure the charge accumulation during electron beam irradiation (20 min).

An electron beam is guided to the sample from the hole of the ring electrode, and a pulsed pressure wave is propagated to the charged sample from the piezoelectric element attached to the back surface of the lower electrode. During the process of the pulsed pressure wave propagating through the charged sample, the displacement current flowing outside is measured as the signal. The waveform of this displacement current shows the accumulated charge distribution.

(3) **Electron accumulation results** (Fig. 11.13)

PI film sample: Negative charge accumulates within the penetration range of high-speed electrons. The red line shows the charge distribution immediately after the

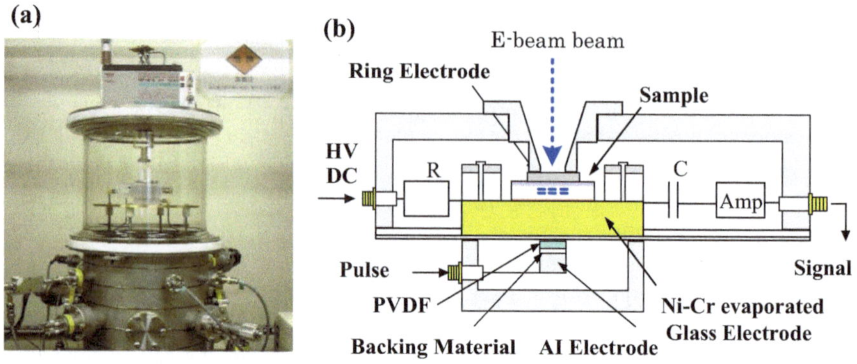

Fig. 11.12 The measurement system for electron beam irradiation conditions

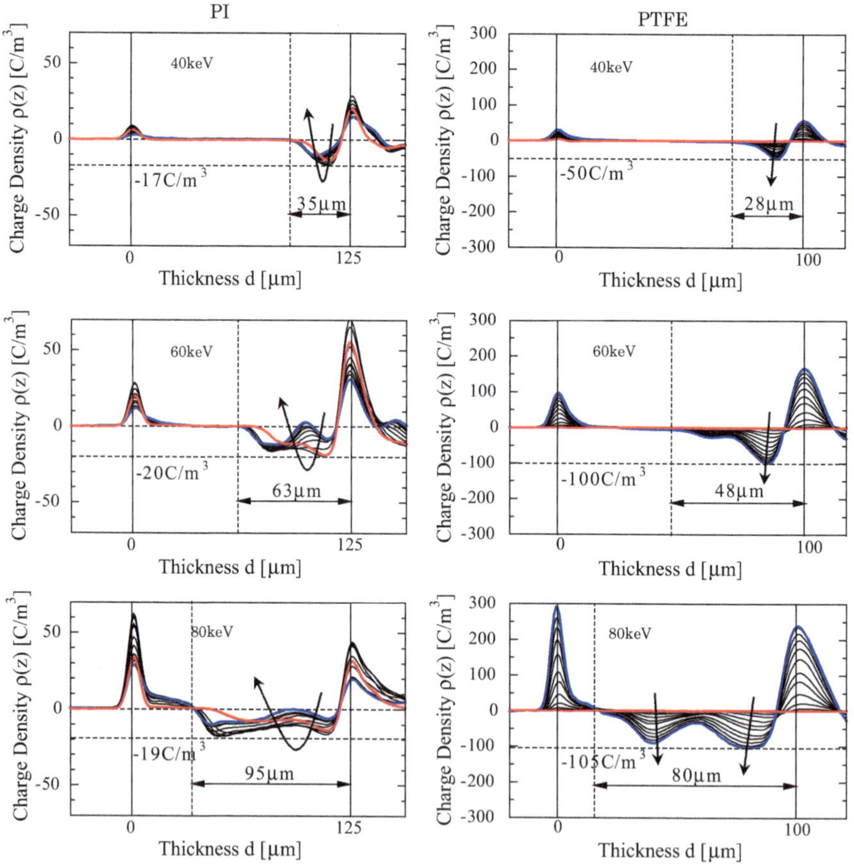

Fig. 11.13 The measured results of different samples under electron beam irradiation

start of electron beam irradiation. When the irradiation energy is high (60, 80 ekV), the amount of positively induced charge on the opposite ground electrode increases, so that the accumulated negative charges (electrons) migrate to the ground electrode. The blue line shows the residual charge distribution when the electron beam irradiation stops.

PTFE film sample: A main feature of the PTFE sample is that negative charge (electrons) does not accumulate immediately after the start of irradiation (red line). Negative charge (electrons) accumulates after a certain time (over 10 minitue). Similarly, it moves toward the positively induced charge on the opposite ground electrode [15, 16].

11.10 Gamma Irradiation and Space Charge Formation

(1) Cobalt 60 gamma ray source

When ^{59}Co in nature absorbs thermal neutrons in a nuclear reactor, it becomes ^{60}Co and spontaneously decays over a half life of 5.26 years to ^{60}Ni. It emits gamma rays with high energy (0.83 to 2.16 meV) in the decay process. ^{60}Co, which has a long half life, is used as a gamma ray source.

(2) Characteristics of space charge accumulation in gamma-irradiated LDPE

Irradiated low-density polyethylene (LDPE; 600 μm) with ^{60}Co γ-rays at a dose of 2.5 kGy (Fig. 11.14a, b). After that, a DC electric field $E = 25$ kV/mm is applied, and the residual accumulated charge distribution $\rho(x,t)$ is measured by the PEA method.

(a) **Unirradiation + applied voltage** (Fig. 11.14c): Since it is unirradiated, electron–hole pairs are not generated in the LDPE. Even if a DC electric field of $E = 25$ kV/mm is applied, no charge accumulation is observed. $\rho(x,t) = 0$.
(b) **γ-ray irradiation + no voltage** (Fig. 11.14d): Electron–hole pairs are generated in the irradiated LDPE. However, since no DC electric field is applied, electron–hole pairs cannot distribute separately. The observation of $\rho(x,t) = 0$ is because positive and negative charges are mixed.
(c) **γ-ray irradiation + voltage application** (Fig. 11.15a–c): Electron–hole pairs are generally generated in the irradiated LDPE. When a DC electric field of $E = 25$ kV/mm is applied there, negative charge (electron) moves to the anode side and positive charge (hole) moves to the negative side, forming hetero charges.
(d) **γ-ray irradiation + after voltage application** (Fig. 11.15d–f): Since positive charge remains predominantly, negative charge is induced on both electrode surfaces at the short circuit state. As a result, the residual positive charge moves toward the both electrode surfaces [17, 18].

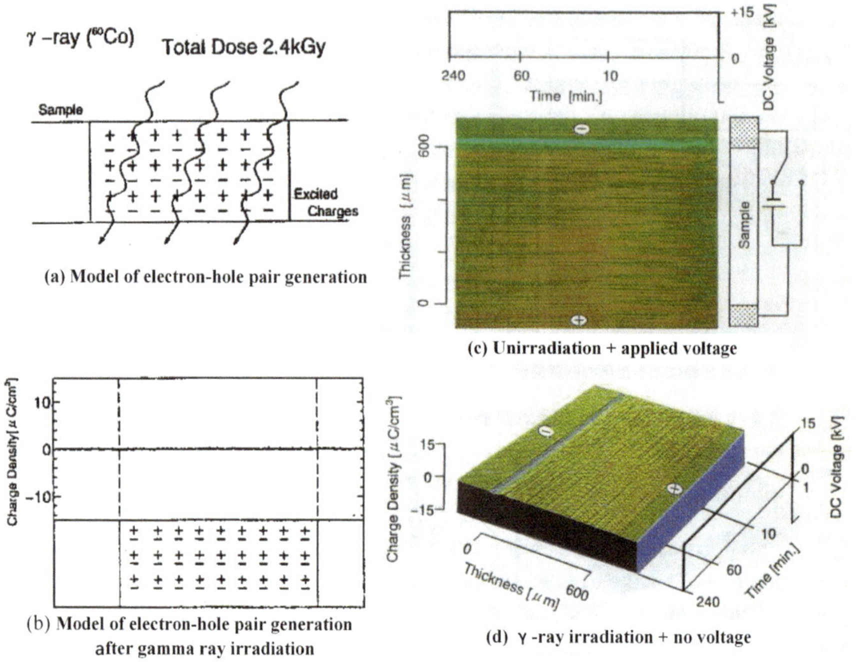

Fig. 11.14 Effect of gamma irradiation on space charge accumulation

11.11 Proton Irradiation and Space Charge Formation

A proton is an ionized hydrogen atom (H+) and is a kind of elementary α-particle. Table 11.1 shows the quantities of other elementary particles (neutrons, electrons). Electrons have both particle and wave characteristics, but protons (protons) only have particle characteristics. < Takada Text Vol. III, Chap. 23>

(1) **Proton irradiation**: During the proton (1.0, 1.5, 2.0 meV) irradiation in vacuum, the accumulated charge distribution $\rho(x,t)$ inside polyimide (PI; 50 μm) is measured by the PEA method.
(2) **Non-iradiation results**, the left in Fig. 11.16: The figure shows the accumulated charge distribution when a DC electric field of 100 kV/mm is applied. Positive and negative charges are injected from both electrodes and their migration distances are very small from the electrodes. They remain as residual charges even after the short circuit, and charges with opposite polarities induce on the electrode surfaces.
(3) **1.0 meV and 1.5 meV iradiation**, the middle in Fig. 11.16: Proton irradiation while applying 100 kV/mm electric field. Positive charges gradually accumulate at the proton penetration range (19 μm and 37 μm). They remain as residual charges even after a short circuit.

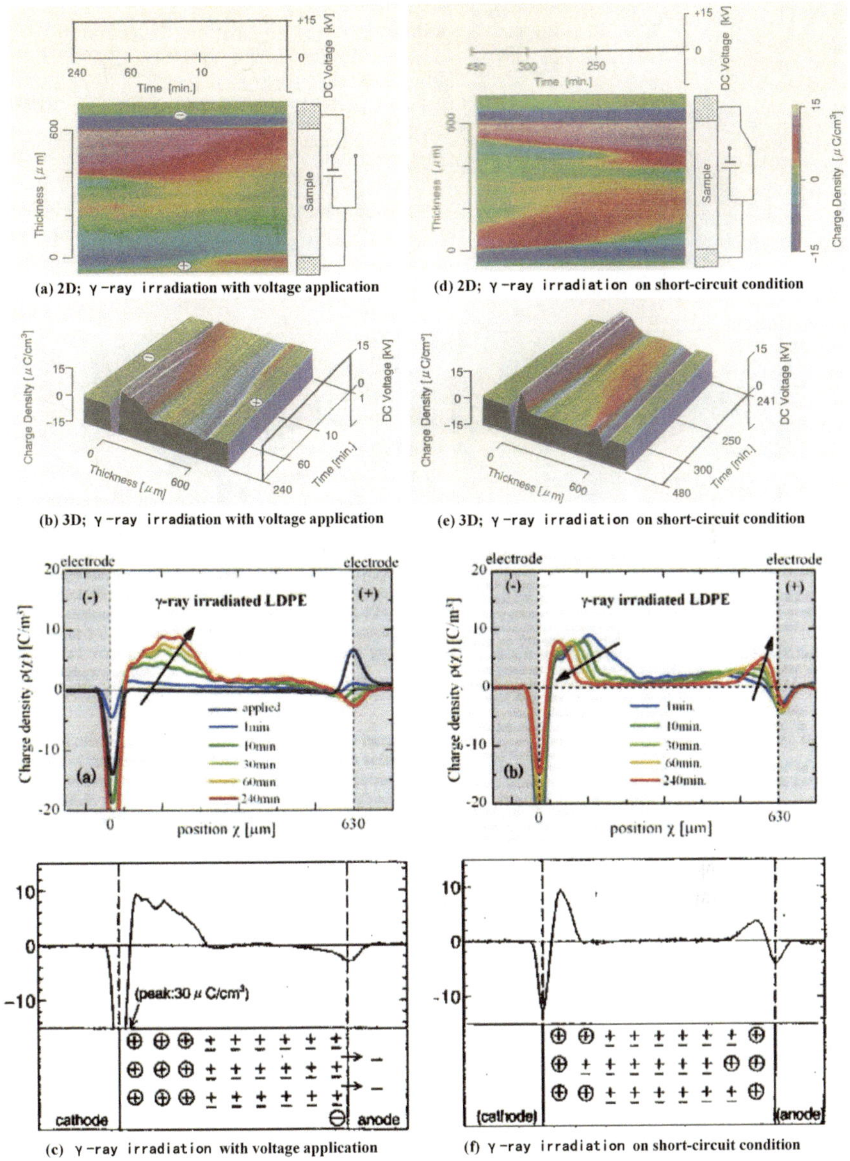

Fig. 11.15 Charge results under gamma irradiation with different conditions

(4) **2.0 meV iradiation,** the right in Fig. 11.16: The proton range (59 μm) is larger than the sample thickness (50 μm), and thus the protons can pass through the sample. Therefore, there is no positive charge accumulation. There is also no residual charge after a short circuit, which is the same as that of non-iradiation one (left in the figure) [19].

Table 11.1 Quantities of other elementary particles

Element	Proton	Neutron	Electron
Mass ratio	1	1	1/1840
Charge [C]	$+1.602 \times 10^{-19}$	0	-1.602×10^{-19}
Maximum penetration range	19 μm/1.0 meV	–	2800 μm
	37 μm/1.5 meV	–	5200 μm
	59 μm/2.0 meV	–	

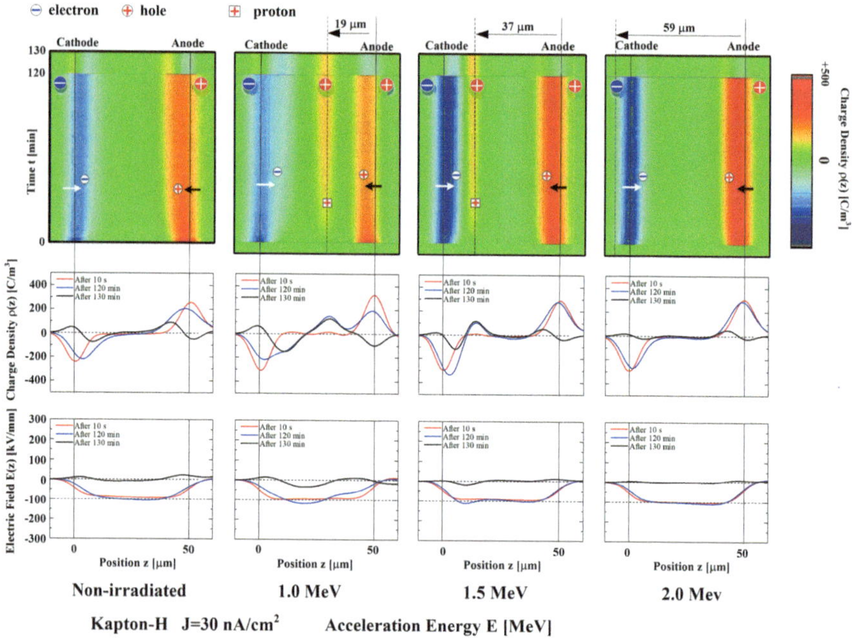

Fig. 11.16 Effect of proton irradiation on charge accumulation

11.12 Comparison of Paraelectric and Ferroelectric PEA Signals

Features: The polarities of the PEA signals of the paraelectric materials (PE and PI) and the ferroelectric materials (PVDF and $LiNbO_3$) are opposite. <see Takada Text Vol. I, Part 2, Chap. 8>

(1) **Paraelectric PEA signal** in Fig. 11.17: The positive pulse voltage $v_p(t)$ is superimposed on the positive or negative DC voltages of $V_{dc} = \pm 5000$ V, which are applied to the paraelectric material (PI). The positive PEA waveform can be observed under the positive DC voltage of $V_{dc} = +5000$ V, and the negative PEA waveform can be observed under the negative DC voltage of $V_{dc} = -5000$ V. The positive PEA waveform generates in the process of shrinking the

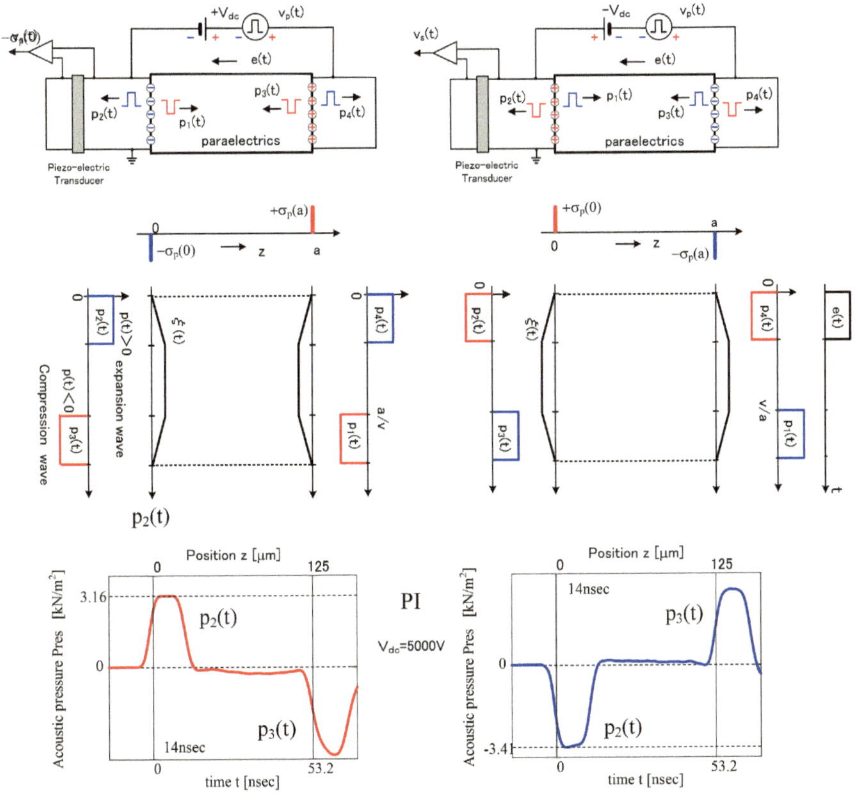

Fig. 11.17 Measured charge accumulations in paraelectric materials

sample, and the negative PEA waveform generates in the process of expanding the sample [20].

(2) **Feroelectric PEA signal** in Fig. 11.18: A positive pulse voltage $v_p(t)$ is applied to ferroelectric PVDF (polyvinylidene fluoride) and $LiNbO_3$ (lithium niobate). DC voltage is not used. Since this ferroelectric substance has already been orientation-polarized, it is not necessary to apply a DC voltage for orientation polarization.

(3) **Nomal polarized PVDF sample,** the left in Fig. 11.18: The ferroelectric PVDF sample is extended by applying a positive pulse voltage $v_p(t)$. In this case, the first negative and second positive PEA signal is observed.

(4) **Reversal of polarized PVDF sample,** the right in Fig. 11.18: The reversed ferroelectric PVDF sample is shrinked under a applied positive pulse voltage $v_p(t)$. In this case, the first positive and second negative PEA signal is observed.

(5) **Future research subjects**: Dielectric/insulating materials include organic materials and inorganic materials. Organic materials only have a very small piezoelectric effect, while it cannot be ignored for inorganic materials. Although there is a problem for inorganic materials to apply under a high electric field

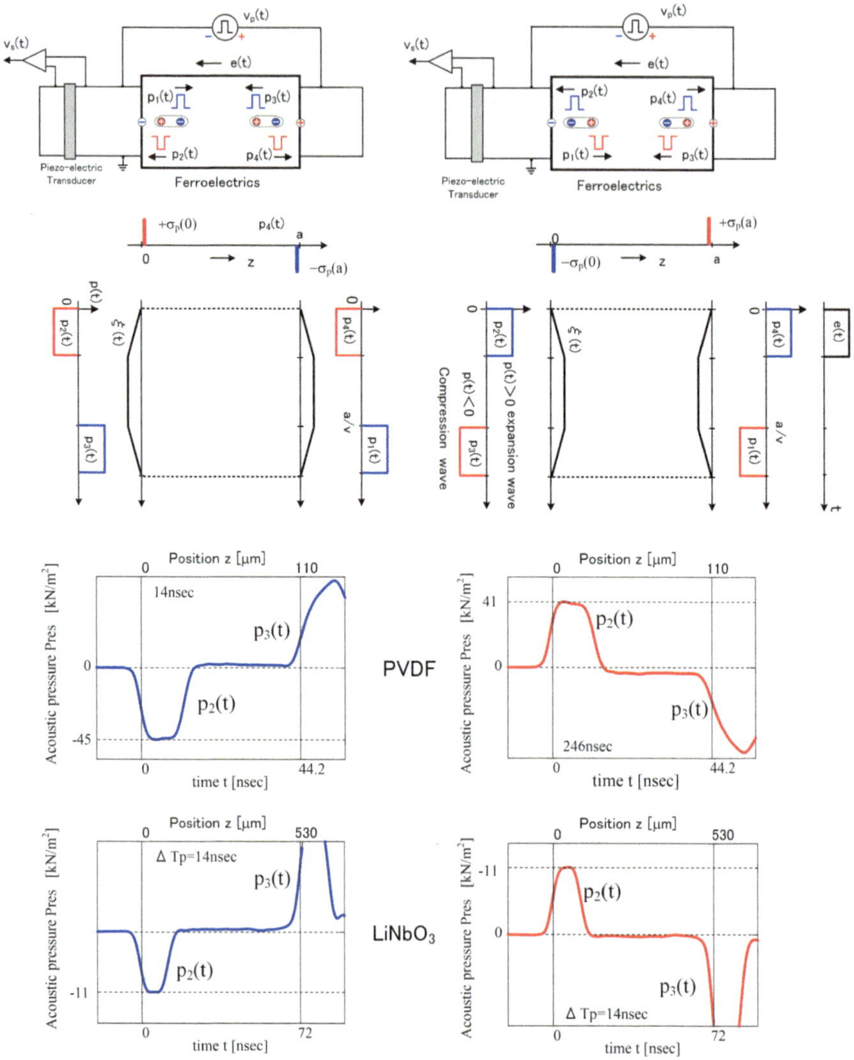

Fig. 11.18 Measured charge accumulations in feroelectric materials

due to charge accumulation, the measurement of them can still be achieved by the PEA method, whose signals include not only accumulated charges and but also polarized charges. Research issues on this area still remain.

11.13 3D Measurement Results of Charge Distribution

The PWP method and the PEA method have been devised to measure the space charge distribution. <see Takada's Text; Vol. I, Part 2, Chap. 7>.

Generally, it is assumed that the sample with stored charge has a flat plate shape, and the stored charge $\rho(z,t)$ changes only on the z direction and is uniform on the x–y plane. Therefore, the pulse pressure wave has also been treated as a longitudinal wave which is uniform on the x–y plane and propagates only on the z direction. However, the actually accumulated charge $\rho(x, y, z, t)$ often has a three-dimensional distribution (x–y-z space). Therefore, an example of trying the PWP method to measure the charge distribution accumulated in three dimensions is introduced below [21–23].

(1) **Measurement device**

Figure 11.19a and b shows the principle for measuring the three-dimensional charge distribution by the pulse pressure wave (PWP: Pressure Wabve Propagation) method and a photograph of the measurement device. Since it is the PWP method, the charged sample is attached to the back (lower) surface of the ground aluminum electrode. Water is attached to the front (upper) surface of the electrode and used as a propagation medium for pulse pressure wave. The pulse pressure wave is narrowed by the acoustic lens so that it has a diameter of 0.1 mm when it reaches the charged sample. This system records the data by measuring the charge distribution $\rho(z, t)$ on the z direction while moving the acoustic lens on the x–y plane by a driving device.

(2) **Measurement resolution**

The acoustic lens head used in the ultrasonic microscope is utilized to generate the pulse pressure wave. A pulse voltage (voltage: 500 V, pulse width: $\Delta T_{\rm p} = 10$ nsec) with a half wavelength time is applied to a piezoelectric element with a resonance frequency of 50 MHz to generate a pulse pressure wave, which usually has one positive/negative cycle/oscillation. An oscillating pressure wave can be used because the impulse response signal processing is later performed.

The applied pulse voltage is 500 V and the pulse width is $\Delta T_{\rm p} = 400$ nsec. Therefore, the positional resolution Δz on the z direction is $\Delta z = v\Delta T_{\rm p} = 1.1$ mm with the PMMA sound velocity of $v = 2680$ m/sec.

(3) **Measured sample and results**

The three-dimensional charge distribution inside a sample partially irradiated with an electron beam is measured. A lead plate (thickness: 200 μm) with a number "7" cut is placed on a PMMA material (thickness: 5 mm), and electron beam irradiation (acceleration energy: 200 keV) is performed to obtain the negative charge distribution with a depth of 100 μm to 250 μm. The charge accumulation is shown in Fig. 11.20. The electron beam can penetrate the part of the lead plate, but it cannot penetrate the PMMA plate, so that the electric charge of "7" is shown (Fig. 11.19d).

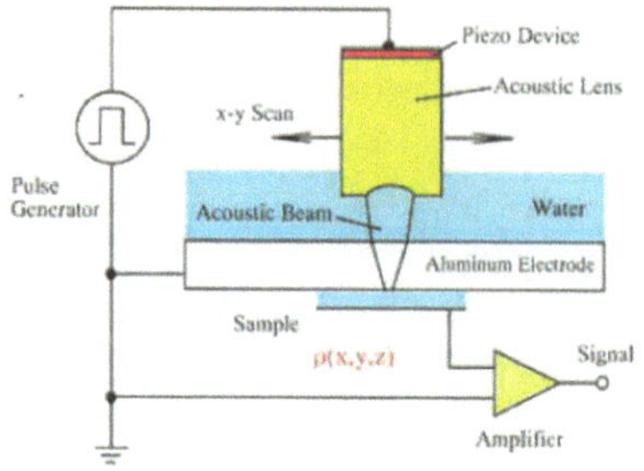

(a) Measurement Principle of three-dimensional charge distribution

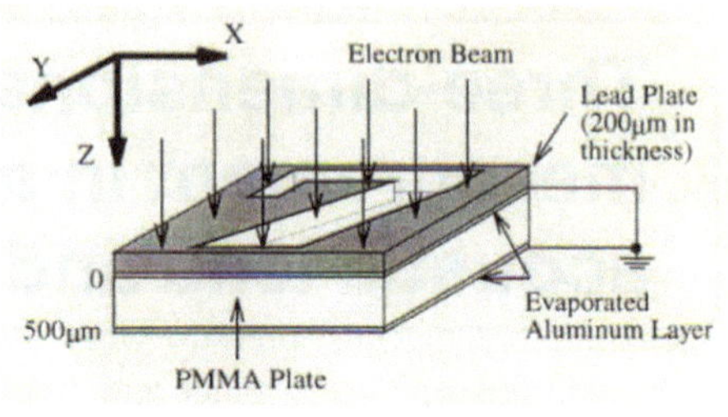

(c) Three-dimensional charge distribution formed by electron beam irradiation

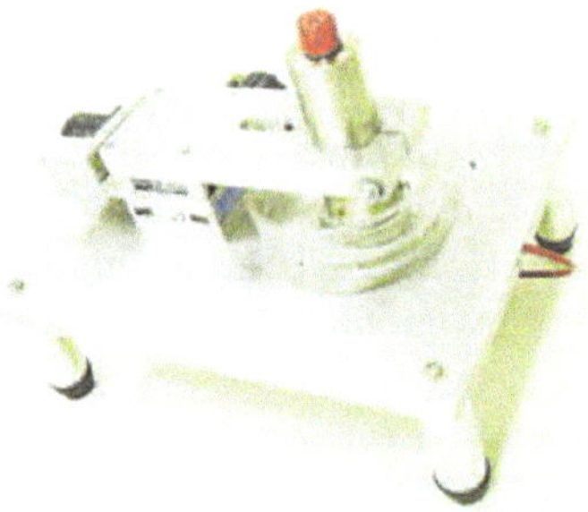

(b) Three-dimensional charge distribution measurement device

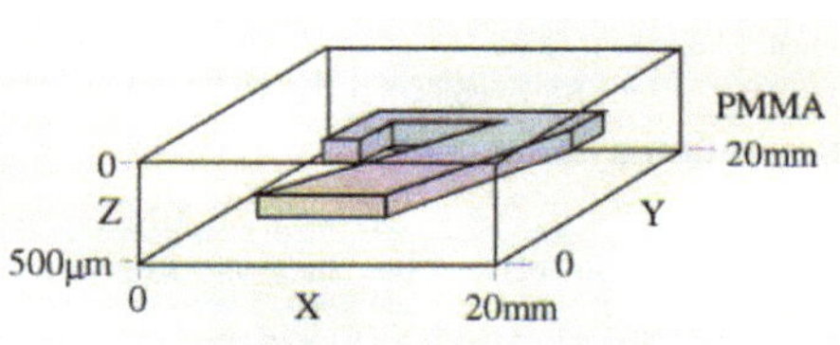

(d) Virtual diagram of three-dimensional charge distribution

Fig. 11.19 Measurement system for three-dimensional charge distribution

11.14 Generation and Recovery of Attenuation and Dispersion of Pressure Wave

(1) Generation and recovery of wave attenuation and dispersion in flat plate samples

Since the polymer insulating material is a kind of viscoelastic body (lossy material), the distortion of pressure wave $p(t)$ occurs when propagating in the material. Figure 11.21b shows the waveform with attenuation and dispersion. When the transfer function $g(z, t)$ cannot keep constant in wave propagation, it means a distortion occurs. In contrast, Fig. 11.21a shows that the distortion does not occur when the transfer function $g(z, t)$ without attenuation and dispersion is constant.

First, the attenuation and dispersion in Fig. 11.21b is discussed. The sheet charge $\sigma(z)$ (green) is located at $z = 0, 1, 2, 3, 4$. When the pulse excitation voltage $v_p(t)$ (pulse electric field $e(t)$) is applied to the sample, a pulse pressure wave $p(z, t)$ (purple) is generated from each sheet charge, as shown in Eq. (11.1). The signal voltage $v_s(t)$ is given in Eq. (11.2), represented by the convolution of the piezo constant of transducer and the pressure wave $p_2(z, t)$.

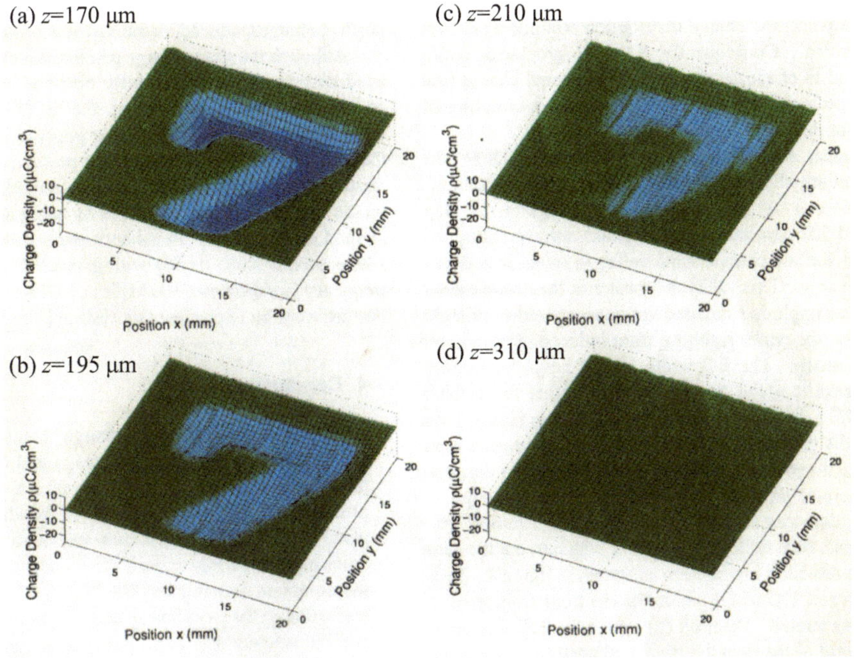

Fig. 11.20 Measured results by the three-dimensional system

$$p_1(t) = C_1 \int_0^t \sigma(\tau)\, e(t-\tau)\, d\tau \tag{11.1}$$

$$v_s(t) = C_2 \int_0^t d(\tau)\, p_1(t-\tau)\, d\tau \tag{11.2}$$

When this wave $p(z, t)$ propagates through the viscoelastic material, the pressure wave $p(t)$ and signal voltage $v_s(t)$ is distorted by the attenuation constant $\alpha(f)$ and dispersion constant $\beta(f)$ of the transfer function $g(z, t)$, as shown in Eqs. (11.3) and (11.4). The waveform broadening due to dispersion also occurs (Fig. 11.21b, red). However, the transfer function $g(z, 0)$ is not attenuated or dispersed (pink series in Fig. 11.21b) when the pulse pressure wave is just generated ($t = 0$). When there is no attenuation and dispersion in Fig. 11.19a, there is no attenuation or dispersion (waveform broadening) phenomena of the pressure wave (Fig. 11.21a, red).

If the sample is very thick, such as the insulation layer of power cable is about 20 mm, the pressure wave in the PEA measurement has a large distortion when the temperature of the insulation material rises. Therefore, the charge distribution $\rho(z, t)$ and electric field distribution $E(z, t)$ cannot be accurately measured. Thus, the attenuation constant $\alpha(f)$ and the dispersion constant $\beta(f)$ of the insulating material

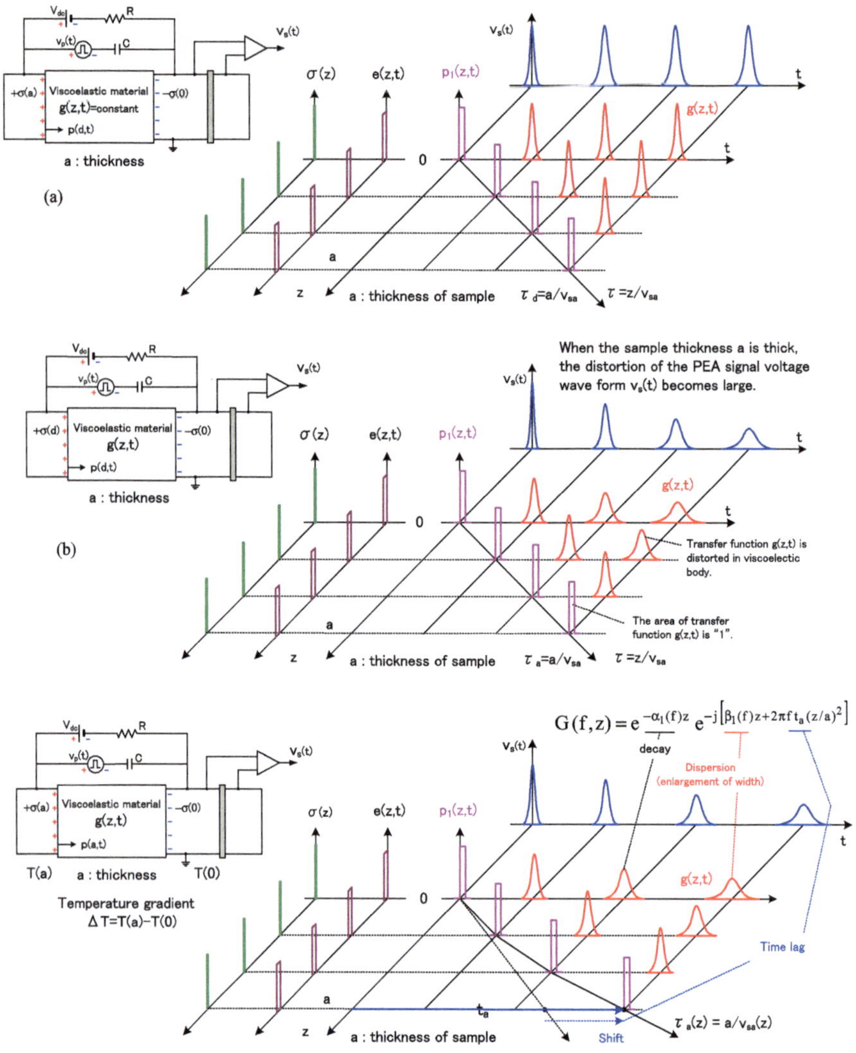

Fig. 11.21 Distortion of pressure wave propagating in the material

are obtained in advance to calibrate the waveform distortion. The algorithm and simulation results are introduced below.

(2) **Recovery example of space charge profile from output signal** [24, 25]

Figure 11.22 shows the PEA signal when an electric field of $E = 5$ kV/mm is applied to a lossy material (LDPE) with a thickness of 2.0 mm. The deformation of the signal waveform in Fig. 11.22 and its recovery process are explained below.

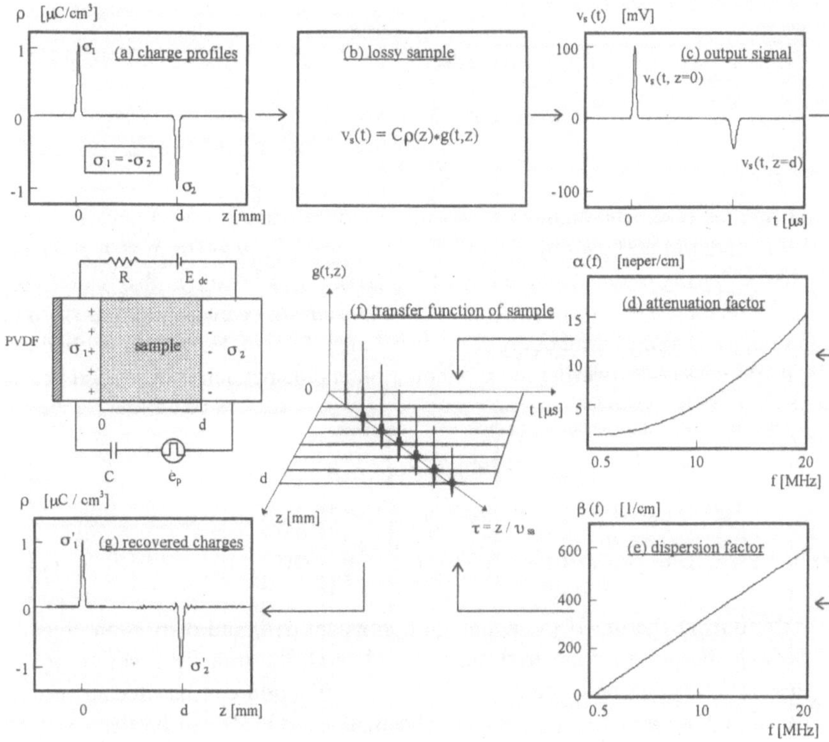

Fig. 11.22 Recovered PEA signal of LDPE with charge

(a) **Charge profile**: When a DC voltage is applied to the sample, surface charge densities σ_1 and σ_2 are induced on the anode and cathode, respectively. When the charge densities σ_1 and σ_2 are measured by the PEA method, the signal waveforms with different polarities and the same magnitude of $\sigma_1 = -\sigma_2$ can be obtained, as depicted in the charge profile of Fig. 11.22a. The result is a signal obtained from an ideal sample without wave attenuation and dispersion. Here, the reason for that the waveforms of charge densities σ_1 and σ_2 are triangular is as follows. As shown in Eq. (11.1), the pressure wave $p_1(t)$ becomes rectangular due to the convolution integral of the charge densities and the pulse electric field $e(t)$ with a finite width (~10 ns). Further, the signal waveform $v_s(t)$ becomes triangular due to the convolution integral of the rectangular piezoelectric function $d(\tau)$ and the rectangular pressure wave $p_1(t)$ in Eq. (11.2).

(b) **Lossy sample**: Generally, polymer insulating materials have viscoelastic properties (lossy material). Thus, attenuation $\alpha(f)$ and dispersion $\beta(f)$ occur when the pressure wave $p(t)$ propagates through the polymer material. Here, the transfer function $g(f, z)$ in Eq. (11.3) is used. The pressure wave $p_2(t)$ propagating in the polymer material can be expressed by the convolution integral of the pressure wave $p_1(t)$ in Eq. (11.4) and the transfer function $g(f, z)$.

$$g(f, z) = e^{-\alpha(f)z} \, e^{-j\beta(f)z} \tag{11.3}$$

$$p_2(t) = C_3 \int_0^t p_1(\tau) \, g(t - \tau) \, d\tau \tag{11.4}$$

(c) **Output signal**: The signal $v_s(t)$ as the pressure wave $p_2(t)$ propagates in the lossy material is shown by Eq. (11.5). As a result, the waveform $v_s(t, z = d)$ of charge density σ_2 is attenuated by $\alpha(f)$, and the signal width is also widened by $\beta(f)$. When $v_s(t, z = 0)$ from σ_1 and $v_s(t, z = d)$ from σ_2 are Fourier transformed and converted into a function in the frequency domain, Eq. (11.6) is obtained. The process of recovering these attenuation and dispersion is described below.

$$v_s(t) = C_4 \int_0^t d(\tau) \, p_2(t - \tau) \, d\tau \tag{11.5}$$

$$V_s(f, 0) = |V_s(f, 0)| \, e^{-j\varphi(f,0)} \, , \, V_s(f, d) = |V_s(f, d)| \, e^{-j\varphi(f,d)} \tag{11.6}$$

(d) **Attenuation factor:** If the attenuation constant $\alpha(f)$ and dispersion constant $\beta(f)$ of the polymer material can be evaluated, the transfer function $g(f, z)$ in Eq. (11.3) can be calculated. This $g(f, z)$ function can be used to recover the signal waveform $v_s(t, z = d)$. The final Eq. (11.7) can be derived from Eqs. 11.3 and (11.6). Equation (11.8) summarizes the real part of Eq. (11.7). From Eq. (11.8), the frequency domain spectrum of the attenuation constant $\alpha(f)$ in Fig. 11.22d can be obtained.

$$\begin{aligned} V_s(f, d) &= V_s(f, 0) \, G(f, z) = V_s(f, 0) \, e^{-\alpha(f)d} \, e^{-j\beta(f)d} \\ |V_s(f, d)| \, e^{-j\varphi(f,d)} &= |V_s(f, 0)| \, e^{-j\varphi(f,0)} \, e^{-\alpha(f)d} \, e^{-j\beta(f)d} \\ |V_s(f, d)| &= |V_s(f, 0)| \, e^{j\varphi(f,d) - j\varphi(f,0)} \, e^{-\alpha(f)d} \, e^{-j\beta(f)d} \end{aligned} \tag{11.7}$$

Attenuation is the real part:

$$|V_s(f, d)| = |V_s(f, 0)| \, e^{-\alpha(f)d} \Rightarrow \alpha(f) = -\frac{1}{d} \ln \frac{|V_s(f, d)|}{|V_s(f, 0)|} \tag{11.8}$$

(e) **Dispersion factor**: Eq. (11.9) is the imaginary part of Eq. (11.7). From Eq. (11.9), the frequency domain spectrum of the dispersion constant $\beta(f)$ can be obtained, as shown in Fig. 11.20e.

Dispersion is the imaginal part.

$$j\varphi(f,0) - j\varphi(f,d) - j\beta(f)d = 0 \Rightarrow \beta(f) = -\frac{1}{d}[\varphi(f,d) - (f,0)] \tag{11.9}$$

(f) **Transfer function of sample**: When the spectrums of the attenuation constant $\alpha(f)$ i n Eq. (11.8) and the dispersion constant $\beta(f)$ of Eq. (11.9) are obtained in the frequency domain, the transfer function $G(f, z)$ in the frequency domain can be represented by Eq. (11.10). The transfer function $g(t, z)$ of Fig. 11.22f in the space and time domains is obtained by the inverse Fourier transform, as shown in Eq. (11.11).

$$G(f,z) = \frac{V_s(f,z)}{V_s(f,0)} = e^{-\alpha(f)z}\, e^{-j\beta(f)z} \tag{11.10}$$

$$g(z,t) = \frac{1}{2\pi}\int_{-\infty}^{+\infty} \left[e^{-\alpha(f)z}\, e^{-j\beta(f)z}\right] e^{j2\pi f}\, df \tag{11.11}$$

(g) **Recovered charge**: $V_s(f, 0)$ in Eq. (11.12) is the spectrum in the frequency domain of $v_s(t, 0)$ without waveform distortion at $z = 0$. $V_s(f, d)$ is the spectrum in the frequency domain of $v_s(t, d)$ after the waveform recovery at $z = d$. Use the inverse Fourier transform to convert $V_s(f, d)$ to $v_s(t, d)$ of Eq. (11.13) in the space and time domains, as shown in Fig. 11.22g. $v_s(t, d)$ is the electrode-induced charge σ'_2 in the figure.

$$G(f,d) = \frac{V_s(f,d)}{V_s(f,0)} = e^{-\alpha(f)d}\, e^{-j\beta(f)d} \tag{11.12}$$

$$V_s(f,d) = V_s(f,0)G(f,d) \quad \Rightarrow \quad v_s(t,d) \tag{11.13}$$

The above is the recovery algorithm of $v_s(t, d)$ drawn in Fig. 11.22. See Takada Text Vol. I, Part 2, Chap. 9.

(3) **Recovery of deformed waveform of flat plate sample under a temperature gradient** [1, 3]

Since a load current flows through the inner conductor of the power cable, the temperature of the inner conductor usually rises up to 90 °C due to the Joule heating. The heat is radiated from the inner conductor to the outer conductor through the insulating layer. Therefore, there is a temperature gradient in the insulating layer. Aiming at this situation, the deformation of pressure wave $p(z, t)$ and PEA voltage signal $v(t)$ of the flat plate sample under a temperature gradient is shown in Fig. 11.21c. The viscoelastic property of the material under a temperature gradient is not uniform in the sample, and thus the sound velocity $v(z)$ and mass density $m(z)$ are functions of space z.

(4) Distorted signal and correction results of a plate sample under a temperature gradient

The red dotted line of Fig. 11.23 shows the original waveforms of the PE material with a thickness of 350 μm measured under the temperature differences of $\Delta T =$ 20 °C and $\Delta T = 40$ °C. When a pulsed pressure wave propagates in a sample with a temperature difference, the sound velocity $v(z)$ becomes a function of the position z. Therefore, the correction calculation is performed using the transfer function $G(f, z)$ as the function of frequency f and position z. The recovered charge waveform is shown by the black solid line in Fig. 11.23 [26, 28]. It shows that the amplitude and shape of the negative charge waveform at the position of 350 μm are the same as those of the positive charge waveform at the position of $z = 0$. Also, the recovered position of the negative charge waveform returns to $z = 350$ μm previously affected by the temperature difference ($\Delta T = 20$ °C and 40 °C).

(5) PEA device for power cable under a temperature gradient

In the previous parts (3) and (4), we introduce the experimental examples of recovered PEA signals of flat plate samples under a temperature gradient. In this part, an example of waveform recovery is introduced when a temperature gradient ($\Delta T =$ 36.8 °C, 52.8 °C) is in the XLPE cable with an insulation layer of 4.5 mm.

Figure 11.24a shows the basic PEA measurement system for a coaxial cable. Concentric fan-shaped electrodes are used on the ground side, and a piezoelectric element film is attached along the electrode surface. When the temperature gradient ΔT is present, the charge signal at the external conductor is also delayed in time.

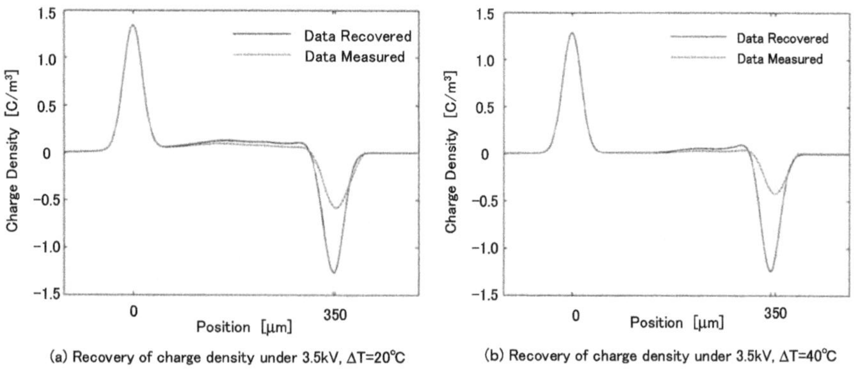

Fig. 11.23 Results of recovery and time-shift correction of charge waveforms of PE with a temperature gradient

Figure 11.24b shows the actual PEA equipment. Since the high voltage and pulse voltage are applied to the outer conductor of the coaxial cable, the digital oscilloscope and other measuring instruments are supplied by the storage battery, which are installed on the high voltage side. It is a very large measurement device [4].

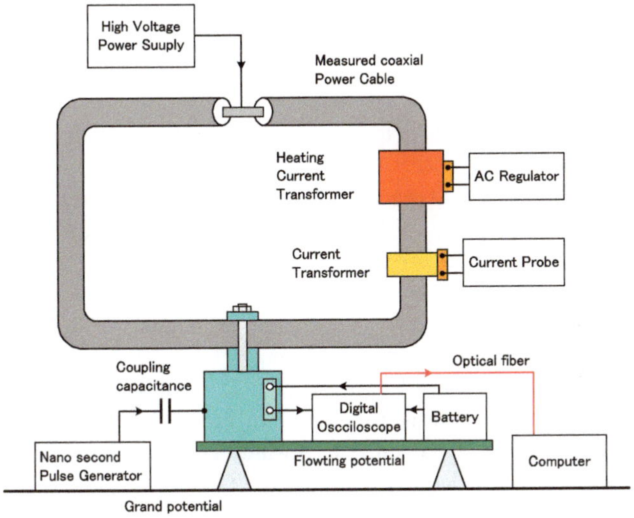

(a) PEA device for coaxial cable with a temperature gradient.

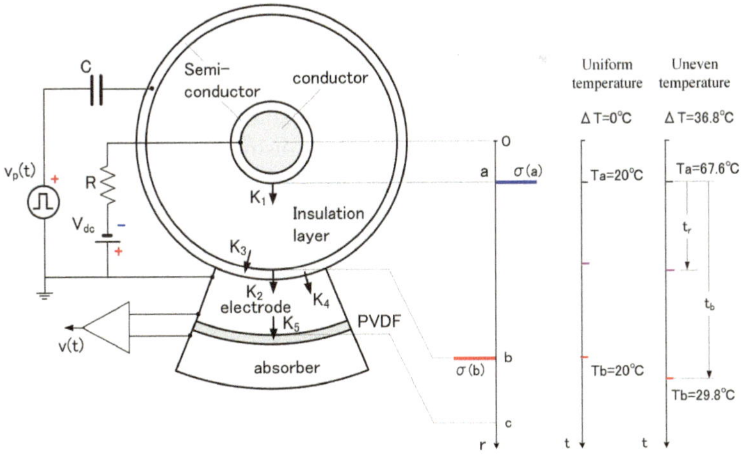

(b) Explanatory diagram of time shift of coaxial cable with a temperature gradient.

Fig. 11.24 The basic circuit of PEA measurement for a coaxial cable

Table 11.2 Conditions of temperature gradient

Load current	90 A (°C)	110 A (°C)
Inner conductor temperature	67.7	88.7
Outer conductor temperature	30.9	35.9
Temperature difference	36.8	52.8

(6) **Distorted signals and correction results of a power cable under a temperature gradient**

The heating of the cable conductor is carried out by the load currents of 90 and 110 A through the center conductor using a current transformer. Table 11.2 shows the temperature of the inner and outer conductors with the temperature difference.

Figure 11.25a shows the PEA signals ($\Delta T = 36.8$ °C: red line, 52.8 °C: blue line) of the insulating layer under a temperature gradient. The PEA signal without a temperature gradient ($\Delta T = 0$ °C) is shown by the black dotted line. Although it is an insulating layer without space charge, the signals of the anode charge and the cathode charge are not the same, showing that the latter is attenuated. Since the insulating layer is as thick as 4.5 mm, the attenuation of pressure waves can be easily recognized. Further, if there is a temperature gradient in the insulating layer, the position of the cathode signal shifts backward. This is because the velocity of the pressure wave is slightly decreased.

Figure 11.25b shows the case where the temperature gradient is $\Delta T = 36.8$ °C. The blue waveform before the correction process changes to the red waveform after the correction process.

Figure 11.25c shows the case where the temperature gradient is $\Delta T = 52.8$ °C. The blue waveform before the correction process becomes the red waveform after the correction process.

For the samples with a thick thickness, temperature rise or temperature gradient, etc., the pressure wave of PEA measurement is distorted and the measured waveform is thus deformed. The correct charge distribution cannot be measured directly. Therefore, it is necessary to recover the signal using the processing algorithm.

Since the algorithm of correction processing is not introduced in detail here, refer to References [24–32].

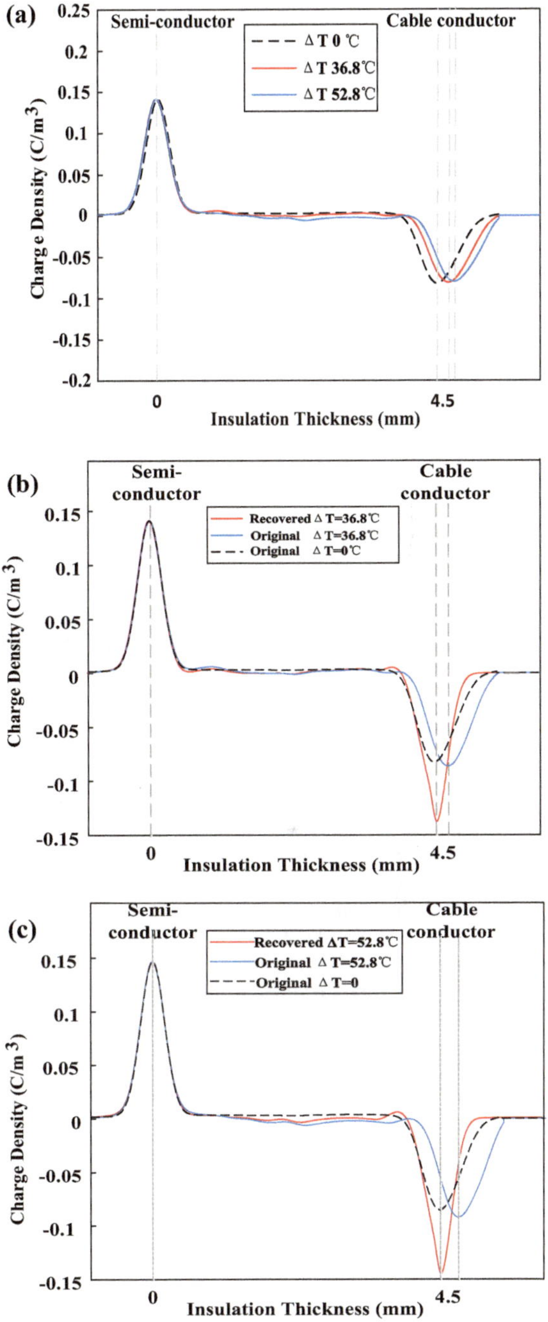

Fig. 11.25 **a** Original PEA signal. **b** and **c** recovery waveforms of coaxial cable in a temperature gradient

References

1. K. Matsui, Y. Tanaka, T. Takada, T. Fukao, K. Fukunaga, T. Maeno, J.M.Alison, Space charge behavior in low-density polyethylene at pre-breakdown, *IEEE Transactions on Dielectrics and Electrical Insulation*, 12(3), 406-415 (2005).
2. T. Takada, Y. Kikuchi, H. Miyake, Y. Tanaka, M. Yoshida, Y. Hayase, Determination of electric charge trapping sites in saturated and aromatic polymers by quantum chemical calculation, *IEEE Transactions on Dielectrics and Electrical Insulation*, 22(2), 1240-1249 (2015).
3. K. Murata, Y. Tanaka, T. Takada, Space charge formation in cross-linked polyethylene under AC voltage, *IEEJ Transactions on Fundamentals and Materials*, 116(12), 1095-1100 (1996).
4. X. Wang, N. Yoshimura, Y. Tanaka, K. Murata, T. Takada, Space charge characteristics in cross-linking polyethylene under electrical stress from dc to power frequency, *Journal of Physics: D Applied Physics*, 31, 2057-2064 (1998).
5. T. Takada, T. Tohmine, Y. Tanaka, J. Li, Space charge accumulation in double-layer dielectric systems—measurement methods and quantum chemical calculations, *IEEE Electrical Insulation Magazine*, 35(5), 36-46 (2019).
6. T. Ito, Y. Tanaka, T. Takada, T. Tanaka, Characteristics of accumulation and decay of space charge in LDPE/EVA laminated, *IEEJ Transactions on Fundamentals and Materials*, 121(2), 129-135 (2001).
7. T. Tanaka, T. Ito, Y. Tanaka, T. Takada, Frequency dependence of interfacial space charge formed in laminated dielectrics under AC voltage application conditions, Paper presented at IEEE Conference on Electrical Insulation and Dielectric Phenomena, 796–799 (2000).
8. T. Takada, Y. Hayase, Y. Tanaka, T. Okamoto, Space charge trapping in electrical potential well caused by permanent and induced dipoles for LDPE/MgO nanocomposite, *IEEE Transactions on Dielectrics and Electrical Insulation*, 15(1), 152-160 (2008).
9. H. Aotama, K. Matsui, Y. Tanaka, T. Takada, T. Maeno, Observation and numerical analysis of space charge behavior in low-density polyethylene formed by ultra-high DC stress, *IEEE Transactions on Dielectrics and Electrical Insulation*, 15(3), 841-850 (2008).
10. R. Liu, T. Takada, N. Takasu, Pulsed electro-acoustic method for measurement of space charge distribution in power cables under both DC and AC electric field", *Journal of Physics D: Applied Physics*, 26(6), 986-993 (1993).
11. K. Suzuki, Y. Tanaka, T. Takada, Y. Ohki, C. Takeya, Correlation between space charge distribution and water-tree location in aged XLPE cable, *IEEE Transactions on Dielectrics and Electrical Insulation*, 8(1), 78-81 (2001).
12. N. Hozumi, H. Suzuki, T. Okamoto, K. Watanabe, A Watanabe, Direct observation of time-dependent space charge profiles in XLPE cable under high electric field, *IEEE Transactions on Dielectrics and Electrical Insulation*, 1(6), 1068-1076 (1994).
13. T. Takeda, N. Hozumi, H. Suzuki, K. Fujii, K. Terashima, M. Hara, Y. Murata, K. Watanabe, M. Yoshida, Space charge measurement in full-size 250 kV DC XLPE cables, *IEEE Transactions on Power Delivery*, 13(1), 28-39 (1998).
14. K. Tadokoro, T. Motoyama, H. Harada, Y. Tanaka, T. Takada, T. Maeno; " Space charge formation by irradiation of visible light in polyimide under DC electric stress ", IEEJ Transactions on Fundamentals and Materials, 129(7), 463-469 (2009).
15. T. Takada, H. Miyake and Y. Tanaka, Pulse acoustic technology for measurement of charge distribution in dielectric materials for spacecraft, *IEEE Transactions on Plasma Science*, 34(5), 2176-2184 (2006).
16. K. Nagasawa, M. Honjoh, H. Miyake, R. Watanabe Y. Tanaka, T. Takada, Charge accumulation in various electron -beam-irradiated polymers", *IEEJ Transactions on Electrical and Electronics Engineering*, 5(4), 410–415 (2010).
17. M. Kojima, Y. Tanaka, T. Takada, Y. Ohoki, Measurement of space charge accumulation in gamma-irradiated polyethylene with DC voltage, *IEEJ Transactions on Fundamentals and Materials*, 115(2), 93-98 (1995).

18. Y. Tanaka, T. Takada, Observation and analysis of charge behavior in gamma-irradiated LDPE, Paper presented at Annual Report Conference on Electrical Insulation and Dielectric Phenomena, Atlanta, USA, 2176–2184 (2006).
19. S. Maruta, H. Miyake, S. Numata, Y. Tanaka, T. Takada, Charge accumulation characteristics in proton beam irradiated polymers, Paper presented at Annual Report Conference on Electrical Insulation and Dielectric Phenomena, Quebec, Canada, 153–156, (2008).
20. T. Takada, Y. Tanaka, Discussion on pulsed electro-acoustic signals caused by accumulated space charge, electrode charge and polarization charge, Paper presented at the papers of Technical Meeting of IEEJ, DEI-18–098 (2018).
21. Y. Imaizumi, K. Suzuki, Y. Tanaka, T. Takada, "Three-dimensional space Charge distribution measurement in electron beam irradiated PMMA, *IEEJ Transactions on Fundamentals and Materials*, 116(8), 684-689 (1996).
22. X. Qin, K. Suzuki, Y. Tanaka, T. Takada, Three-dimensional Space Charge Measurement in a Dielectric using the Acoustic Lens and PWP Method, *Journal of Physics D: Applied Physics*, 32(1), 157-160 (1999).
23. T. Maeno, Three-dimensional PEA charge measurement system, *IEEE Transactions on Dielectrics and Electrical Insulation*, 8(5), 845-848 (2001).
24. Y. Li, Space charge measurement in lossy solid dielectric materials by pulsed electroacoustic method, *Ph.D. Dissertation*, Musashi Institute of Technology, 1994.
25. Y. Li, M. Aihara, K. Murata, Y. Tanaka, T. Takada, Space charge measurement in thick dielectric materials by pulsed electroacoustic method, *Review of Scientific Instruments*, 66(7), 3909-3916 (1995).
26. K. Nagashima, X. Qin, Y. Tanaka, T. Takada, Calibration of accumulated space charge in power cable using PEA measurement, Paper presented at 6th International Conference on Conduction and Breakdown in Solid Dielectrics, 60–64 (1998).
27. X. Wang, C. Chen, J. Hao, K. Wu, X. Chen, W. Li, C. Zhang, Recovery algorithm for space charge waveform in coaxial cable under temperature gradient, *Sensors and Materials*, 22(2), 1213-1219 (2015).
28. B. Vissouvanadin, T. Vu, L. Berquez, S. Roy, G. Teyssedre, C. Laurent, Deconvolution techniques for space charge recovery using pulsed electroacoustic method in coaxial geometry, *IEEE Transactions on Dielectrics and Electrical Insulation*, 21(2), 821-828 (1998).
29. H. Wang, K. Wu, Q. Zhu, X. Wang, Recovery algorithm for space charge waveform under temperature gradient in PEA method, *IEEE Transactions on Dielectrics and Electrical Insulation*, 22(2), 1213-1219 (2015).
30. Q. Zhu, X. Wang, K. Wu, Y. Cheng, Z. Lv, H. Wang, Space charge distribution in oil impregnated papers under temperature gradient, *IEEE Transactions on Dielectrics and Electrical Insulation*, 22(1), 142-151 (2015).
31. J. Dong, Z. Shao, Y. Wang, Z. Lv, X. Wang, K. Wu, W. Li, C. Zhang, Effect of temperature gradient on space charge behavior in epoxy resin and its nanocomposites, *IEEE Transactions on Dielectrics and Electrical Insulation*, 24(3), 1537-1546 (2017).
32. C. Chen, Y. Bu, X. Wang, C. Cheng, K. Wu, A comparison of space charge behaviors in coaxial cable and film sample under temperature gradient, *IEEE Transactions on Dielectrics and Electrical Insulation*, 26(6), 1941-1948 (2019).

Part III
Utilization of Quantum Chemical Calculation Analysis

Chapter 12
Basics of Quantum Chemical Calculation

12.1 Terms in the Electron Energy Level Diagram

The terms in the energy level diagram used in the Quantum Chemical Calculation/DFT calculation are explained based on the example of the ethane molecule (C_2H_6), as shown in Fig. 12.1.

(a) DFT: The abbreviation of Density Functional Theory, which is an algorithm for calculating the orbital electron wave function $\psi(x, y, z)$ and the energy level E of the Schrödinger equation.

(b) Valence band: The energy level filled with orbital electrons from ① to ⑥ in the energy level diagram belongs to the valence band. The two electrons of α spin and β spin can exist in an energy level (Pauli Exclusion Principle). ① At the −276.6 eV level, there are four 1S orbital electrons in two carbons C. ② At the −20.39 eV level, orbital electrons exist in the entire C_2H_6 molecule. Moreover, orbital electrons exist in each energy level E of ③ to ⑥. Since the carbon C has six orbital electrons and the hydrogen H has one, the total orbital electrons of the ethane molecule are ($6 \times 2 + 1 \times 6 =$) 18.

(c) Empty band: Empty band inculdes ⑦ ⑧ ⑨ and so on where orbital electrons do not exist. When an external electron comes into the ethane C_2H_6 and the molecule is negatively charged, the electron enters the LUMO level of empty band. Electrons in the ⑥ HOMO level can be excited by light energy $h\nu$[eV] or similar energies. Then they move to the ⑦ LUMO level in the empty band. At this time, the electrons at the ⑥ HOMO level leave and the holes remain there.

(d) HOMO level: The abbreviation of Highest Occupied Molecular Orbital. It is the highest energy level in the valence band.

(e) LUMO level: The abbreviation of Lowest Unoccupied Molecular Orbital. It is the lowest energy level in the empty band.

T. Takada et al., *Electric Charge Accumulation in Dielectrics: Measurement and Analysis*,
https://doi.org/10.1007/978-981-19-6156-4_12

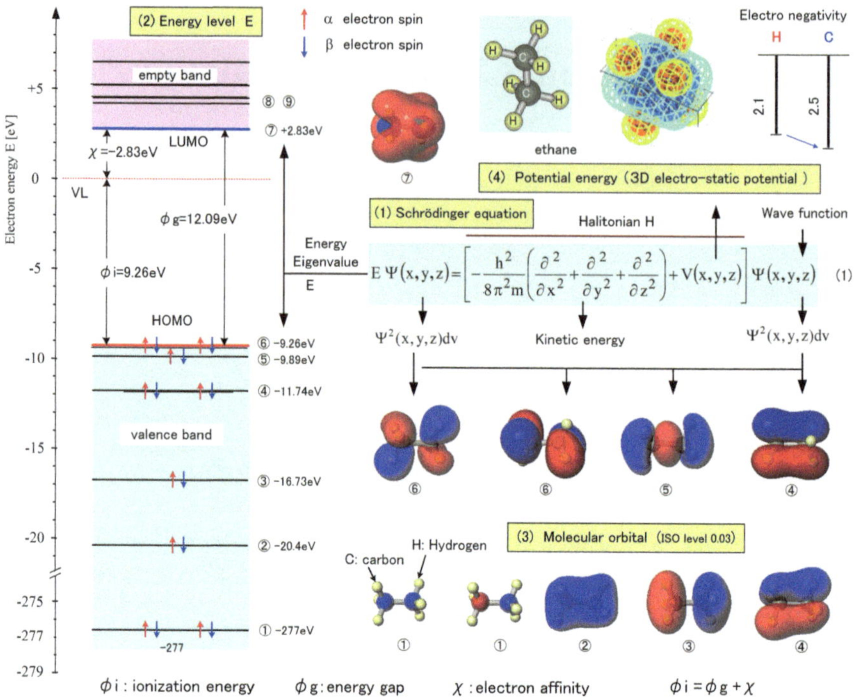

$$E\,\Psi(x,y,z)=\left[-\frac{h^2}{8\pi^2 m}\left(\frac{\partial^2}{\partial x^2}+\frac{\partial^2}{\partial y^2}+\frac{\partial^2}{\partial z^2}\right)+V(x,y,z)\right]\Psi(x,y,z) \quad (1)$$

Fig. 12.1 The example of the ethane molecule in the Quantum Chemical Calculation/DFT calculation

(f) Energy gap ϕ_g [eV]: The energy difference between the LUMO level and the HOMO level. ϕ_g [eV] = LUMO–HOMO.

(g) Vacuum level VL: The energy level o f$E = 0$ is defined as the vacuum level VL. When the electron in the HOMO level (negative charge) gains energy and jumps out of the molecule, holes (positive charge) remain in the original molecule. The electric field of the remaining holes (positive charges) still attracts the electrons (negative charges) in jumping process. If the electron has enough energy, it cannot be pulled back to the molecule. The energy level in which the electron cannot be pulled back is the vacuum level VL.

(h) Ionization energy ϕ_i [eV]: The difference between the VL level and the HOMO level. When the electron at the HOMO level (negative charge) gains energy and jumps out of the original molecule (ethane C_2H_6 molecule), the hole (positive charge) remains in the original molecule and generates due to this ionization.

(i) Electron affinity χ [eV]: The difference between VL level and LUMO level. The electron affinity is $\chi > 0$ when VL level > LUMO level, and $\chi < 0$ when VL level < LUMO level.

See Sect. 12.6 for the explantion of the polarity of electron affinity.

(j) Schrödinger equation: The Schrödinger equation is shown in the figure, where the time dependence is not added in it. Hamiltonian H is in the bracket [] on the right side and given by the sum of the kinetic energy of the electron wave (first term) and the potential of the electron wave space (second term). Then, at some electron energy levels E, the wave function $\psi(x, y, z)$ exists as a standing wave. The potential component of the Hamiltonian term is created by the atomic protons (positive charges) and extranuclear electrons (negative charges) of an atom. Also, the potential $V(x, y, z)$ is determined by the bonding state of atoms. The DFT is the software that can find the solution of the intrinsic electron energy E and the wave function $\psi(x, y, z)$ of the Schrödinger equation.

(k) Molecular orbitals: The molecular orbitals are shown in ① to ⑦ on the right side of the energy levels in Fig. 12.1. ① The molecular orbital of −276.6 eV level indicates the existence of 1S orbital electron wave of carbon C. This molecular orbital is represented by the product $\psi^2(x, y, z)dv$ of the square of electron wave function $\psi(x, y, z)$ and its infinitesimal space dv. Specifically, it refers to a region within a molecule where electrons are likely to be found.

12.2 Atomic and Molecular Orbitals

Here, it is explained that molecular orbitals are composed of atomic orbitals. Figure 12.2 shows the DFT calculation results of the electron energy level and the molecular orbital of water molecule H_2O. The Density Functional Theory (DFT) is a calculation method for solving the Schrödinger equation in Sect. 12.1. The simplest STO 3G is adopted as the basic function (Basis) to explain the DFT concept, though its calculation accuracy is low.

(1) Molecular orbital

The molecular orbital is shown in the center of Fig. 12.2. The shape of this molecular orbital differs depending on the electron energy level. Focus on the molecular orbital 5H of −4.08 eV (−0.15 Hartree) as the HOMO level, and it is found that electron waves exist on the both sides of the H_2O plane. The molecular orbital 1H of −513 eV (−18.85 Hartree) is 1S orbital (spherical) of oxygen O. The molecular orbital 2H of −25 eV (−0.91 Hartree) consists of 1S and 2S orbitals of oxygen O (spherical) and 1S orbital of hydrogen H (spherical). Furthermore, the molecular orbital 3H of −11 eV(−0.40 Hartree) consists of 1S orbital of hydrogen H and 2PX orbital of oxygen O. These 1S and 2PX orbits are called atomic orbitals.

(2) Create molecular orbitals from atomic orbitals and electron wave functions

The molecular orbital is represented by the product ψ^2 (x, y, z) dv of the square of the electron wave function $\psi(x, y, z)$ and its infinitesimal space dv. Therefore, $\psi(x, y, z)$ is obtained by the linear superposition of the atomic orbitals of the atoms that make up the molecule (equation in Fig. 12.3). As for the atomic orbital function χ,

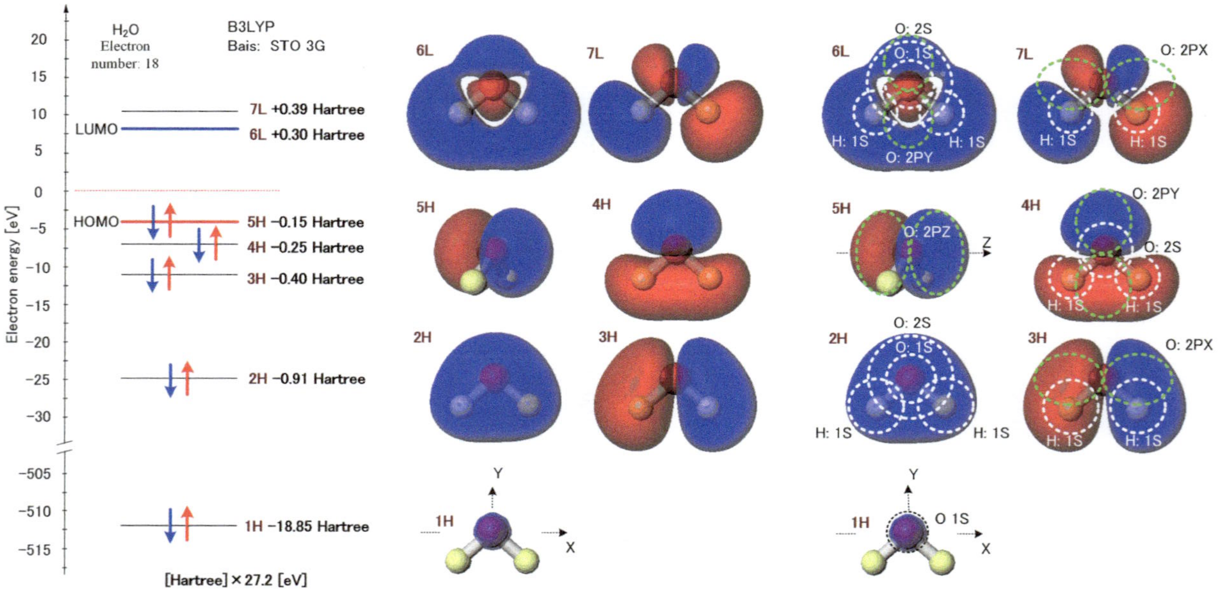

Fig. 12.2 The DFT calculation results of the electron energy level and the molecular orbital of H_2O

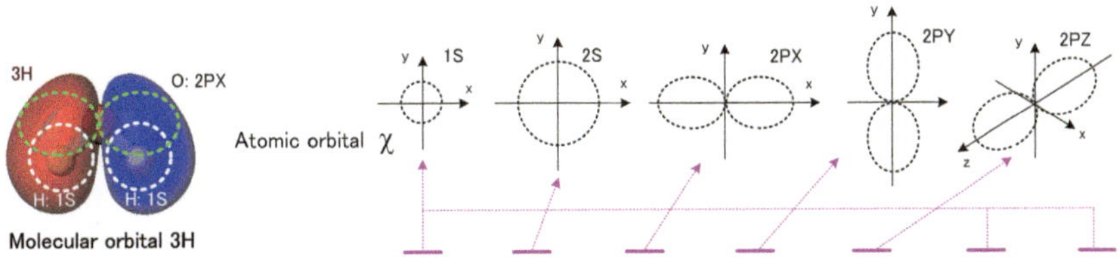

The electron wave function of molecular orbital 3H (Linear superposition of electron orbitals) →

$$\Psi(3H) = c_1\chi_1 + c_2\chi_2 + c_3\chi_3 + c_4\chi_4 + c_5\chi_5 + c_6\chi_6 + c_7\chi_7$$

C_1=0.00 C_2=0.00 C_3=0.62 C_4=0.00 C_5=0.00 C_5=0.45 C_7=−0.45

F: full band
V: empty band

Eigenvalues of electron orbital energy Unit: Hartree

Gaussian calculation's log file Search: "Molecular orbital coefficient" then get the table.

b3lyp/sto-3g			1	2	3	4	5	6	7
			F	F	F	F	F	V	V
		Eigenvalues	−18.85	−0.9122	−0.4046	−0.2548	−0.1461	0.3043	0.38694
1	O	1S	0.9936	−0.2301	0.00005	−0.1233	0	0.1161	−0.0002
2		2S	0.0284	0.82462	−0.0003	0.60362	0	−0.737	0.00158
3		2PX	0	0.00013	0.61973	−9E−05	0	0.0015	0.93175
4		2PY	−0.004	−0.1749	0.00006	0.71633	0	0.7843	−0.0013
5		2PZ	0	0	0	0	1	0	0
6	H	1S	−0.006	0.16649	0.45037	−0.312	0	0.7366	−0.7909
7	H	2S	−0.006	0.16634	−0.4504	−0.3121	0	0.7392	0.78807

Coefficients *C* of molecular orbital 3H

Fig. 12.3 Different molecular orbitals and calculation coefficients

spherical 1S orbital and 2S orbital, gourd type 2PX orbital, 2PY orbital, 2PZ orbital, and other orbitals have been drawn, as shown at the top of Fig. 12.3. Then, $\psi(x, y, z)$ is the sum of the products of the atomic orbital χ and the coefficient C (equation in Fig. 12.3).

The table in Fig. 12.3 shows the eigenvalues E (Hartree units) of the electron energy and the values of the coefficient C obtained from the Schrödinger equation. For example, focus on $E = -18.85$ Hartree ($= -18.85 \times 27.2 = -513$ eV), which is the eigenvalue of the lowest energy level of O oxygen (leftmost up in the table), the coefficient $C_1 = 0.9936$ (yellow) of the 1S orbit of oxygen O (yellow mark) is the largest, and the coefficients of the other orbits can be ignored. Thus, its wave function becomes 1S orbit of oxygen with $\psi(x, y, z) = C_1\chi_1$. This is the molecular orbital only in oxygen shown by the bottom of Fig. 12.3. Also, look at $E = -0.4046$ Hartree ($= -0.4046 \times 27.2 = -11$ eV) of the eigenvalue of molecular orbital 3H (center row of the table).

The coefficient $C_3 = 0.61973$ of 2P orbital of oxygen with the coefficients $C_6 = 0.45037$ and $C_7 = -0.4504$ (yellow mark) of 1S and 2S orbitals of hydrogen are large values, while the coefficients of the other orbits can be ignored. So the wavefunction becomes $\psi(x, y, z) = C_3\chi_3 + C_6\chi_6 + C_7\chi_7$, which are composed of 2PX orbit of oxygen O and 1S orbits of two hydrogen H. Therefore, the molecular orbitals at the upper left corner of Fig. 12.3 are drawn by dotted lines to indicate 2PX orbitals of oxygen O and 1 s orbitals of hydrogen H. Furthermore, the molecular orbitals displayed by this $\psi(3\mathrm{H}, x, y, z) = C_3\chi_3 + C_6\chi_6 + C_7\chi_7$ become the product of the square of the electron wave function and its infinitesimal space dv, i.e. $\psi^2\,(x, y, z)\,\mathrm{d}v$ is shown by red and blue color in Fig. 12.3. Finally, the molecular orbital 3H of ψ^2 (3H, x, y, z) dv in Fig. 12.2 is also the result obtained by this way. Further, for other eigenvalues, each $\psi(x, y, z)$ can be obtained from the coefficients C numbered 1 to 7 in the table of Fig. 12.3, and the molecular orbitals for each energy eigenvalue E are similarly obtained. The other results are shown by the 2H, 4H and 5H molecular orbitals in Fig. 12.2.

12.3 Diagram of Electron Energy Level

(1) Electron energy level and molecular orbital of propane C_4H_{10}

Before discussing the actual polymer materials, the electron energy levels and molecular orbitals of small molecule butane C_4H_{10} are outlined in Fig. 12.4. The diagram of electron energy level is shown on the left of the figure. The lowest level is −277 eV in the 1S orbit of carbon C of No. 1, No. 2, No. 3 and No. 4. The next level is the No. 5 of −21.46 eV, which is far from the lowest level. There are 17 levels under the highest HOMO level in the valence band, which is the No. 17 of −8.64 eV. The level above the forbidden zone is the LUMO in the empty band, which is No.18 of + 2.55 eV, and there are also levels above it. There are no electrons in the empty band. Thus, the 1S orbital of carbon C is as low as −277 eV, while the LUMO level is + 3 eV. The range of electron energy level is very wide. These electron energy levels are the eigenvalues E of the Schrödinger equation in the figure.

The valence band is composed of 17 levels of eigenvalues. The electron number of hydrogen H is 1, the electron number of carbon C is 6, and the total electron number of C_4H_{10} is ($6 \times 4 + 1 \times 10 =$) 34. Two electrons of α-spin and β-spin can exist in one energy level according to "Pauli Exclusion Principle". Therefore, the valence band becomes 17 levels. When excess electrons come from the outside, they enter into the empty band. There is a band gap between the LUMO level and the HOMO level. At the energy level of −277 eV, four 1S molecular orbitals of carbon C are formed. This molecular orbital is given by the product of the square of the wave function $\psi(x, y, z)$ of the Schrödinger equation and the infinitesimal space dv, i.e. $\psi^2(x, y, z)\mathrm{d}v$. The No. 5 molecular orbital (−21.46 eV) is distributed within the whole C_4H_{10} molecule. The HOMO level as No.17 molecular orbital (−8.64 eV) crosses hydrogen H and carbon C. The LUMO level as No. 18 molecular orbital (+2.55 eV) expands inside and outside the C_4H_{10} molecule. When discussing the charge accumulation and transfer in polymer materials, the electrons in the LUMO level and the HOMO level play a major role, and thus the electron energy level in the range of −15 eV to +5 eV across the forbidden band is targeted.

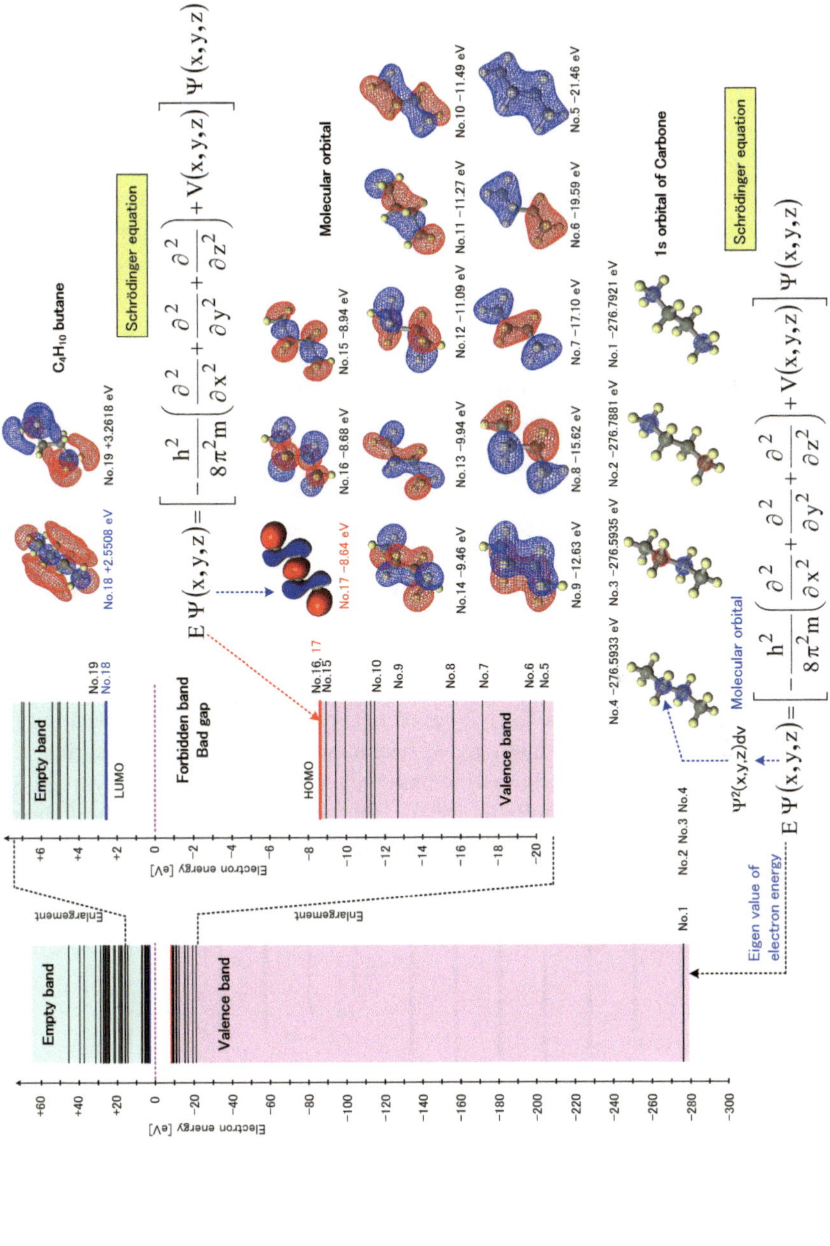

Fig. 12.4 The electron energy levels and corresponding molecular orbitals

12.4 Mulliken Atomic Charges and Electrostatic Potential Distribution

(1) Electronegativity

When atoms covalently bond to form a molecule, the waves of electrons overlap and the center of electron wave is biased. The electron bias depends on the electronegativity of each atom. The atom's electronegativity is shown in Fig. 12.5. The electronegativities of Li(1.0) or Na(0.9) are small, while those of H(2.1), F(4.0) and Cl(3.0) are large. When the electronegativities between the bonded atoms are different, the electrons are attracted to the atom with the higher electronegativity. For example, consider the electronegativity of the covalently bonded molecule of hydrogen H and carbon C in hexane C_6H_{14} shown by Fig. 12.6. Electron shift occurs from hydrogen H with a small electronegativity of 2.1 to carbon C with a large electronegativity of 2.5, and hydrogen H^+ is positively charged and carbon C^- is negatively charged. By observing the calculated electrostatic potential distribution $V(x, y, z)$ of C_6H_{14} in Fig. 12.6, carbon C^- is certainly negatively charged (blue), and hydrogen H+ is positively charged (orange).

(2) Mulliken atomic charges

The electrostatic potential distribution $V(x, y, z)$ of the ethane C_2H_6 molecule is introduced in Sect. 12.1. Here, the hexane C_6H_{14} is taken as a example and the method of obtaining $V(x, y, z)$ using the Mulliken atomic charge $\Delta q(a, b, c)$ of each bond atom from the DFT calculation result is introduced.

Figure 12.6 illustrates a method for obtaining $V(x, y, z)$ of the n-Hexane C_6H_{14} molecule. The left side of the figure shows the molecular structures of Mulliken atomic charge $\Delta q(a, b, c)$ and C_6H_{14}, respectively. Due to the electronegativity, the Mulliken atomic charge of hydrogen H becomes positive and that of carbon C becomes negative. Since the carbons C at both ends are bonded to three hydrogens H, $\Delta q(a, b, c)$ is large. The Mulliken atomic charge $\Delta q(a, b, c)$ of each atom is displayed in the three-dimensional coordinates (a, b, c).

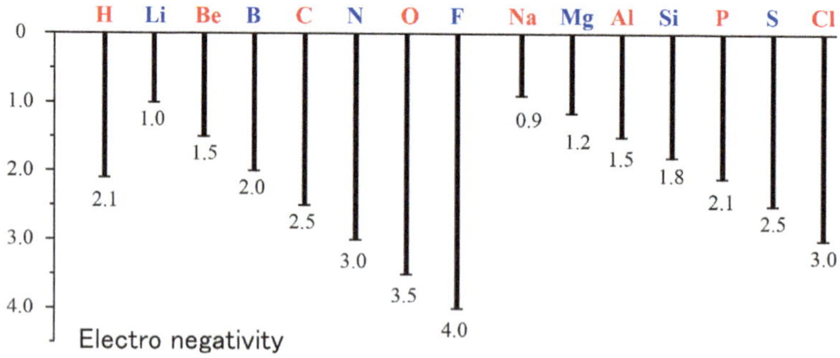

Fig. 12.5 The electronegativities of different atoms

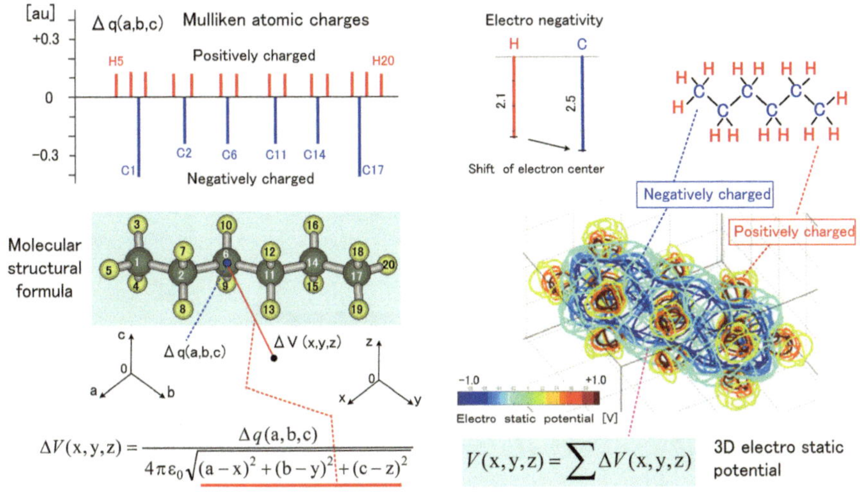

Fig. 12.6 The Mulliken atomic charges and the calculated electrostatic potential distribution

(3) Electrostatic potential distribution

The electric potential $\Delta V(x, y, z)$ at the space coordinates (x, y, z) in one area $\Delta(a, b, c)$ is given by Eq. (12.1). The potential $V(x, y, z)$ at space coordinates (x, y, z) due to Mulliken atomic charge of all bonded atoms is given by Eq. (12.2).The lower right part of Fig. 12.6 shows the calculating result of the electric potential distribution $V(x, y, z)$ of the C_6H_{14} molecule by using Eq. (12.2). It is confirmed that there is a negative potential distribution (blue) near the main chain carbon C, and a positive potential distribution (yellow orange) is near the side chain hydrogen H.

$$\Delta V(x, y, z) = \frac{\Delta q(a, b, c)}{4\pi \varepsilon_0 \sqrt{(a-x)^2 + (b-y)^2 + (c-z)^2}} \tag{12.1}$$

$$V(x, y, z) = \sum \Delta V(x, y, z) \tag{12.2}$$

(4) Electrostatic potential distribution of double bond (>C=O)

Figure 12.6 shows the n-Hexane C_6H_{14} molecule, and here we introduce a comparison of the n-Decane $C_{10}H_{22}$ molecule and the Oxide center $C_{10}H_{20}O$ molecule in Fig. 12.7. The HOMO level in the energy level diagram of the $C_{10}H_{20}O$ molecule in which oxygen O in the center of the molecule is present locally rises and the LUMO level falls. This is due to the localization of the electrostatic potential distribution $V(x, y, z)$ of the central oxygen-carbon double bond (>C = O). As a result, the π electron wave of the double bond of >C = O makes the HOMO level and the LUMO level closer to each other.

The Mulliken atomic charges $\Delta q(a, b, c)$ of both molecules are compared on the lower left and right sides of the figure. $+\Delta q(a, b, c)$ of hydrogen H in n-Hexane

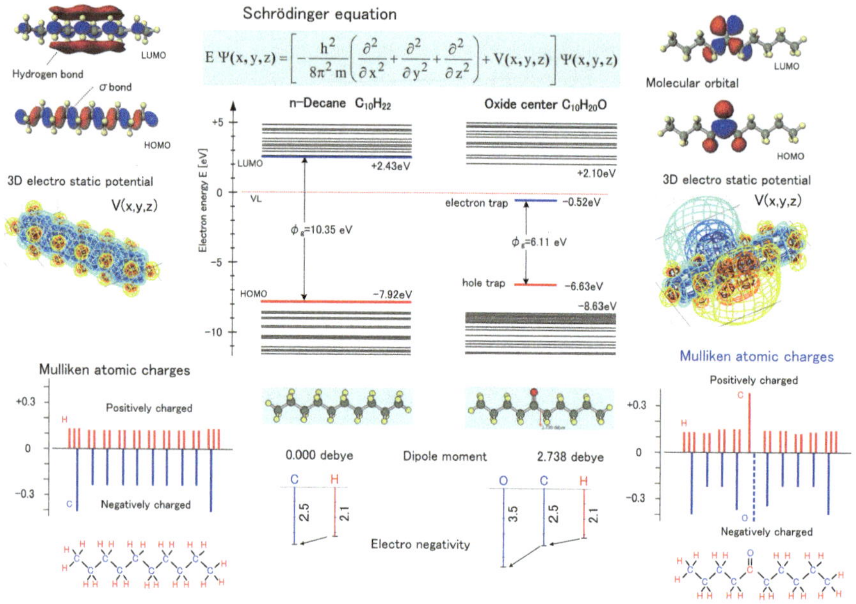

Fig. 12.7 The calculated results of the n-Hexane C_6H_{14} molecule

C_6H_{14} molecule has a positive polarity, and $-\Delta q(a, b, c)$of carbon C has a negative polarity. In particular, $-\Delta q(a, b, c)$ of carbon C at both ends is large because it bonds with three hydrogens H.

The carbon C of the double bond (>C = O) of the oxidized $C_{10}H_{20}O$ molecule becomes $+\Delta q(a, b, c)$ with positive polarity and oxygen O becomes $-\Delta q(a, b, c)$ with negative polarity. The value is large. The adjacent carbon C has a negative polarity of $-\Delta q(a, b, c)$, while the carbon C bonded to the oxygen O has a positive polarity of $+\Delta q(a, b, c)$. Therefore, the range of the electrostatic potential distribution $V(x, y, z)$ in the double bond (>C = O) portion is expanded.

If an electron carrier (negative polarity) comes close to it, it will be trapped by $+V(x, y, z)$ distribution. Further, if the hole carriers (positive polarity) come close to them, they can also be attracted and trapped in the $-V(x, y, z)$ distribution. The localized HOMO level of the energy level of the $C_{10}H_{20}O$ molecule in Fig. 12.7 becomes the trap site for holes, and the LUMO level becomes the trap site for electrons. By understanding the electric potential distributions $V(x, y, z)$ of each molecule, it is possible to deduce the origin of the formation of trap sites in polymer materials and the glass transition temperature due to intermolecular integration caused by the electric potential distributions $V(x, y, z)$ of each molecule.

12.5 Chemical Structure and Shape of Molecular Orbitals

In order to consider a model for electronic properties of molecules, the basics of Quantum Chemical Calculation/DFT calculation are introduced. Here, Fig. 12.8 compares the DFT calculation results of various molecules [1].

(a) Saturated hydrocarbon (n-Decane $C_{10}H_{22}$): LUMO level is +2.43 eV, HOMO level is −7.92 eV with the energy gap of $\phi_g = 10.35$ eV. The potential of hydrogen H is protruding and looks like a Rahotsu (curly hair of Buddha head).

(b) Oxidized hydrocarbon ($C_{10}H_{20}O$): There is a double bond (>C = O) due to oxidation at the center. This double bond becomes the LUMO level of −0.52 eV and the HOMO level of −6.63 eV, which is narrowed to $\phi_g = 6.11$ eV. Positive and negative potentials extend to the double bond. The dipole moment is as large as $m =$ 2.738 Debye.

(c) Unsaturated hydrocarbon (Conjugated DB $C_{10}H_{18}$): When a conjugated double bond with two double bonds is provided, the energy gap becomes narrower

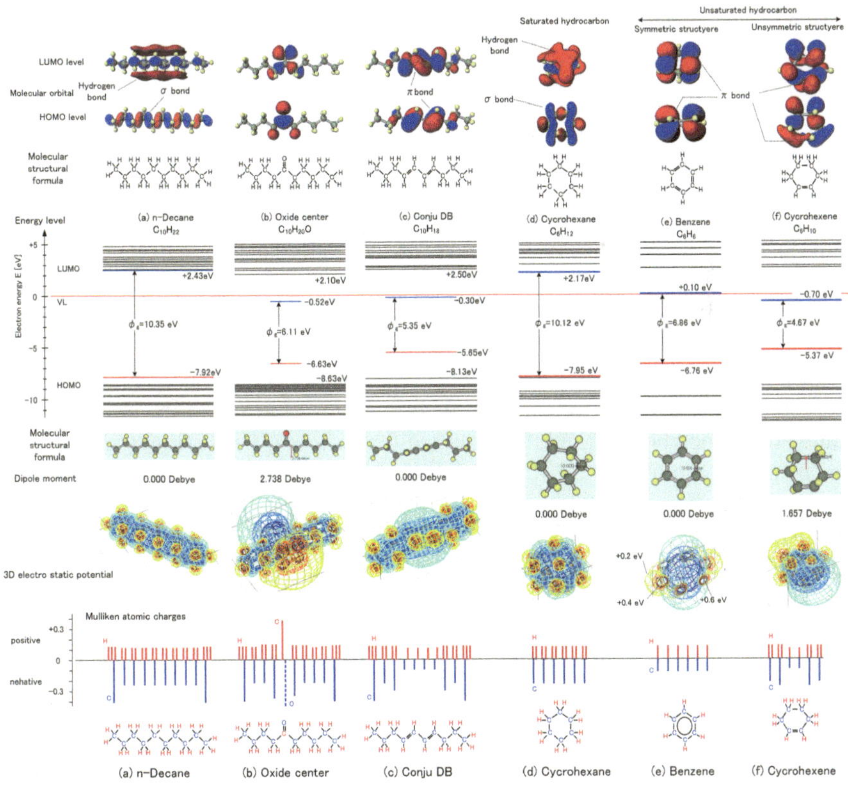

Fig. 12.8 The DFT calculation results of various molecules

with the value of $\phi_g = 5.35$ eV. Negative potential spreads to two double bonds (π electron orbits).

(d) Cyclic saturated hydrocarbon (Cycrohexane C_6H_{12}): Cyclic hydrocarbon without double bond has an energy gap of $\phi_g = 10.12$ eV, which is slightly narrower than that of (a) saturated hydrocarbon. The potential of hydrogen H is protruding like Rahotsu of Buda head.

(e) Cyclic hydrocarbon (Benzene C_6H_6): A cyclic hydrocarbon has three conjugated double bonds. The energy gap is $\phi_g = 6.66$ eV. Negative potential spreads vertically above and below the benzene ring plane. The dipole moment of benzene is $m = 0$. Hydrogen H is positively charged, then the positive potential is cyclic. The π electrons of carbon C are negatively charged, then the negative potential is distributed above and below the molecular surface. Benzene is quadrupole polarized. The potential of this quadrupole polarization extends far, and the intermolecular interaction is strong.

(f) Cyclic unsaturated hydrocarbon (Cycrohexene C_6H_{10}): A cyclic hydrocarbon has one double bond. The energy gap is narrower at $\phi_g = 4.67$ eV. The negative potential spreads to the double bond side, and the positive potential spreads to the hydrogen H side. Therefore, the dipole moment is as large as $m = 1.657$ Debye [2].

12.6 Positive Polarity and Negative Polarity of Electron Affinity

In the explanation of electron affinity in Sect. 12.1, it is stated that when the LUMO level is higher than the vacuum level VL, the electron affinity χ is negative polarity, i.e. $\chi < 0$, and it is $\chi > 0$ when the LUMO level is lower. The representative example of the former affinity is PE and that of the latter is PTFE [3, 4]. Here, we observe how the electronic behavior in the molecule changes depending on the polarity of the electron affinity. In order to compare the electronic behaviors in the molecule, the saturated hydrocarbon PE and the saturated fluorocarbon PTFE shown in Fig. 12.9 are taken as examples.

(1) Comparison of energy levels: (a) The LUMO level of PE molecules in which five $C_{24}H_{50}$ are arranged in parallel is higher than the VL. Therefore, the electron affinity is the negative polarity of $\chi < 0$ (Fig. 12.9, left). On the other hand, (b) the LUMO level of the PTFE molecule in which four $C_{16}F_{34}$ are arranged in parallel is lower than the VL, and thus the electron affinity is the positive polarity of $\chi > 0$ (right in Fig. 12.9).

(2) Comparison of molecular orbitals: Fig. 12.9c shows the molecular orbitals of the LUMO and HOMO levels of PE molecules and Fig. 12.9d shows the orbitals of PTFE molecules, respectively. The molecular orbital (electron wave) of the HOMO level of both the PE molecule and the PTFE molecule exists in the intra-chain. On

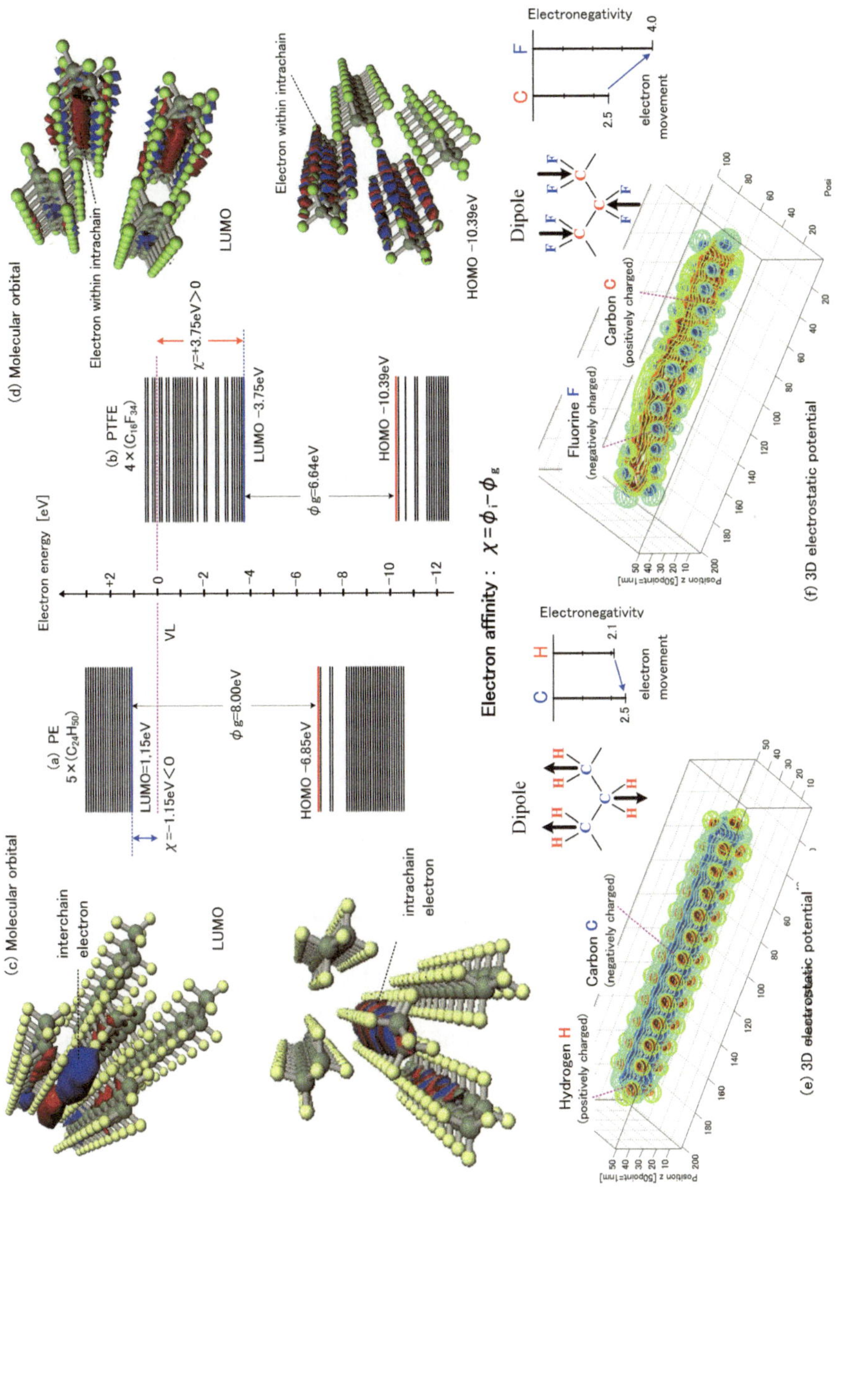

Fig. 12.9 The calculation results of the PE and PTFE molecules

the other hand, the LUMO level of the PE molecule exists in the inter-chain, while the LUMO level of the PTFE molecule exists in the intra-chain. Therefore, when the electron wave exists in the inter-chain, the electron affinity becomes the negative polarity of $\chi < 0$.

(3) Comparison of electrostatic potential: Fig. 12.9e shows the electrostatic potential $V(x, y, z)$ of PE molecule and Fig. 12.9f shows the potential of PTFE molecule, respectively. The carbon C at the main chain of the PE molecule is negatively charged (blue) and the hydrogen H at the side chain is positively charged (orange). On the other hand, the carbon C at the main chain of the PTFE molecule is positively charged (orange), and the fluorine F a tthe side chain is negatively charged (blue). The polarities of the potential distributions of PE and PTFE molecules are completely opposite. This charge polarity depends on the electronegativity of the bonded atom. See the part (1) of Sect. 12.4 for more information.

(4) Model of the electron affinity of $\chi < 0$ for PE: The LUMO level is a level in which excess electrons from the outside can enter. When the excess electrons in the vacuum level VL try to enter the PE molecule from the outside, they are repelled by the negative charge (blue) of the carbon C at the main chain. Therefore, the excess electrons need extra energy (electron affinity χ). Excess electrons with the energy of electron affinity χ cannot exist in the carbon C region at the main chain shown by Fig. 12.9e, but they can exist in the inter-chain of Fig. 12.9c.

On the other hand, when the excess electrons try to enter the PTFE molecule, the carbon C at the main chain in Fig. 12.9f is positively charged (orange). Thus, the excess electrons release energy and have an electron affinity of $\chi > 0$, and they can exist in the PTFE molecule. Therefore, the model with the electron affinity of $\chi < 0$ is that the excess electrons cannot enter into the molecule unless they get extra energy.

12.7 Application of Fermi–Dirac Distribution Function

Free electrons exist in the Fermi level (work function) of metal. The electrons in the metal electrode are emitted from the electrode when they get sufficient energy. Further, when the orbital electrons of the HOMO level of the polymer material (PE, PTFE, PI) get sufficient energy, they are excited to the LUMO level and electron–hole pairs are generated. In this way, the Fermi–Dirac Distribution Function is applied to analyze the rate of electron emission from the electrode and the generation rate of electron–hole pair.

(1) Fermi-Dirac Distribution Function

The probability of electron emission from the electrode is shown in Fig. 12.10 and the generation probability of electron–hole pair in polymer materials are examined using the Fermi–Dirac Distribution Function (FD) in Eq. (12.3). FD is drawn on the right side of the figure. The vertical axis is the difference between the electron energy

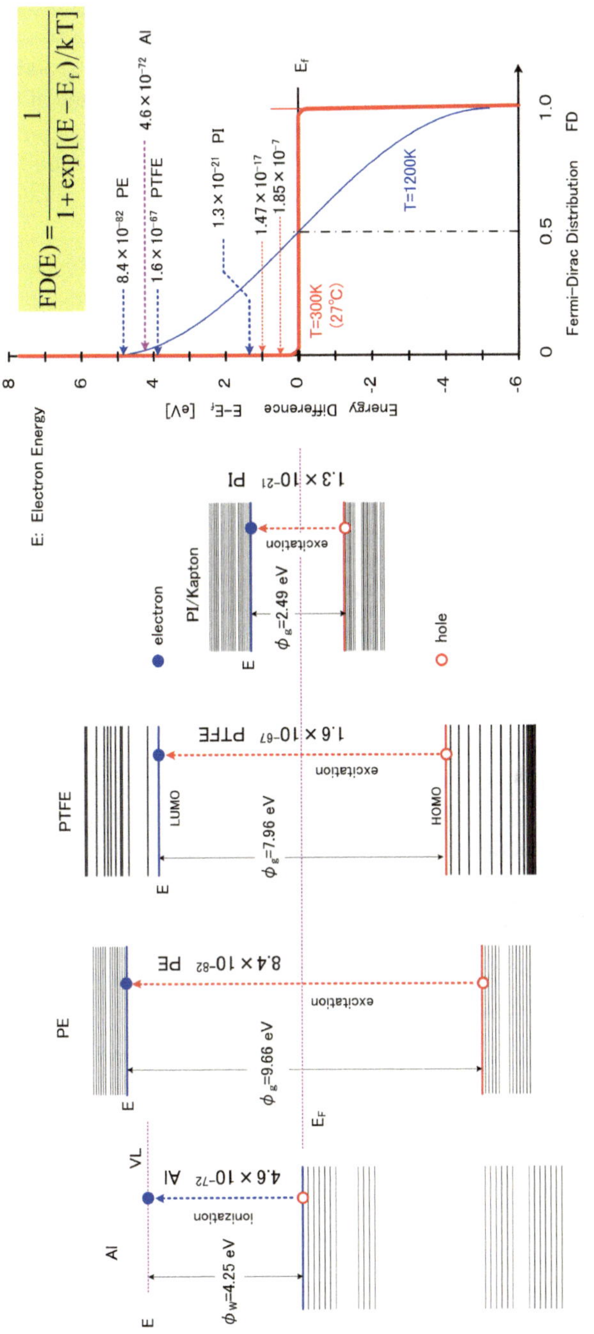

Fig. 12.10 The calculation results based on the Fermi–Dirac Distribution Function

E and the Fermi level E_F ($E-E_F$), and the horizontal axis is the probability FD. The shape of this FD is FD = 50% = 0.5 when $E = E_F$, which approaches FD = 1.0 in the $E < E_F$ region and FD = 0 in the $E > E_F$ region. Comparing the shapes when the temperature is $T = 1200$ K (blue line on the right side of the figure) and $T = 300$ K (red line on the right side of the figure), it can be seen that the shape of FD strongly depends on the temperature T. At the room temperature of 27 °C ($T = 300$ K), the electron energy E rises slightly above the Fermi level E_F, and FD becomes extremely small. On the other hand, FD becomes 1.0 when the electron energy is slightly lower than the Fermi level.

$$\begin{aligned} FD &= \frac{1}{1+\exp\left[\frac{E-E_F}{kT}\right]} \approx \exp\left[-\frac{E-E_F}{kT}\right] \\ &= \exp\left[-\frac{0.5\,\varphi_g}{kT}\right] \quad \text{where} \quad (E-E_F) = 0.5\,\varphi_g \end{aligned} \tag{12.3}$$

(2) **Electronic excitation on insulating material and metal electrode**

The characteristics of the probability FD of polymer materials and metal electrodes at room temperature are the target here, which are shown in the red line on the right side of the figure. The difference between the Fermi level E_F [eV] and the LUMO level E is the half of the energy gap ($E_{(LUMO)}-E_F$) = $0.5\phi_g$[eV]. Now consider the probability FD at which an electron in the HOMO level at the temperature of $T = 300$ K (red) is excited to the LUMO level. Focus on the shape of FD in Eq. (12.3), FD = 0.5 = 50% at ($E-E_F$) = 0. The region of ($E-E_F$) < 0 approaches FD ≈ 1.0, and conversely the region of ($E-E_F$) > 0 shows the FD is near zero (ultra small value). When T is 300 K (red), the thermal energy is $kT = 0.025$ eV. In general, ($E-E_F$) >> $kT = 0.025$ eV holds, and thus the region of ($E-E_F$) > 0 can be approximated to the Maxwell-Boltzman equation of the two items in Eq. (12.3).

For PI/Kapton (Fig. 12.10): The energy gap is $\phi_g = 2.49$ eV. Therefore, from Eq. (12.3), it can be calculated as FD = 1.3×10^{-21} at $T = 300$ K.

For PTFE (Fig. 12.10): The energy gap is $\phi_g = 7.96$ eV. Therefore, from Eq. (12.3), at $T = 300$ K, FD becomes the value of 1.6×10^{-67}.

For PE (Fig. 12.10): The energy gap is $\phi_g = 9.66$ eV. Therefore, from Eq. (12.3), at $T = 300$ K, FD becomes the value of 8.4×10^{-82}.

In the case of Al metal electrode (Fig. 12.10): The work function of the difference between the Fermi level E_F of Al metal and the vacuum level E_{VL} is $\phi_W = 4.25$ eV. In the case of metal, Eq. (12.4) is used. At T=300 K, the value becomes FD=4.64×10^{-72}.

$$\begin{aligned} FD &= \frac{1}{1+\exp\left[\frac{E_{VL}-E_F}{kT}\right]} \approx \exp\left[-\frac{\varphi_W}{kT}\right] \\ &\text{where} \quad (E_{VL}-E_{FM}) = \varphi_W \end{aligned} \tag{12.4}$$

(3) Comparison of electronic excitation between insulating material and impurity semiconductor

The insulating material may contain impurities. Therefore, the probability FD of electronic excitation between an insulating material containing impurities and an impurity semiconductor (n-type & p-type) is compared at $T = 300$ K. In the case of the intrinsic insulating material as shown in Fig. 12.11a, the material does not contain impurities. The energy gap is assumed to be $\phi_g = 8.0$ eV. Therefore, from Eq. (12.3), the value is FD = 7.2×10^{-68}.

Insulating material with impurities (Fig. 12.11b): Assuming that the HOMO level of impurities is the impurity donor and the donor gap is $\phi_d = 4.0$ eV, FD becomes the value of 2.7×10^{-34}.

In the case of n-type semiconductor (Fig. 12.11c): Generally, the donor level is in the range of 0.03 to 0.06 eV, so here we assume that $\phi_d = 0.06$ eV. Similarly, Eq. (12.3) can be applied and the value of FD is 0.31. It is difficult to draw a small value of $\phi_d = 0.06$ eV in the figure, so be careful because the donor level is enlarged in the figure.

In the case of p-type semiconductor (Fig. 12.11d): The acceptor level is also assumed to be $\phi_a = 0.06$ eV. Similarly, the value of FD can be caulculted as 0.31. Please note that the acceptor level is also enlarged in the figure.

In the case of impurity semiconductors, $\phi_d = 0.06$ eV and $\phi_a = 0.06$ eV, and thus the generation probability of electron–hole pairs is as large as 0.31. On the other hand, the generation probability of electron–hole pairs in the insulating material is extremely small, as in the range of 10^{-34} to 10^{-68}. Due to this large difference of the FD probability, semiconductors have been developed into electronics, while insulating materials are used to ensure the safety of electrical products.

(4) Apply FD probability to the formed interface of double-layer material

When different dielectrics are contacted, the interface becomes positively and negatively charged. When they are separated from each other, the positive charge and the negative charge are separated and the voltage between them becomes high, and spark discharge may be observed. Further, when the metal electrode and the insulating material come into contact with each other, an electric double layer is formed at the interface. Morikawa et al. have performed computational chemical simulations for the growth of electric double layers in the process of contacting Cu electrodes and pentacene molecules [5]. Here, we apply this simulation to the formation of the double layer between the SC electrode and the LDPE material (Fig. 12.12).

When the SC electrode and the LDPE are separated from each other: The work function of the SC electrode is 5.56 eV. At $T = 300$ K, the probability of the electrons on the SC electrode being emitted to the outside beyond the vacuum level VL is $FD_1 = 4.6 \times 10^{-94}$. On the contrary, the probability that the electrons in the HOMO level of LDPE are emitted to the outside beyond the vacuum level VL is $FD_2 = 3.2 \times 10^{-46}$. The relationship is clear: $FD_1 = 4.6 \times 10^{-94} << FD_2 = 3.2 \times 10^{-46}$. In this state, the electrons in the HOMO level of LDPE move to the SC electrode in the process of bringing them into close contact with each other ($Z_c = 3.7$ Å at the closest

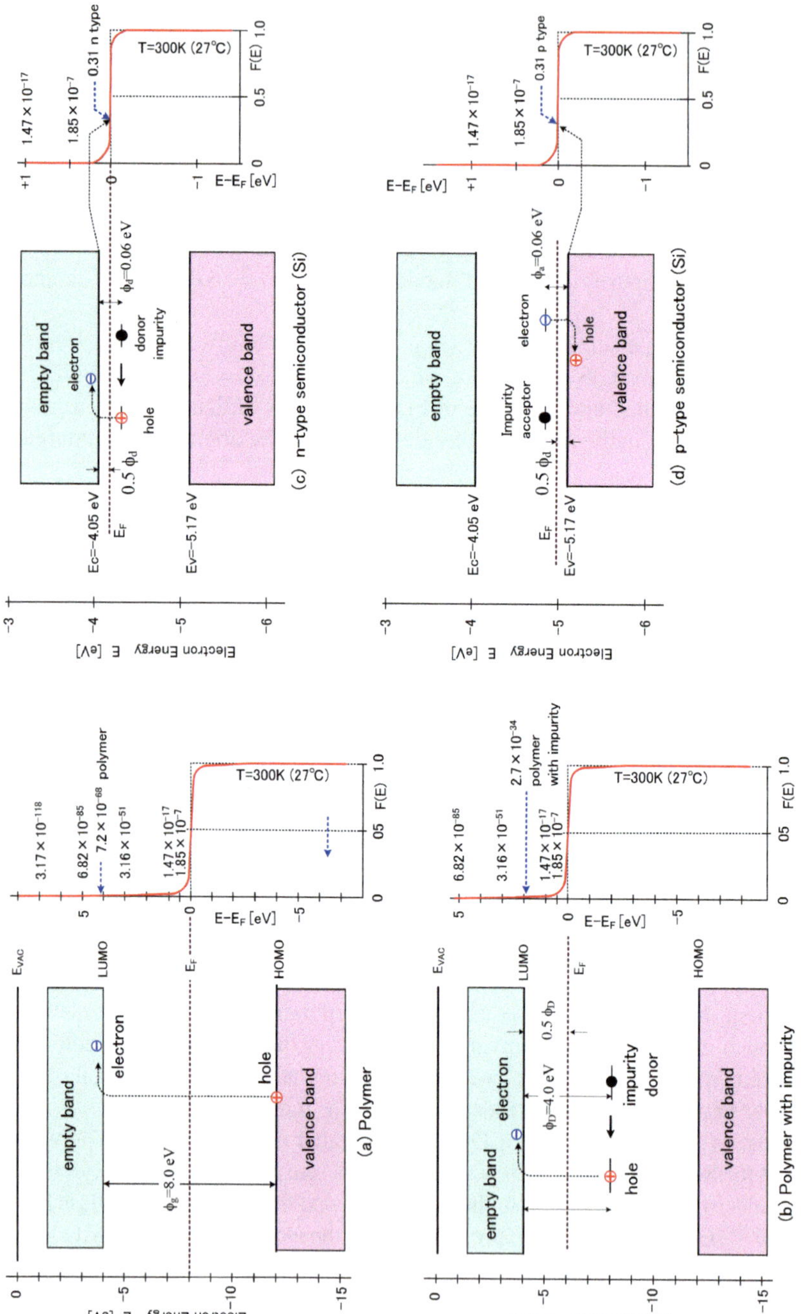

Fig. 12.11 The probability of electronic excitation of the two materials at different states

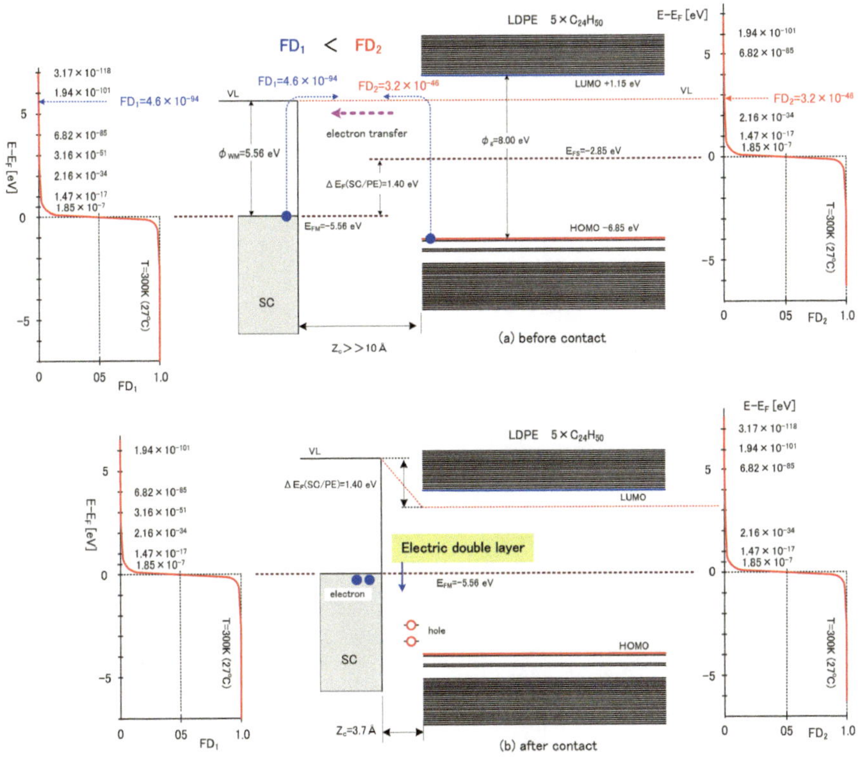

Fig. 12.12 The probability of the double layer between the SC electrode and the LDPE material

approach state). As a result, the Fermi levels of both are consistent, negative charges (electrons) transfer to the surface of the SC electrode, and positive charges (holes) remain on the LDPE surface. An electric double layer is formed at the interface. In this double layer, a potential difference of $\Delta E_F(\text{SC/PE}) = 1.40$ eV of the Fermi level between the two is generated. In the case of LDPE, this electric double layer can be used to explain the measurement results in which positive charges are injected and transferred from the anode.

(5) **Electrode charge injection barrier**

It is described in part (4) above by the Fermi–Dirac Distribution Function that the electric double layer is formed at the interface between the electrode and the dielectric. The charge injection barrier ϕ_B can also be explained from the above process, and we introduce it here. Figure 12.13a shows the state before the contact between the electrode and the dielectric, and Fig. 12.13b represents the state after contact. In the center, the Fermi level of $E_F = E_M$ indicates the Fermi level of the dielectric match, and $E_F < E_M$ means that the Fermi level of the dielectric is lower than that of the electrode. Moreover, $E_F > E_M$ means that the Fermi level of the dielectric is higher than that of the electrode.

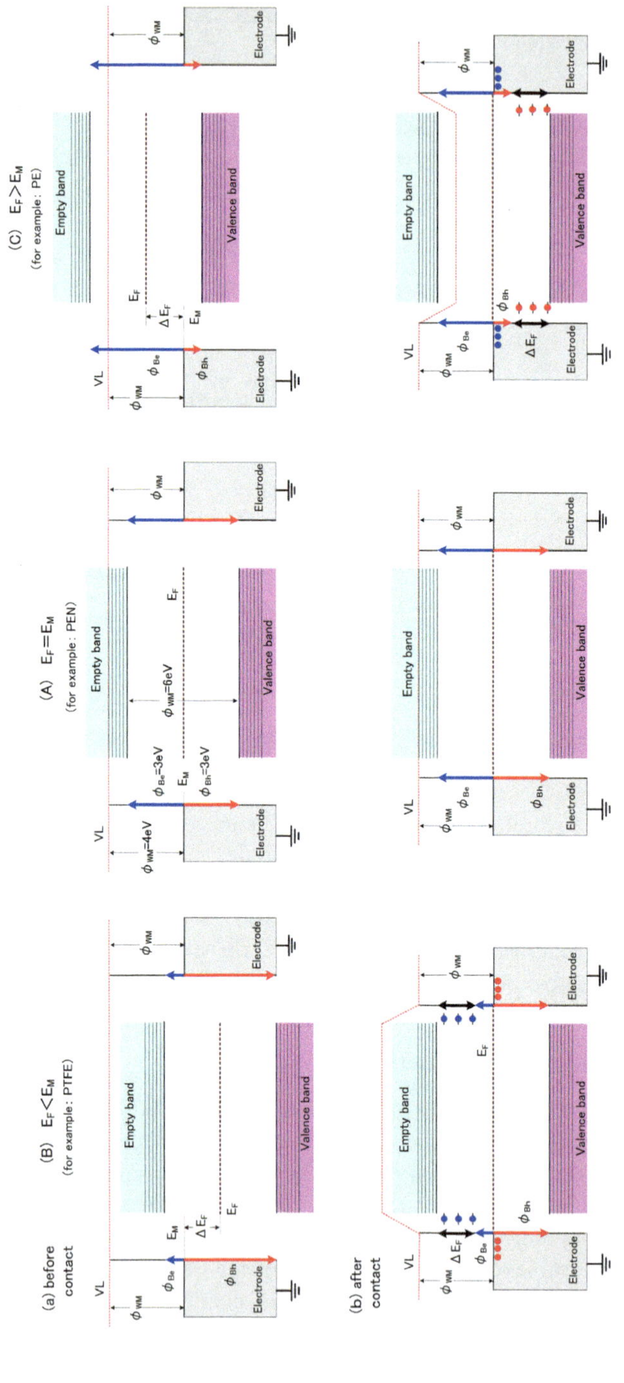

Fig. 12.13 The calculated results of the electrode and the dielectric at different contact states

The electron injection barrier ϕ_{Be} and the hole injection barrier ϕ_{Bh} before contact are drawn in the upper part of the figure. When (A) $E_F = E_M$, $\phi_{Be} = \phi_{Bh}$. When (B) $E_F < E_M$, $\phi_{Be} < \phi_{Bh}$, and when (C) $E_F > E_M$, $\phi_{Be} > \phi_{Bh}$.

The electron injection barrier and the hole injection barrier after contact are drawn in the lower part. In case of (A) $E_F = E_M$, $\phi_{Be} = \phi_{Bh}$, but in case of (B) $E_F < E_M$, ϕ_{Be} decreases and electron injection becomes easy because negative charge exists on the dielectric surface. On the other hand, in the case of (C) $E_F > E_M$, there are positive charges on the surface of the dielectric, so ϕ_{Be} is reduced and hole injection becomes easier.

Generally, when the electrode and the dielectric are in contact, the Schottky effect of charge injection model is discussed, but this effect is small. The charge injection of the electric double layer formed on the contact surface is more dominant than that, and the experimental results can be explained well based on these analysis.

See the part (4) of Sect. 3.1 for more explanation.

12.8 Observation of Macromolecules by Molecular Dynamics and Density Functional Calculations

Up to the previous chapter, we have dealt with the energy level of a small single molecule with the carbon number of 10. The actual polymer is in an amorphous state where many macromolecules with the carbon number of $C = 1000$ or more are entangled. It is observed how the energy levels of many molecules differ from those of single molecule by Molecular Dynamics and Density Functional Calculations.

(1) **Molecular Dynamics Calculation of aPE and aPEN**

The molecules to be calculated are the following aPE and aPEN, and their chemical structure formulas are shown in Fig. 12.14.

aPE (amorphous polyethylene): Condense fifty $C_{120}H_{242}$ molecules into a pseudo polymer with a density (0.83 g/cm^3) close to that of LDPE.

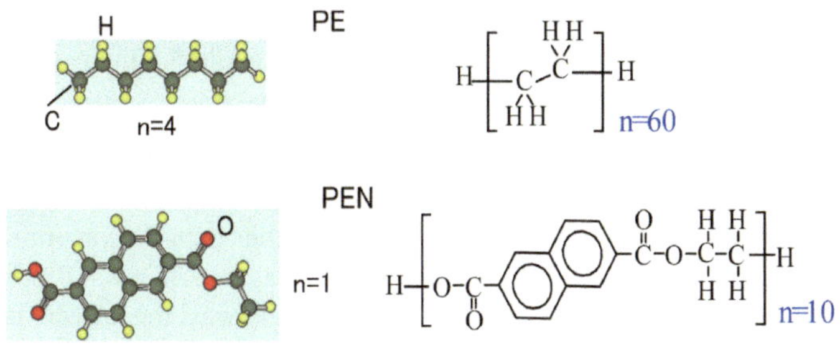

Fig. 12.14 The chemical structure formulas

aPEN (amorphous polyethylenenaphthrate): Condense thirty-five H-$(C_{14}H_{10}O_4)_{n=10}$-H molecules into a pseudo polymer with a density (1.26 g/cm^3) close to that of PEN.

(2) **Calculation results of molecular dynamics**

Figure 12.15 shows the results of the molecular dynamics calculation of $C_{120}H_{242}$ molecules of aPE and H-$(C_{14}H_{10}O_4)_{n=10}$-H molecules of aPEN. Each molecule is depicted by a line model. When 50 and 35 linear molecules are put in the calculation pixel and the molecular dynamics calculation is performed, the molecular shape changes to an energetically stable configuration due to the temperature and pressure of the system. Then, its density (specific gravity) rises and becomes close to the value of the actual polymer resin. NPT calculation is performed at 300 K, and the calculation is stopped when the aPE molecular density is about 0.83 g/cm^3 and that of aPEN is about 1.26 g/cm^3, which are close to the actual densities.

The molecular dynamics calculation is a method for calculating the trajectory of particles (molecules and atoms) based on Newton's equation of motion. GROMACS 5.0 [6, 7] is used as the calculation software for molecular dynamics, and the calculation procedure is as follows.

Step 1 Energy minimization: The target molecule is randomly placed in the calculation cell and the energy minimization calculation is performed. A one-nanosecond canonical ensemble (NVT) calculation at 298 K and an isothermal isobaric ensemble (NPT) calculation at 400 K are performed at 101 kPa.

Step 2 Force field calculation: GAFF (General Amber Force Field) [8] is used as the force field. Nosé-Hoover thermostat [9, 10] and Parrinello- Rahman barostat [11] are used for temperature control and pressure control, respectively. The PME (Particle-Mesh Ewald) method [12] is used for the interaction.

(3) **Comparison of energy levels**

The Quantum Chemical Calculation by DFT is performed using Gaussian 09. The DFT calculation can derive the molecular orbital (electron wave) of the electron energy level and evaluate the electronic states of atoms and molecules. The function is B3LYP and the basis set is 6-31 g. From the calculation results of Fig. 12.15, three adjacent molecules of aPE-2 M (Fig. 12.16c) and aPEN-2 M (Fig. 12.16d) are extracted. After that, the energy level of each molecule is obtained by the DFT calculation. The energy levels are shown in Fig. 12.16a, b.

For aPE 2 M/3 M molecule: The HOMO level is in the range of −7.08 eV to −7.09 eV, and the LUMO level is in the range of +1.05 eV to +1.30 eV. In addition, the LUMO level of the aPE 2 M/3 M molecule is higher than the vacuum level VL, and the electron affinity is negative. The band gap is $\phi_g = 8.38$ eV ~ 8.14 eV. These values depend on the closeness of the molecules. By the way, the LUMO level of PE is higher than the vacuum level, and the electron affinity is negative.

In the case of aPEN 2 M/3 M molecule: The HOMO level is in the range of −5.59 eV to −5.60 eV, and the LUMO level is in the range of −2.49 eV to −2.56 eV. These

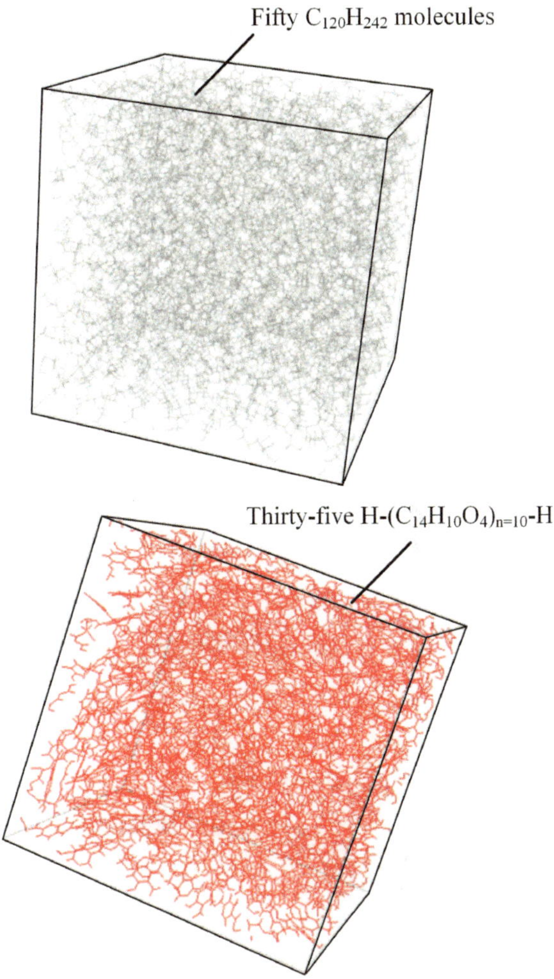

Fig. 12.15 Results of the molecular dynamics calculation

values also depend on the closeness of molecules. The LUMO level is lower than the vacuum level VL, and the electron affinity is positive. The polarity is opposite to that of the aPE molecule described above. The band gap is $\phi_g = 3.03$ eV to 3.11 eV, which is about 38% narrower than that of aPE molecule. Consider why the band gap of aPE is wide and that of aPEN is narrow. The major difference between the two molecules is that PEN has a π-electron double bond, while PE has only an σ-electron single bond. This π-electron bond exists in naphthalene and the carbonyl group. It raises the HOMO level, lowers the LUMO level, and narrows the band gap. The details are described in Sect. 2.2.

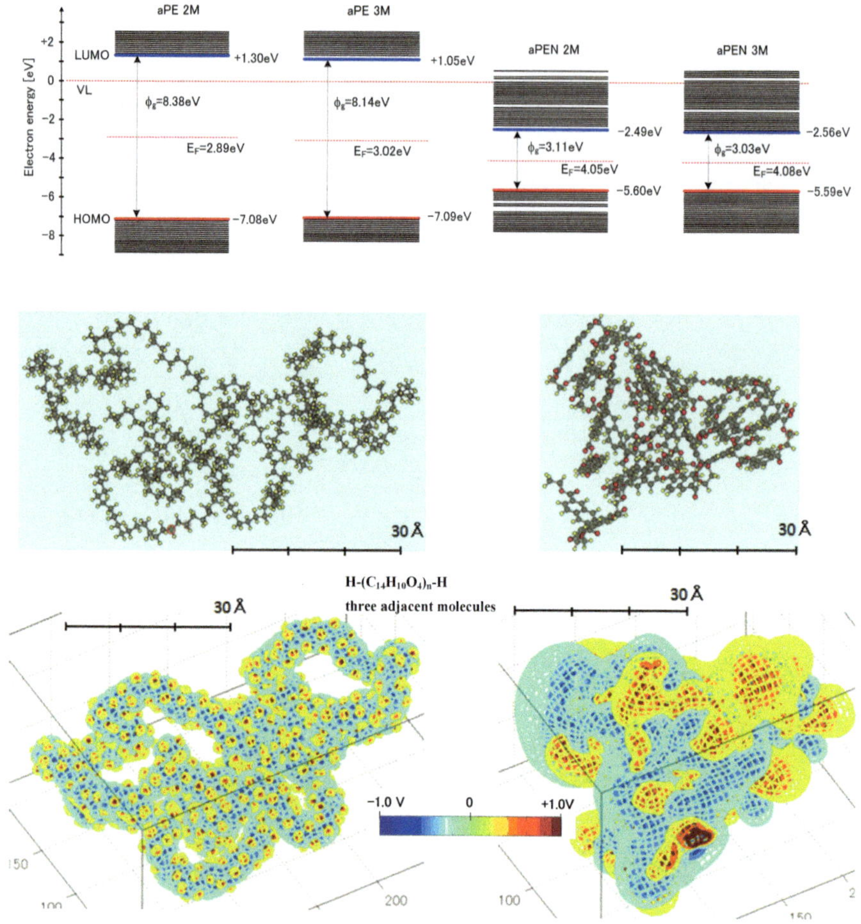

Fig. 12.16 The calculated results of the three adjacent molecules of the two materials

(4) Comparison of three-dimensional electrostatic potential distribution

Figure 12.16c, d respectively show the shapes of the molecules obtained by extracting only three adjacent molecules from the MD calculation results of aPE and aPEN. The shapes of the extracted molecules are randomly arranged. What can be seen from Fig. 12.16c, d is that the molecules are not crystalline but are arranged in an amorphous form. That is, there are places where the molecules are close to each other, and there are places where the molecules are separated from each other, and both are mixed. In addition, the molecule has straight and bent parts. In PEN, the electrostatic force of the electric dipole created by naphthalene and the carbonyl groups becomes attractive and repulsive, and they are entwined with each other. In fact, PE is a soft material and PEN is a hard material. The calculated glass transition temperature of aPE is as low as Tg = −38 °C and that of aPEN is as high as Tg = +103 °C.

Therefore, we have calculated the three-dimensional electrostatic potential distribution $V(x, y, z)$ of the aPE and aPEN molecules obtained from the Mulliken charges, from which the DFT calculation results are obtained (Fig. 12.16e, f). The potentials are displayed in color, with warm colors showing positive potentials and cold colors showing negative potentials. The positive and negative potentials of the aPE molecule are localized without spreading. On the other hand, the positive and negative potentials of the aPEN molecule spread and the potential overlaps within the molecule and between adjacent molecules. That is, it can be inferred that the interaction between PEN molecules is strong, but that of PE molecules is weak. If the interaction is strong, it can be said that the material has a high glass transition temperature and becomes a hard material. On the contrary, if the interaction is weak, the material has a low glass transition temperature and becomes a soft material.

References

1. Y. Hayase, M. Tahara, Y. Tanaka, T. Takada, M. Yoshida, Relationship between electric potential distribution and trap depth in polymeric materials, *IEEJ Transaction on Fundamental and Materials*, 129(11), 455-462 (2009).
2. W. Wang, Y. Tanaka, T. Takada, Space charge of polyethylene and electronic structures analysis of trapping site by common chemical group, *Sensors and Materials*, 29(8), 1223-1231 (2017).
3. S. Serra, E. Tosatti, S. Iarlori, S. Scandolo, G. Santoro, Interchain electron states in polyethylene, *Physical Review B*, 62(7), 4389-4393 (2000).
4. W. Wang, T. Takada, Y. Tanaka, S. Li, Space charge mechanism of polyethylene and polytetrafluoroethylene by electrode/dielectrics interface study using quantum chemical method, *IEEE Transactions on Dielectrics and Electrical Insulation*, 24(4), 2599-2606, (2017).
5. K. Toyoda, Y. Nakano, I. Hamada, K. Lee, S. Yanagisawa, Y. Morikawa, First-principles study of the pentacene/Cu(111) interface: absorption states and vacuum level shifts, *Journal of Electron Spectroscopy and Related Phenomena*, 174(1), 78-84 (2009).
6. H. Berendsen, D. Spoel, R.V. Drunen, GROMACS: A message-passing parallel molecular dynamics implementation, *Computer Physics Communications*, 91(1), 43-56 (1995).
7. J. Abrahama, T. Murtolad, R. Schulz, S. Pall, J.C. Smith, B. Hess, E. Lindahl, GROMACS: High performance molecular simulations through multi-level parallelism from laptops to supercomputers, *SoftwareX*, 1, 19-25 (2015).
8. J. Wang, R.M. Wolf, J.W. Caldwell, P.A. Kollman, D.A. Case, Development and testing of a general amber force field, *Journal of Computational Chemistry*, 25(9), 1157-1174 (2004).
9. S. Nose, A molecular dynamics method for simulations in the canonical ensemble, *Molecular Physics*, 52(2), 255-268 (1984).
10. W.G. Hoover, Canonical dynamics: Equilibrium phase-space distributions, *Physics Review A*, 31(3), 1695-1697 (1985).
11. M. Parrinello, A. Rahman, Polymorphic transitions in single crystals: A new molecular dynamics method, *Journal of Applied Physics*, 52, 12, 7182-7190 (1981).
12. T. Darden, D. York, L. Pedersen, Particle mesh Ewald an $N \cdot \log(N)$ method for Ewald sums in large systems, *The Journal of Chemical Physics*, 98(12), 10089-10092 (1993).

Chapter 13
Application Examples of Quantum Chemical Calculation

13.1 Molecular Chain Approaching and Trap Site Formation

(1) XLPE Molecule Approaching and Trap Site

The formation of trap sites of cross-linked polyethylene (XLPE) is observed. The XLPE molecules are created by cross-linking multiple linear and branched molecules (Fig. 13.1a). Energy distribution and spatial distribution of molecular orbitals of XLPE molecules are observed by Quantum Chemical Calculation/DFT calculation (Fig. 13.1b, c) [1]. The results are summarized below.

- Electron waves are generated between approaching molecules, and molecular orbitals of localized levels appear (Fig. 13.1b).
- The LUMO level is reduced, and the level forms an electron trap with the depth of 0.3–0.8 eV (Fig. 13.1c).
- The HOMO level rises slightly and the level forms a hole trap with the depth about 0.1 eV (Fig. 13.1c).

(2) Molecular Orbitals and Traps of Two Approaching Polyethylene PE ($C_{12}H_{26}$) Molecules

The result in Fig. 13.1 shows that an electron wave is generated between the molecules that are close to each other and a molecular orbital appears in the localized level. Therefore, two PE ($C_{12}H_{26}$) molecules are further arranged in parallel under different molecular spacings, and then the molecular orbital (electron wave) is observed by Quantum Chemical Calculation/DFT calculation, as shown in Fig. 13.2.

Range of molecular spacing in Fig. 13.2c: It shows the two linear PE ($C_{12}H_{26}$) molecules with the varied spacing R from 4 Å to 9 Å.

Electron energy level in Fig. 13.2a: It shows the energy levels for different molecular spacings. The LUMO level decreases from +2.31 eV ($R = 10.0$ Å), reaches a

T. Takada et al., *Electric Charge Accumulation in Dielectrics: Measurement and Analysis*,
https://doi.org/10.1007/978-981-19-6156-4_13

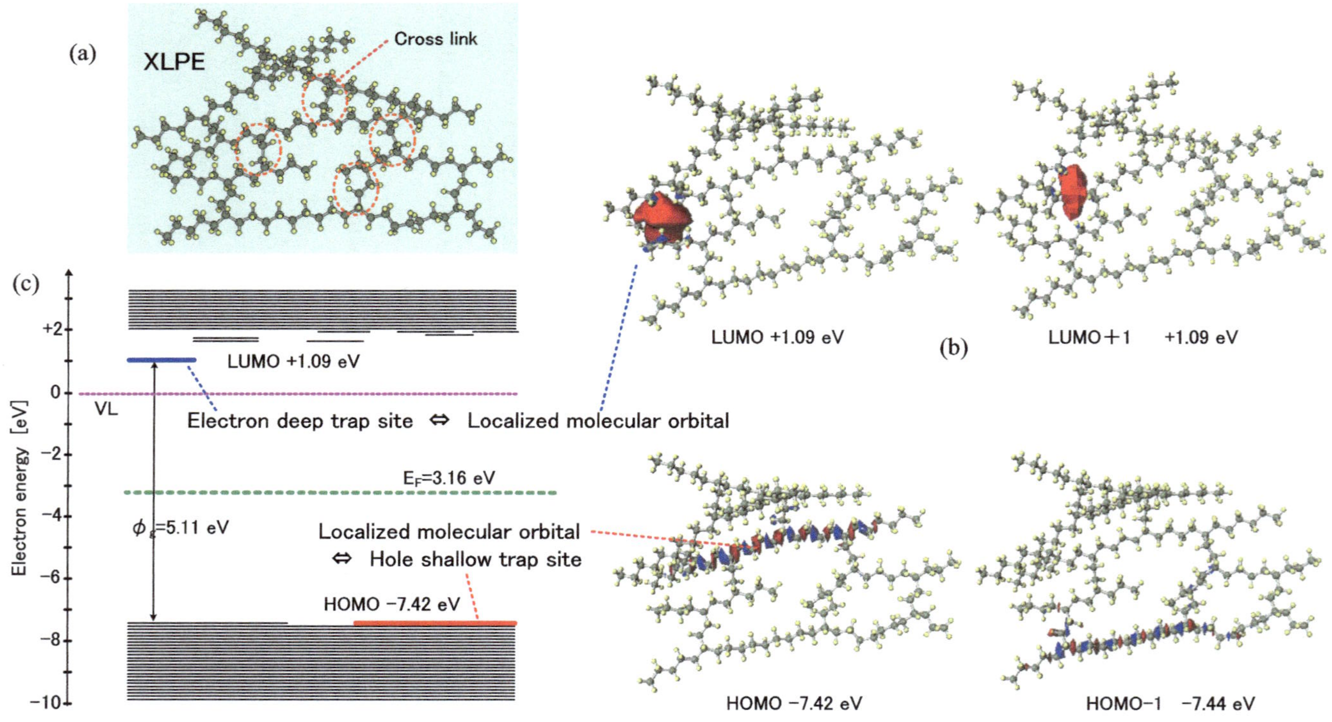

Fig. 13.1 Approaching XLPE molecules and trap sites

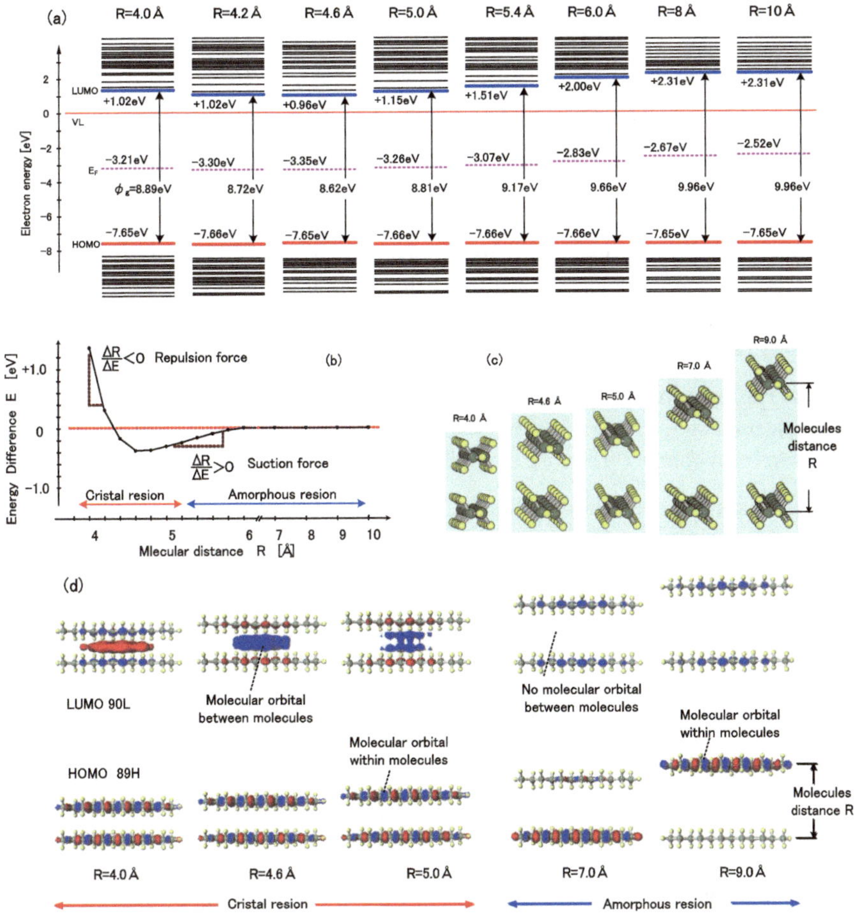

Fig. 13.2 Observation of molecular orbitals when the molecular spacing of PE $(C_{12}H_{26})_2$ is changed

minimum of +0.96 eV ($R = 4.6$ Å), and rises again (+1.02 eV) at $R = 4.0$ Å. The HOMO level remains almost constant. The energy gaps ϕ_g are 9.96 eV, 8.62 eV and 8.89 eV, respectively.

Total Kinetic Energy Difference in Fig. 13.2b: The Total Kinetic Energy can be obtained from DFT calculation. The vertical axis is the Energy Difference value based on the Total Kinetic Energy when the two linear molecules are sufficiently separated. The horizontal axis is the molecular spacing R from 4 Å to 10 Å. The energy difference decreases as the molecular spacing R becomes smaller, and the minimum value is at $R = 4.6$ Å. The energy rises sharply when the molecules approach further. The two linear molecules are the closest and stable at $R = 4.6$ Å. The crystal region of PE material is considered to be within the closest distance of 4.6 Å between molecules. This is supposed.

(3) **HDPE Molecular Orbitals and Traps**

Low-density polyethylene (LDPE) has an amorphous structure in which linear molecules are disordered, while cross-linked polyethylene (XLPE) has an entangled amorphous structure in which linear molecules are cross-linked with branched molecules (Fig. 13.3d), and high-density polyethylene (HDPE) has linear molecules. Here we take one model of molecular orbitals and traps of HDPE as an example, which is composed of a crystalline region and an amorphous region arranged in parallel (Fig. 13.3d).

Crystal region and non-crystalline region in Fig. 13.3a: HDPE molecule is a model in which four molecules of PE ($C_{12}H_{16}$) are arranged in parallel, and PE ($C_{10}H_{22}$) molecules that connect them are combined in the non-crystalline region.

3D electrostatic potential distribution in Fig. 13.3d: As described in the part (2) of Sect. 12.4, hydrogen H is positively charged and carbon C i s negatively charged. The electrostatic potential distribution in the crystalline region has regularity, while it becomes irregular at the boundary in the non-crystalline region.

Electron energy level and molecular orbital in Fig. 13.3b, c: The LUMO level is +0.88 eV, and molecular orbitals exist between molecules in the crystalline region. The LUMO +37 level which is 37 levels above LUMO exists in the amorphous region at +2.72 eV. The HOMO level is −7.55 eV in the crystalline region. The HOMO-8 level, which is eight levels below HOMO, is in the amorphous region at −7.88 eV. In both cases, the molecular orbital exists in the main chain of the molecule.

Simulated HDPE trap in Fig. 13.3e: An electron energy level of the pseudo HDPE model is drawn by connecting 4 crystalline regions and 3 amorphous regions. From this energy level, it can be inferred that the LUMO level of the crystalline region is the electron trap region and the amorphous region is the electron conduction region. Similarly, the HOMO level of the crystalline region can be assumed as the hole trap region, and the non-crystalline region is the hole conduction region. Their trap depths are shallow.

(4) **Charge and Electrostatic Potential Distribution of Approaching PE Molecules**

In the above part (1), it is described that electron waves exist and molecular orbitals are created where PE molecules are close to each other. We speculate that the location of the molecular orbital becomes the center site of charge trap. Here, two long PE molecules of $C_{60}H_{122}$ are orthogonalized, and the intervals Δz of their intersections are set as 4.2 Å, 3.2 Å and 2.0 Å. Then, the electron energy level and molecular orbital are observed on the left of Fig. 13.4. The molecular orbital of LUMO levels is formed near the intersections where the intervals are 3.2 Å and 2.0 Å. We assume there is an electron trap site near this place.

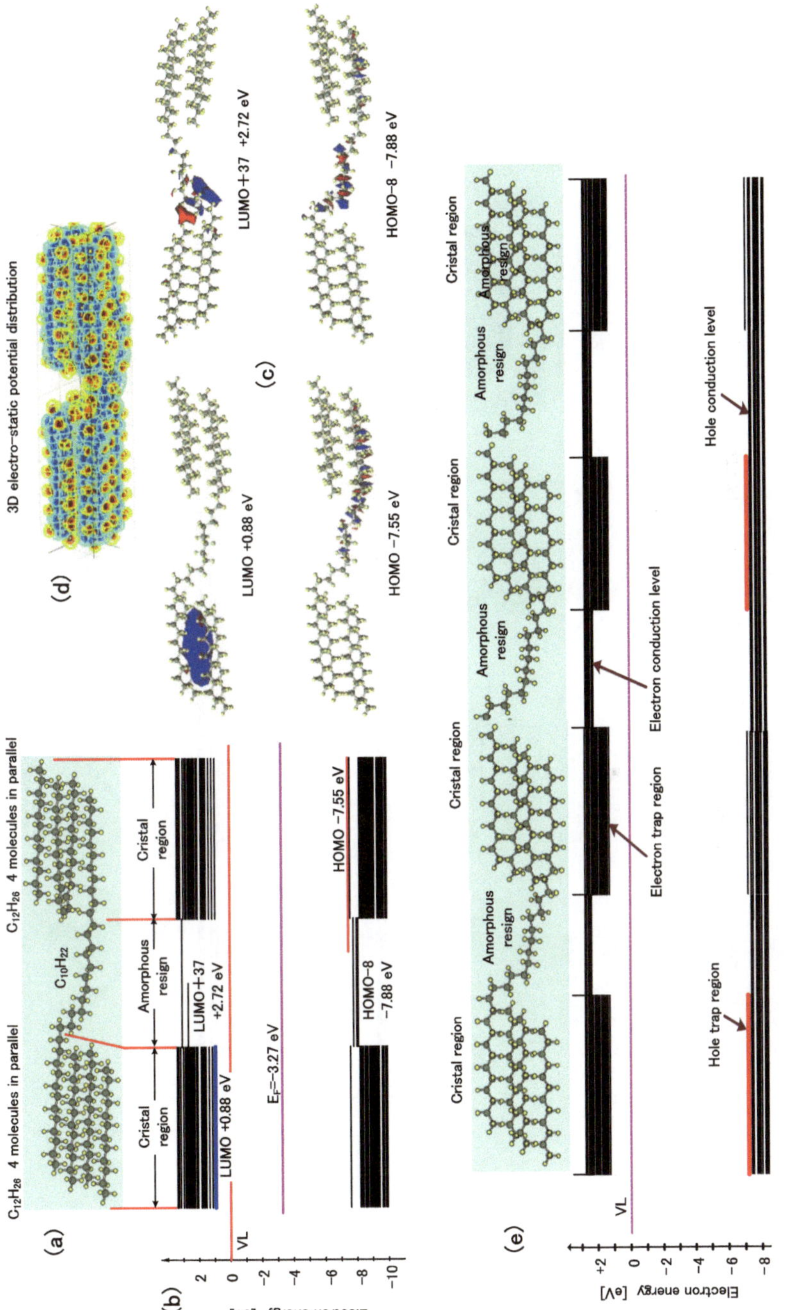

Fig. 13.3 Electron energy levels and trap formation in the pseudo-HDPE model

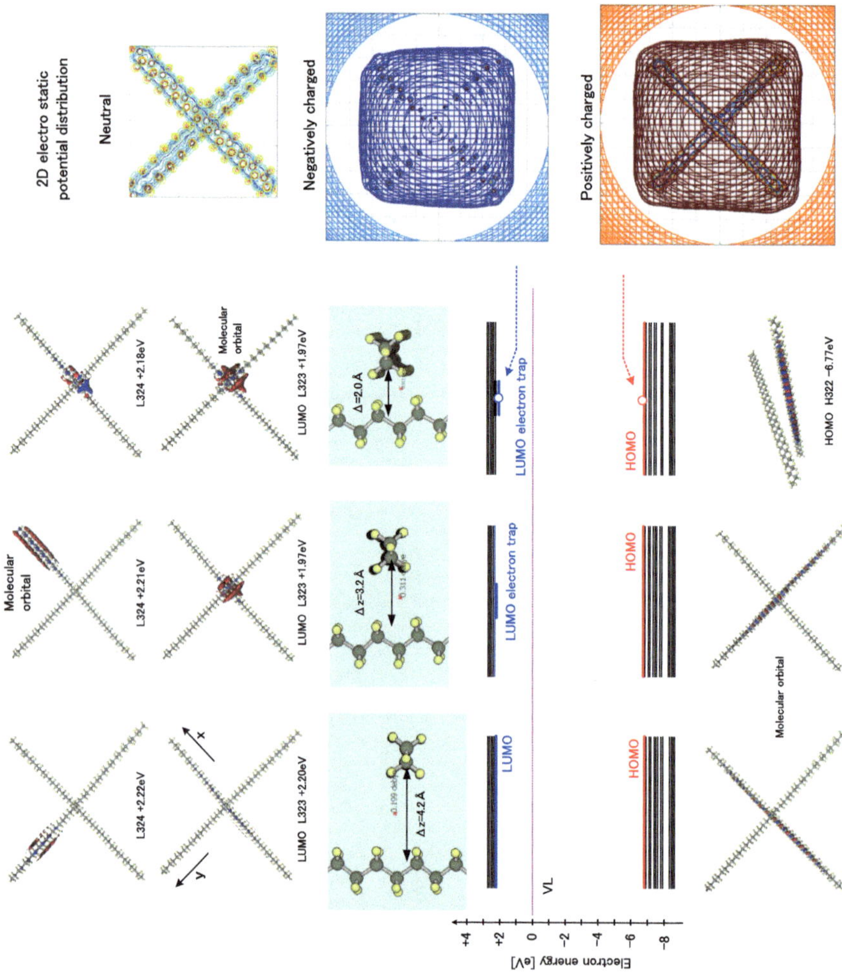

Fig. 13.4 Charge and electrostatic potential distribution of close PE molecules

When one electron is removed from the two $C_{60}H_{122}$ molecules, they are positively charged. When one electron is added, they are negatively charged. The electrostatic potential distribution $V(x, y, z)$ under these two situations is calculated from Milliken atomic charges. This $V(x, y, z)$ is shown on the right side of Fig. 13.4. Positively charged $V(x, y, z)$ draws the potential distribution around the intersection of two PE molecules. Therefore, it is confirmed that the positive charge is trapped at the close intersection. However, the positive charges are distributed not only at the intersections but also along the molecular chains. It is thus certain that the intersection is the center site of the trap. The center of negatively charged $V(x, y, z)$ also has an intersection.

13.2 Polymerization of Monomers and Polymers

"Electret earphones" are being developed by irradiating a PET film with an electron beam. A "thermal electret" has also been developed by applying high electric field/high temperature. These can be achieved because a PET film can trap electrons for a long time. Then, we want to know where the electron is trapped in the PET molecular structure. With this motivation, we began to study the PET trap mechanism. Macromolecules are the synthesis of various monomers. The LUMO and HOMO levels of this monomer are strongly dependent on its molecular structure. It is noticed that when the monomers with different levels are polymerized, they become the polymers with different trap depths.

This section presents a comparison of the LUMO and HOMO levels of various monomers.

(1) **Observation of Energy Level and Molecular Structure of PET**

Figure 13.5 shows the molecular structure of PET. In addition, the reaction formula of monomer polymerization of the PET is shown on the right side of Fig. 13.5, which is produced by synthesizing ⑤ ethylene glycol and ⑮ terephthalic acid. Therefore, the molecular structures of ethylene glycol and terephthalic acid are shown, and the electron energy levels of the molecules of the PET are calculated by DFT calculation (Fig. 13.5, right). Then, the electron trap is considered from the observation of each energy level and molecular orbital. The results is described below. Observe the relationship between the LUMO (134L) level of PET and the upper levels including the LUMO+1(135L) level, LUMO+2(136L) level and so on with their respective molecular orbitals and molecular structures. The molecular orbital up to LUMO+8(144L) level ⑮ is localized in terephthalic acid. The molecular structure above the LUMO+9(145L) level distributes throughout the whole PET molecule including ⑤ ethylene. Therefore, the upper levels above LUMO can be electron trap region and the higher levels LUMO+9(145L) can be electron conduction level. Similarly, observe the relationship between the HOMO (133H) level and the HOMO-2 (132L) level with their molecular orbitals and structures. As a result, the high levels below HOMO can be the hole trap and the low levels below HOMO (125H) can be the hole conduction level.

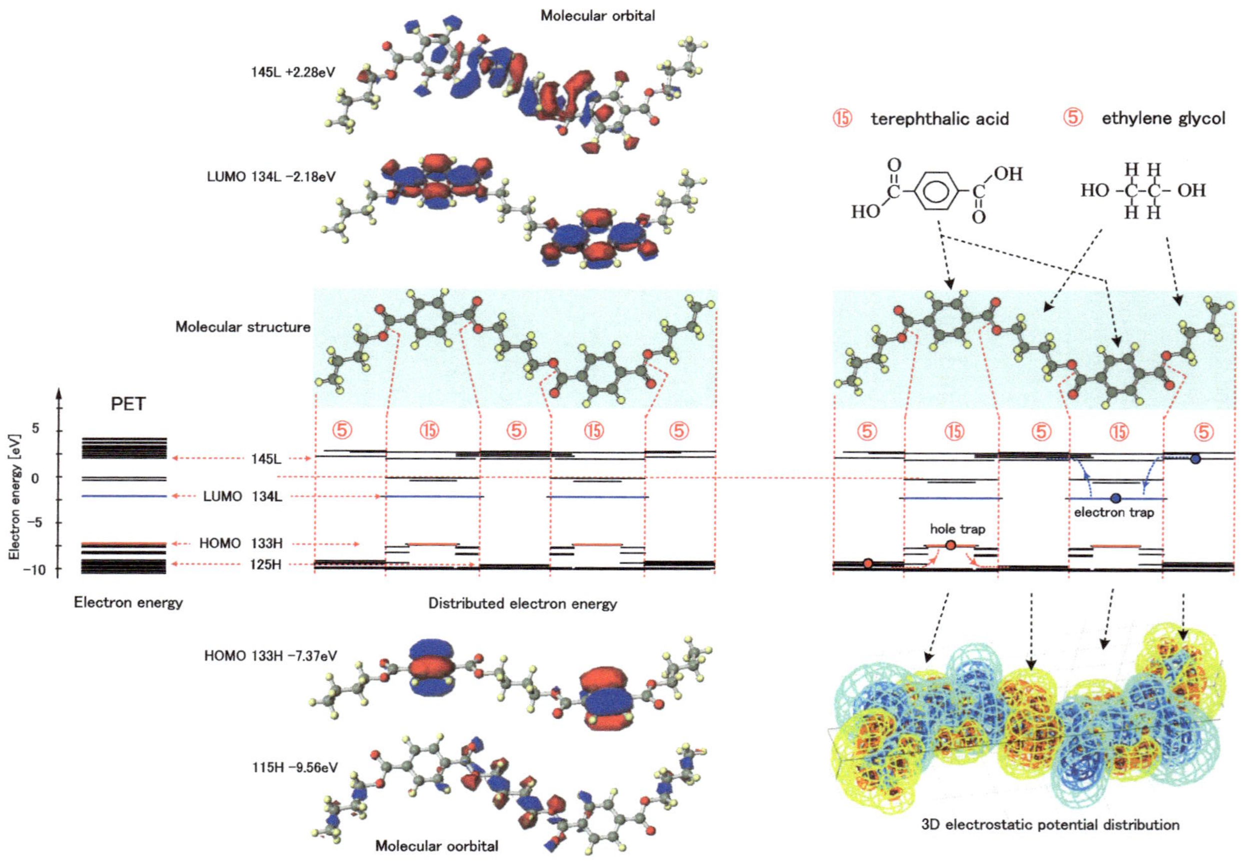

Fig. 13.5 Observation of PET molecular orbitals, electrostatic potential distribution, and energy levels

(2) **Monomer Polymerization of Various Polymers**

In the part (1) above, the relationship between the electron energy level of PET and the molecular orbital is observed, and the trap sites for electrons and holes are considered. Therefore, Fig. 13.6 shows the monomer polymerization process of various polymer insulating materials. The polymer after polymerization remains the chemical structure of a part of the monomer. Therefore, the DFT calculation results of the electron energy levels and molecular orbitals of the monomer with the molecular structure after synthesis are summarized in Figs. 13.7, 13.8, and 13.9. The numbers in red circles in Fig. 13.6, Figs. 13.7, 13.8, and 13.9 correspond to each other.

(3) **Grouping of Monomer Molecules**

The characteristics of the electron energy levels of various monomers in Fig. 13.6 are roughly divided into the following 3 groups.

Group A LUMO > VL (Fig. 13.7): The LUMO levels of the molecules ② to ⑤ are classified into the groups with negative electron affinity ($\chi < 0$) higher than the vacuum level VL. These molecules are σ bonds (single bonds). The molecular orbital of HOMO level is in the molecule, and the electron wave is in the molecular chain. On the contrary, the molecular orbital of LUMO level is outside the molecule, and the electron wave is around hydrogen H in the side chain. The only exception is that the LUMO level of PTFE is lower than the vacuum level VL under the case of the positive electron affinity ($\chi > 0$). See Sect. 13.1–13.6 for the difference between the positive and negative polarities of the electron affinity of PTFE and PE. Here, PTFE is indistinguishable to Group A.

Group B LUMO≒VL (Fig. 13.8): The LUMO levels of the molecules ⑥ to ⑩ are close to the vacuum level VL. The molecules with one or two benzene rings are in this group. The π-electron wave of the conjugated double bond of the benzene ring forms the LUMO and HOMO levels in molecular orbitals. Therefore, the π-electron wave of the conjugated double bond narrows the LUMO and HOMO levels.

Group C LUMO < VL (Fig. 13.9): The LUMO levels of the molecules of ⑪ to ⑮ are 3–4 eV lower than the vacuum level VL. The molecules consisting of a benzene ring and a ketone (>C = O) in the main chain are in this group. In this group, the π-electron wave of the conjugated double bond forms a molecular orbit across the benzene ring and the ketone (>C = O). As a result, the LUMO level decreases.

(4) **Relationship Between Localized Molecular Orbitals and Traps**

In the above part (3), it is found that the LUMO and HOMO levels of the molecules of Group A, B, C are different. If a polymer is composed of the monomer molecules from Group A and B, Group A and C, or Group B and C, the intervals of LUMO and HOMO levels can be combined as "narrow" and "wide" characteristics. We have already described the PET polymer in Fig. 13.5 of the part (1) and observed the relationship between electron energy levels, localized molecular orbitals, and trap formation. Then, the trap formation process is described. As for other polymer

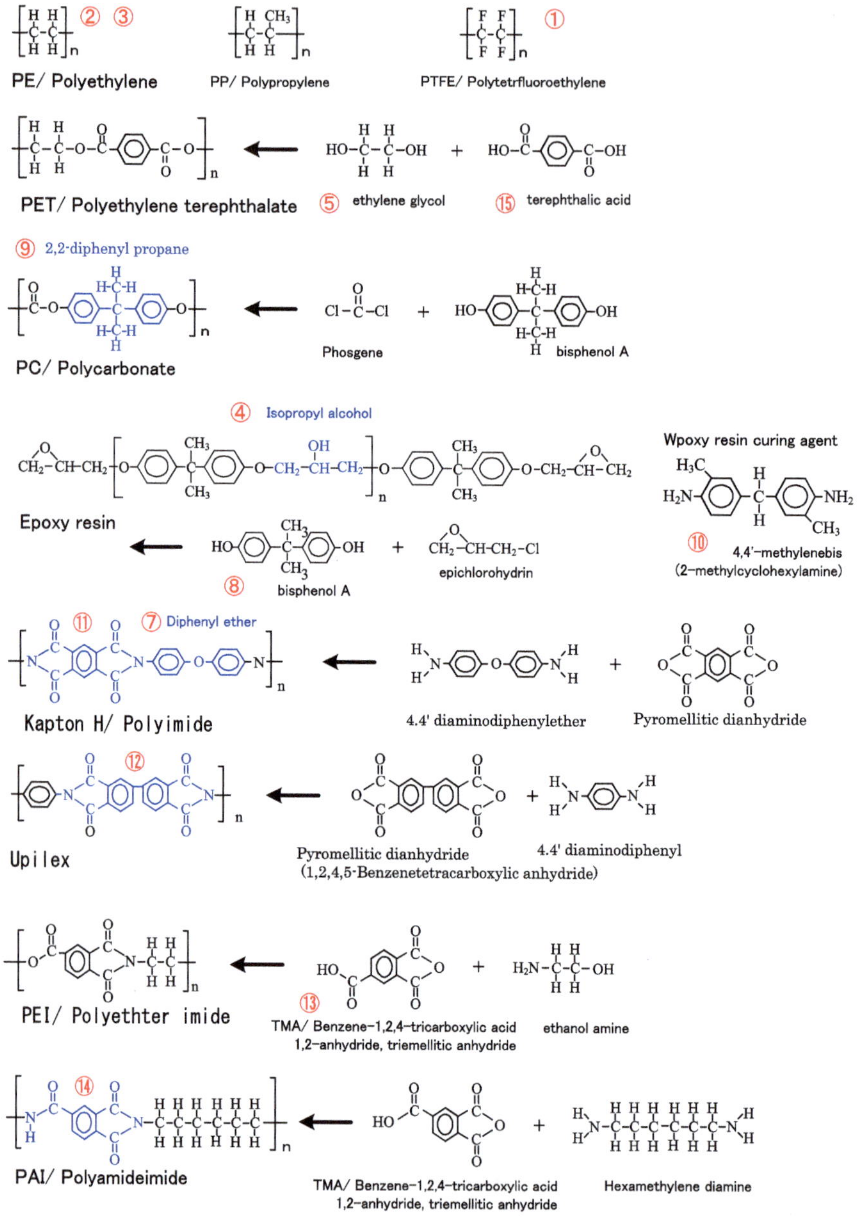

Fig. 13.6 Monomer polymerization process of various polymer insulating materials

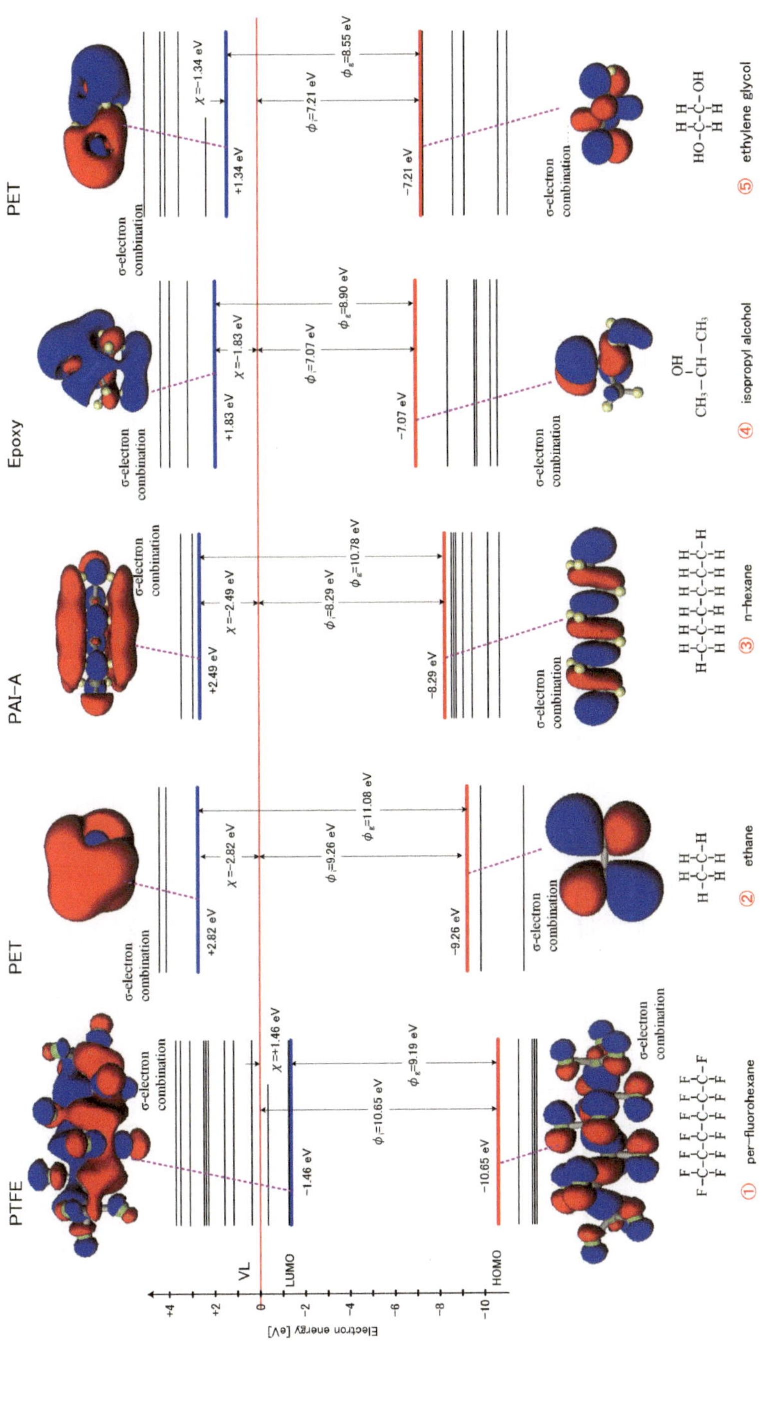

Fig. 13.7 Electron energy levels and molecular orbitals of various monomers of Group A

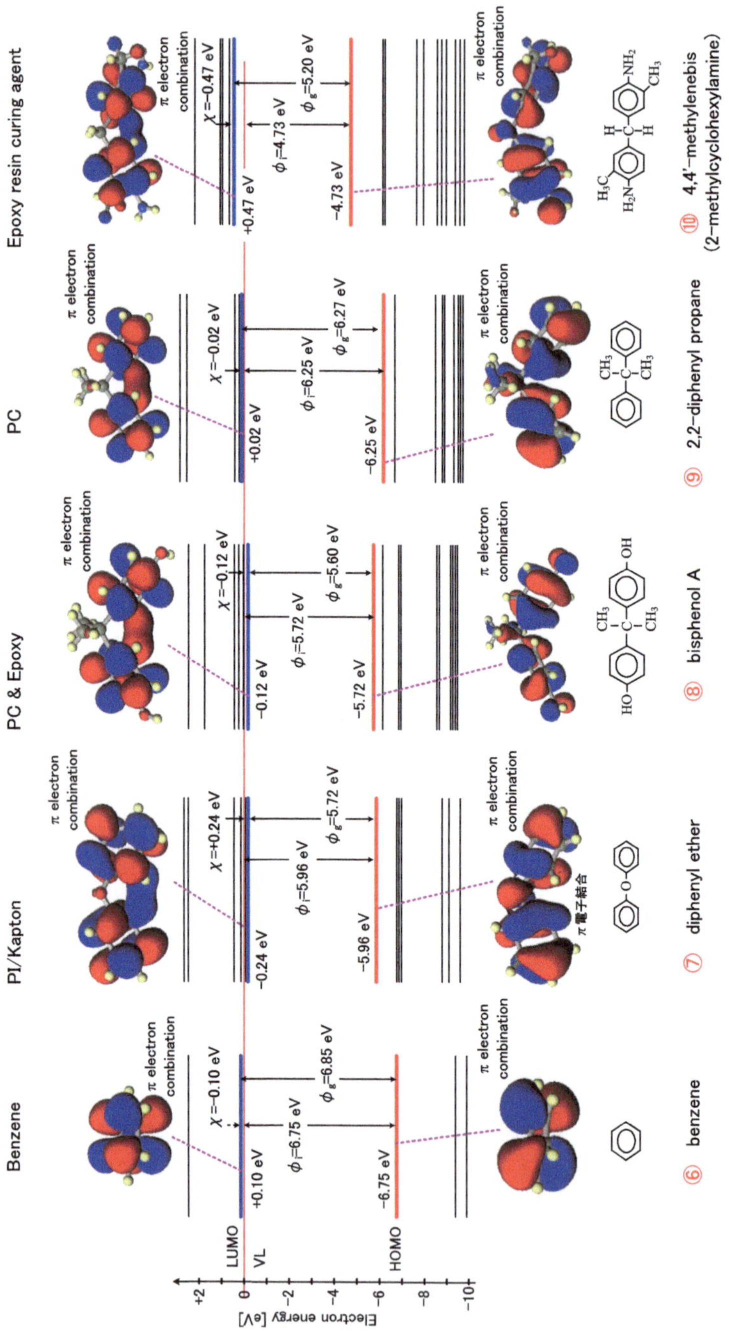

Fig. 13.8 Electron energy levels and molecular orbitals of various monomers of Group B

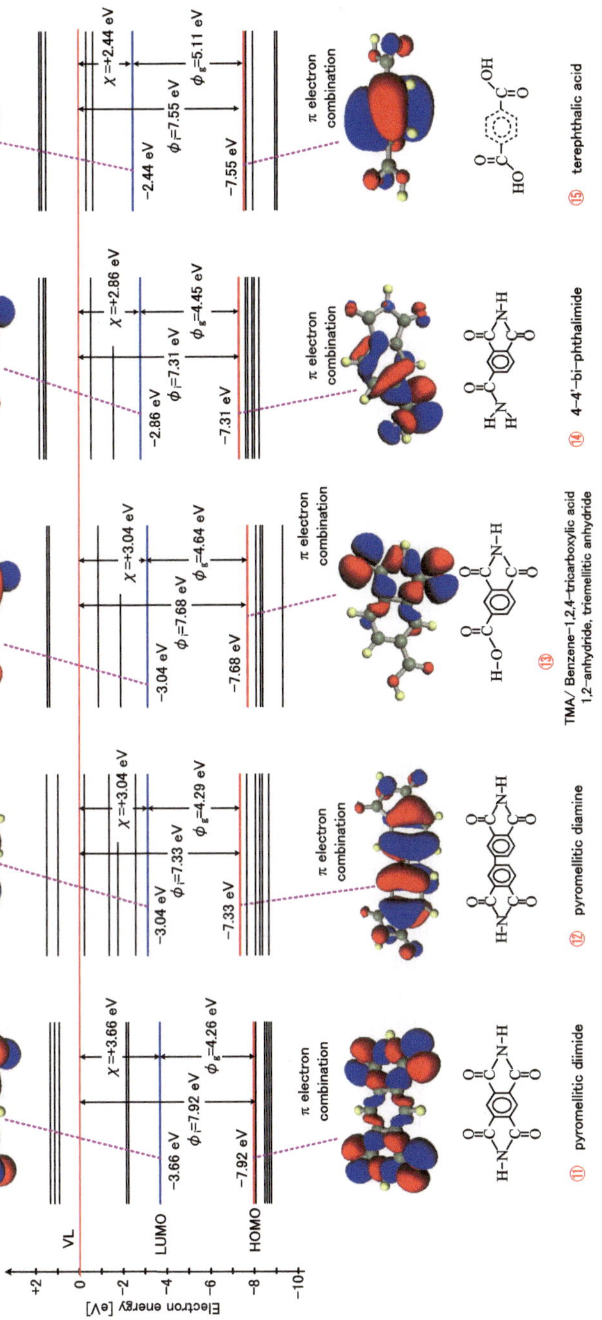

Fig. 13.9 Electron energy levels and molecular orbitals of various monomers of Group C

materials, the formation of traps can be analyzed similarly by observing the electron energy level and the localized molecular orbitals.

The examples of Polyimide/Kapton in Sect. 14.3 and PEN in Sect. 14.4 are introduced to discuss the relationship between charge accumulation and trap formation.

13.3 Chemical Structure and Formation of Trap Sites for Charge

In Sect. 13.2, we find that the double bond (π electron), conjugated double bond (2 π electrons), and benzene ring (3 π electrons) have narrow LUMO and HOMO levels. Therefore, the mechanism of trap site formation by introducing these double bonds (>C = O) into linear PE molecules is investigated by DFT calculation.

(1) **Charge Trap of Oxidized Double Bond (>C = O)**

Figure 13.10 shows the DFT calculation results of the energy levels and molecular orbitals of the PE molecule $C_{24}H_{50}$ and the oxidized PE molecule $C_{24}H_{48}O$. The gap between LUMO and HOMO levels of the $C_{24}H_{48}O$ molecule is narrow, and the molecular orbitals are also concentrated in the oxidized part (>C = O). The electrons moving in the PE molecule are trapped in the LUMO level with low energy. Therefore, this LUMO level becomes a trap site. On the contrary, in the case of holes, the HOMO level becomes a trap site.

In order to consider the mechanism of trap site formation, we focus on the Schrödinger equation in the figure. In the case of $C_{24}H_{50}$ molecule (the left in the figure), the potential distribution $V(x, y, z)$ is regularly distributed in positive (hydrogen H) and negative potentials (carbon C) in the left figure "C". This $V(x, y, z)$ is the $V(x, y, z)$ of the Schrödinger equation. Since $V(x, y, z)$ has positive and negative potentials regularly distributed, the eigenvalue E of electron energy is regularly distributed almost throughout the whole molecule in the left figure "C".

On the other hand, in the case of $C_{24}H_{48}O$ molecule, $V(x, y, z)$ is concentrated and changed in the oxidized part (>C = O) ("A" in the right figure). Therefore, the eigenvalue of LUMO and HOMO levels is determined by this changed $V(x, y, z)$. In addition, the characteristic value (Eigen value) besides the oxidized portion (>C = O) is also determined by $V(x, y, z)$ of other positions. This eigenvalue is distributed almost throughout the molecule ("B" in the figure).

(2) **Trap Site with Charging Process**

In the above part (1), it is described that the trap sites for charge are created in the oxidized part (>C = O) of $C_{24}H_{48}O$/oxidized PE molecule. Here, the electrostatic potential distribution $V(x, y, z)$ is observed by positively or negatively charging the $C_{24}H_{48}O$ molecule by DFT calculation. Figure 13.11 shows the two-dimensional $V(x, y)$ and three-dimensional $V(x, y, z)$ distributions, and their polarities are (a)

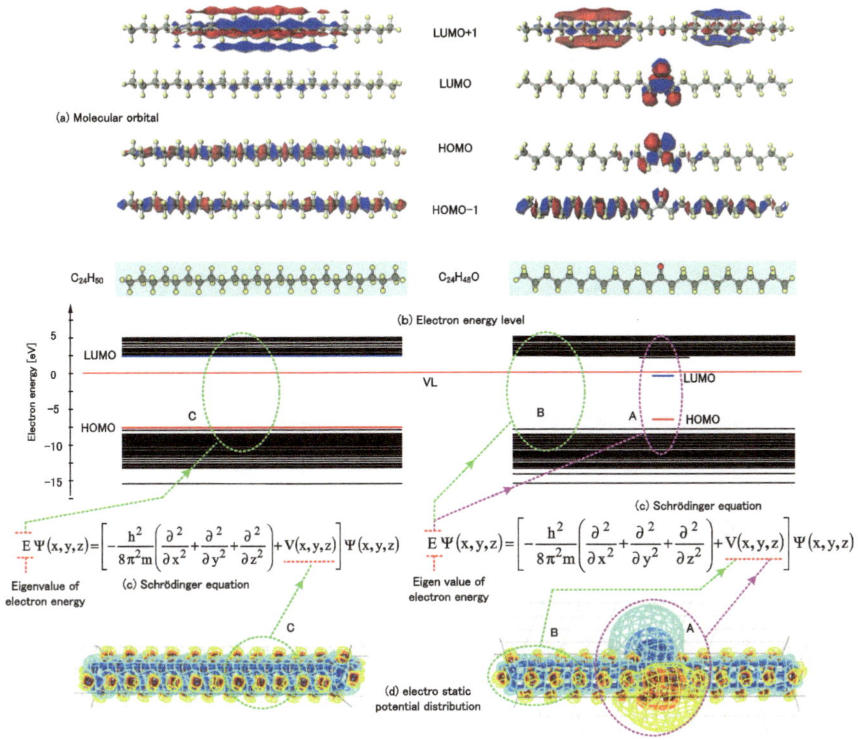

Fig. 13.10 Relationship between trap site formation and Schrödinger equation

neutral, (b) positively charged, (c) negatively charged. The neutral one (a) shows the $V(x, y, z)$ distribution of the electric dipole of $> C^+ = O^-$ in the oxidized part. The $V(x, y, z)$ distributions of the (b) positively charged (+q) and (c) negatively charged (−q) are almost spherical with the oxidized part (>C = O) as the center. The equipotential surface with $V(x, y, z) = \pm 1.0$ V is close to a sphere with a radius of 1.44 nm.

As described above, the HOMO level and the LUMO level, which are locally present in the band gap of forbidden band, are the hole and electron trap sites, respectively. This can be explained by the fact that they coincide with the center of the charged $V(x, y, z)$ distribution.

(3) **Chemical Structure Trap**

The energy levels of various monomers are compared in Sect. 13.2. Here, the polymer molecules of (a) narrow carboxyl group (-COOH) or (b) narrower pyromellitic diimide with the energy between LUMO/HOMO levels and the wider $C_{24}H_{46}$/PE molecule between LUMO/HOMO levels are taken as examples. Their molecular orbitals are investigated. Observe the relationship between molecular orbital, electrostatic potential $V(x, y, z)$ distribution and LUMO/HOMO levels (Fig. 13.12).

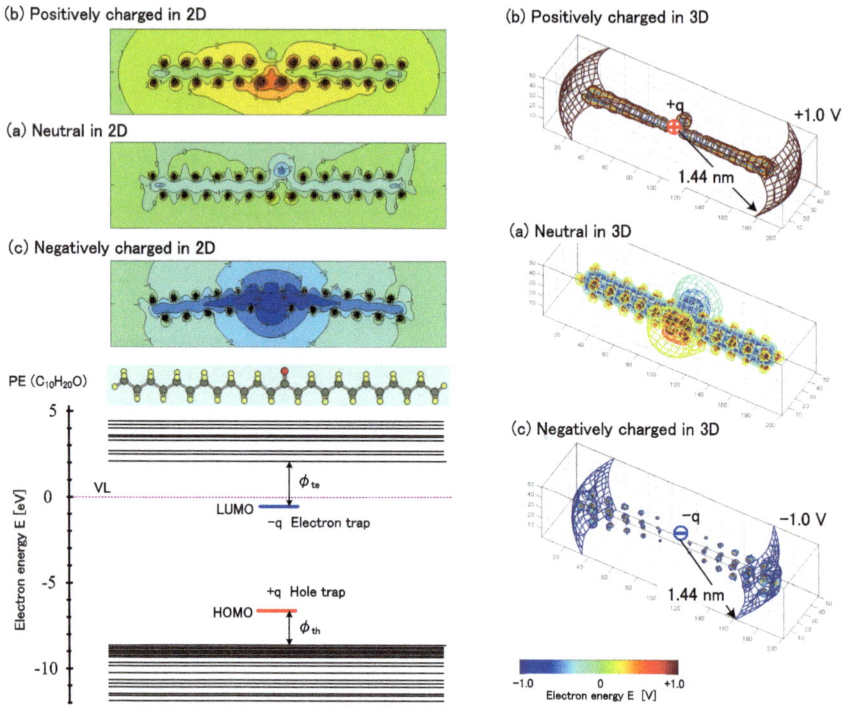

Fig. 13.11 Electrostatic potential distribution $V(x, y, z)$ and electron energy level of $C_{24}H_{48}O$ molecule charged positively, neutrally and negatively

(a) **Carboxyl group (-COOH)**: Since the carboxyl group has a ketone group ($>C^+ = O^-$) and a hydroxyl group ($-O^-H^+$), oxygen O is negatively charged which has a high electronegativity. Carbon C and hydrogen H in this chain are positively charged. As a result, a new $V(x, y, z)$ distribution can be created near it, and the LUMO/HOMO level becomes narrow by the solution of the Schrödinger equation. The LUMO level becomes an electron trap, and the HOMO level becomes a hole trap.

(b) **Pyromellitic Diimide**: It has one benzene ring (C_6H_6) and four ketone groups ($>C = O$) on both sides (see ⑪ in Fig. 13.9). This portion has 7 conjugated double bonds (π electron bonds). By observing the electrostatic potential $V(x, y, z)$ distribution of this part, it can be found that the negative potential of oxygen O of the ketone group ($>C = O$) appears strongly, and the positive potential appears near hydrogen H and carbon C of the benzene ring. The solution of the Schrödinger equation by this $V(x, y, z)$ distribution also shows narrow LUMO/HOMO levels. Similarly, this LUMO level becomes an electron trap and the HOMO level becomes a hole trap.

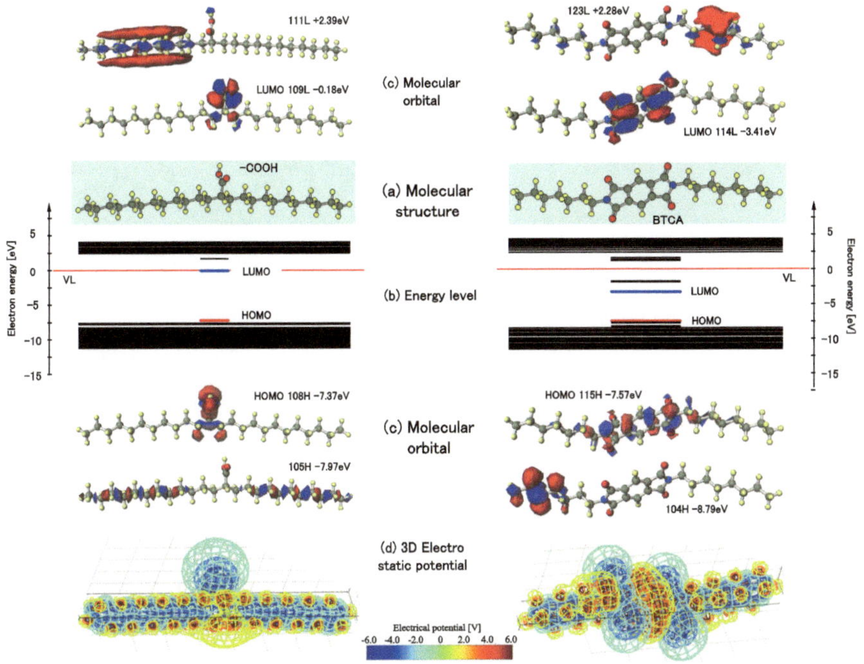

Fig. 13.12 Molecular orbital and electrostatic potential of PE molecule with carboxyl group and pyromellitic diimide

13.4 Formation of Trap Sites in Macromolecular Systems (PE and PEN)

Section 13.1 has described the approaching molecular chain, and Section 13.3 has described the formation of trap sites in chemical structures. Here, the formation of trap sites in macromolecular system (PE and PEN) is observed by molecular dynamics and DFT calculation. The molecular dynamics calculation of macromolecular system is described in Sect. 12.8.

(1) Molecular Dynamics Calculation of Macromolecular System (PE and PEN)

The calculated molecules are PE and PEN, and their chemical formulas are shown in Figs. 13.13 and 13.14. PE (polyethylene) is a pseudo polymer with a density (0.83 g/cm^3) close to LDPE, which is obtained by condensing 50 $C_{120}H_{242}$ molecules by molecular dynamics calculation. PEN (polyethylenenaphthrate) is a pseudo polymer that condenses 35 H-$(C_{14}H_{10}O_4)_{n=10}$-H molecules with a density (1.26 g/cm^3) close to PEN. After the calculation of molecular dynamics, one $C_{120}H_{242}$ $\times$ 1 molecule (1M), two adjacent $C_{120}H_{242}$ $\times$ 2 molecules (2M), and three adjacent $C_{120}H_{242}$ $\times$ 3 molecules (3M) are extracted and shown on the left side of Fig. 13.13. Similarly, in the case of PEN, the extracted part is shown in Fig. 13.14.

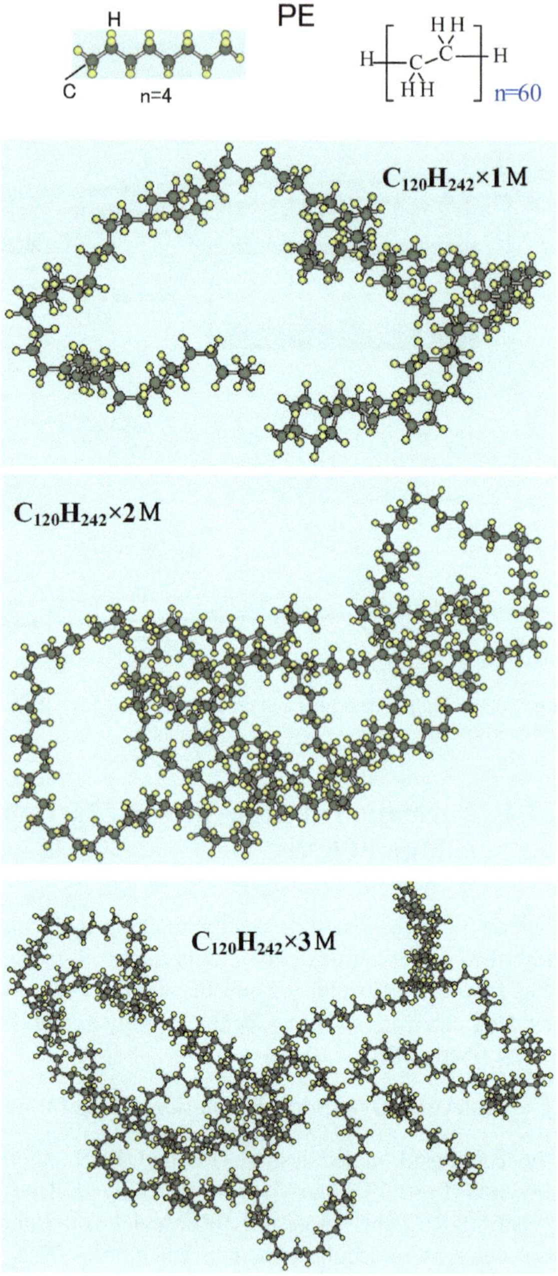

Fig. 13.13 Results of PE chemical structure and molecular dynamics calculations

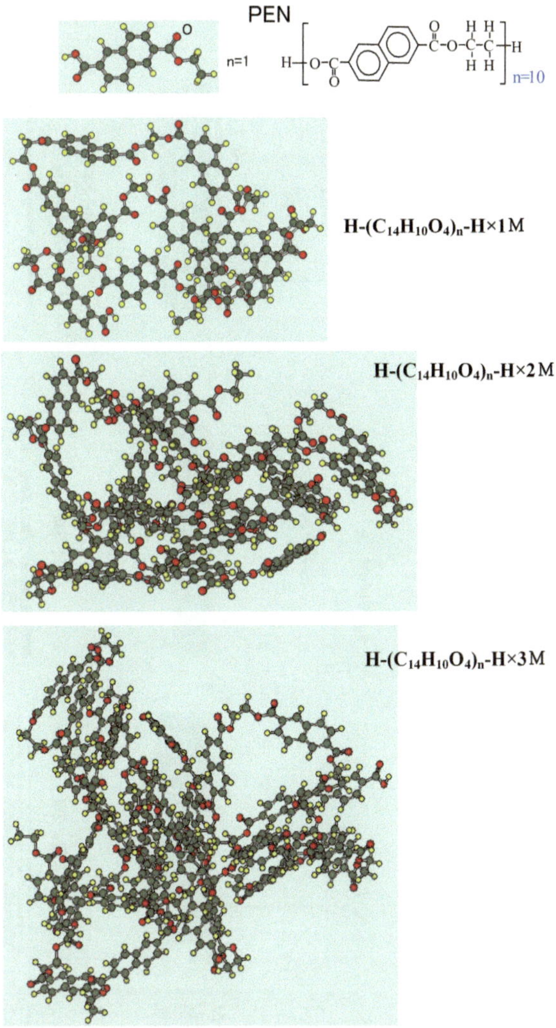

Fig. 13.14 Results of PEN chemical structure and molecular dynamics calculations

(2) Electron Energy Level of Macromolecular System (PE and PEN)

Figure 13.15 shows the DFT-calculated electron energy levels of the 1M, 2M, and 3M molecules extracted from PE and PEN in the amorphous state of the macromolecular systems in Figs. 13.13 and 13.14. The values of HOMO level, LUMO level, Fermi level E_F, and band gap ϕ_g of each molecule are compared. There is no large difference in these values between 1M, 2M, and 3M molecules in the amorphous state.

The right side of Fig. 13.15a shows the electron energy levels of the straight PE 1M- $C_{120}H_{242}$ molecule and the right side of Fig. 13.15b shows that of the straight PEN 1M- $C_{42}H_{14}O_{12}$ molecule. This straight molecule is single and not affected by nearby molecules. From the comparison between the electron energy levels of

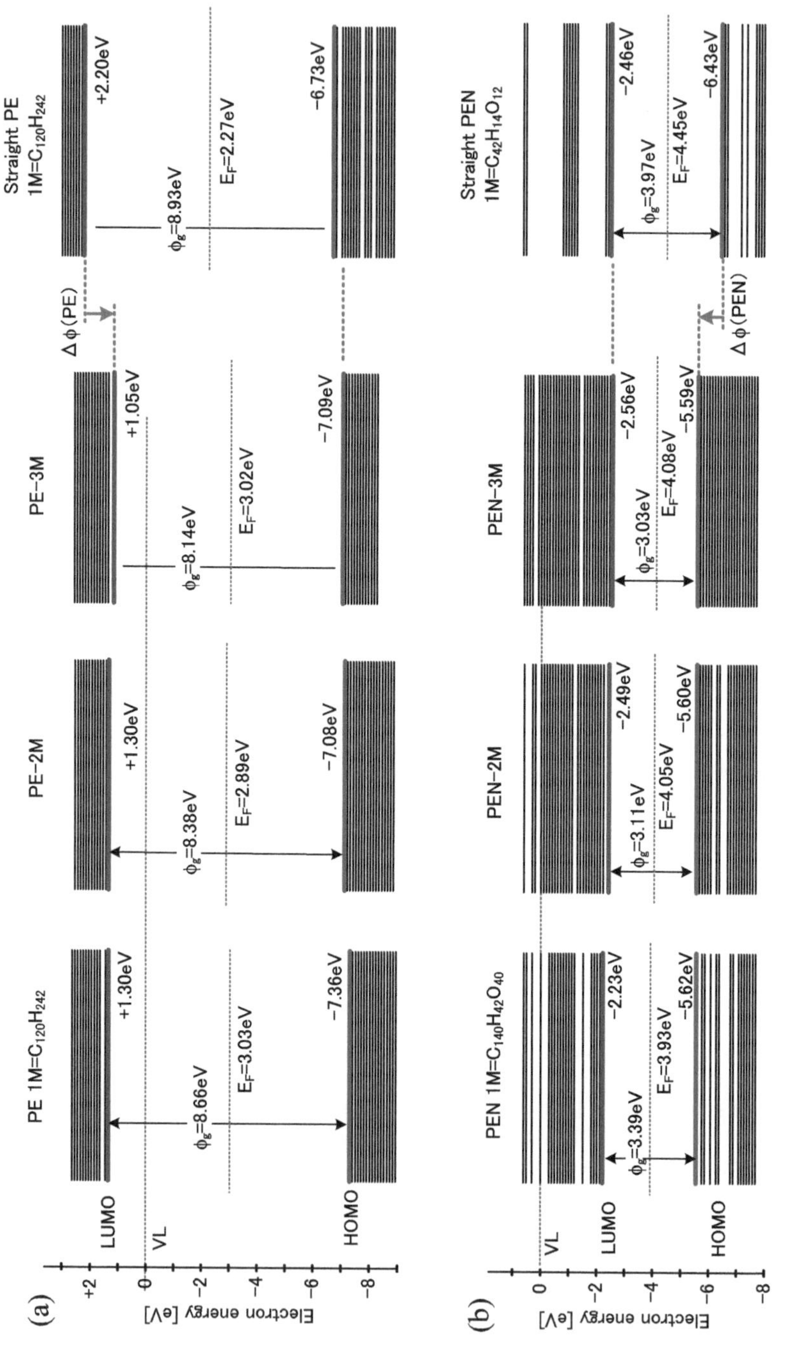

Fig. 13.15 Electron energy levels of PE and PEN 1M, 2M, 3M molecules

PE-1M, 2M, 3M molecules and straight PE, the LUMO level decreases from +2.20 eV to +1.05 eV with $\Delta\phi$(PE) = −1.15 eV. On the other hand, when the PEN-1M, 2M, 3M molecules and straight PEN are compared, the HOMO level increases from −6.43 eV to −5.59 eV with $\Delta\phi$(PEN) = +0.84 eV.

What causes the decrease in the LUMO level of PE with $\Delta\phi$(PE) = −1.15 eV and the increase in the HOMO level of PEN with $\Delta\phi$(PEN) = +0.84 eV is considered next.

(3) Observation of Molecular Orbital of Macromolecular System (PE and PEN)

In order to consider the reason of the decrease in the LUMO level of PE and the increase in the HOMO level of PEN, the energy levels of macromolecular system (PE, PEN) with the molecular orbitals are calculated. Figures 13.16 and 13.17 show the results.

Observation of HOMO level of PE: Even if the molecular chains approach, the molecular orbital is formed in the molecular chain. The HOMO level is the same as that of straight PE.

Observation of LUMO level of PE: At the trap site, the molecular orbitals of electron waves of hydrogen H are observed in the (b) closely approaching molecules of the molecular chain and (c) bend molecule. The LUMO level is lower than that of straight PE. For details, see the part (1) of Sect. 13.1.

Observation of HOMO level of PEN: (b) HOMO level of close molecules is as high as −5.59 eV, and (a) HOMO level of single molecule is as low as −5.60 eV.

Observe the LUMO level of PEN: (a) LUMO level of single molecular part changes to −2.49 eV and LUMO+1 level changes to −2.28 eV.

(4) Observation of Decrease of LUMO Level of PE and Increase of HOMO Level of PEN

It was difficult to explain the change of LUMO level and HOMO level in the macromolecular system (PE, PEN). Therefore, we observe changes in the LUMO and HOMO levels during the parallel approach of two molecules (Figs. 13.18 and 13.19).

Decrease of LUMO level of PE (Fig. 13.18): When PE molecules is approached to each other, the LUMO level is significantly reduced and the HOMO level is almost unchanged. The decrease in the LUMO level of PE is due to the electron wave of hydrogen H in adjacent molecules, as already explained in Sect. 13.1(1).

Elevation of HOMO level of PEN (Fig. 13.19): When the PEN molecule is approached, the increase of HOMO level is seen more strongly than the decrease of LUMO level. It is speculated that the closeness of the two Naphthalene surfaces causes the rise of the HOMO level due to the overlapping of π-electron waves.

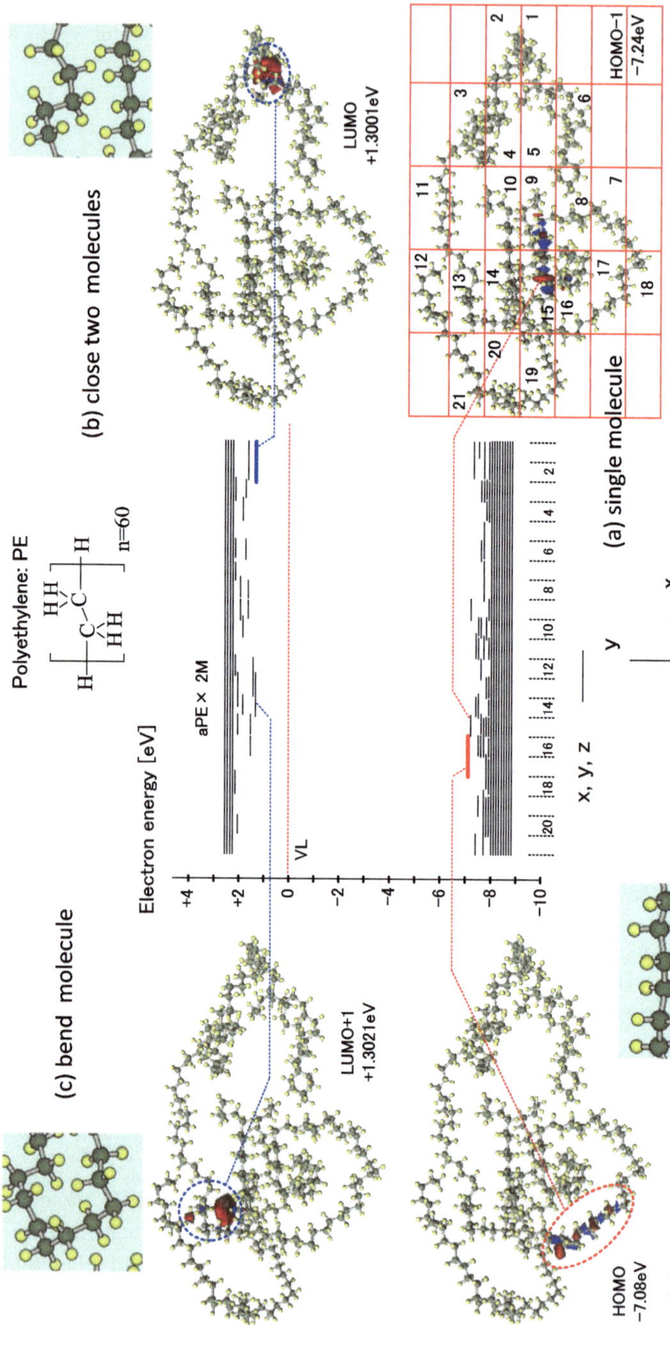

Fig. 13.16 Observation of molecular orbital and distribution of trap site of PE

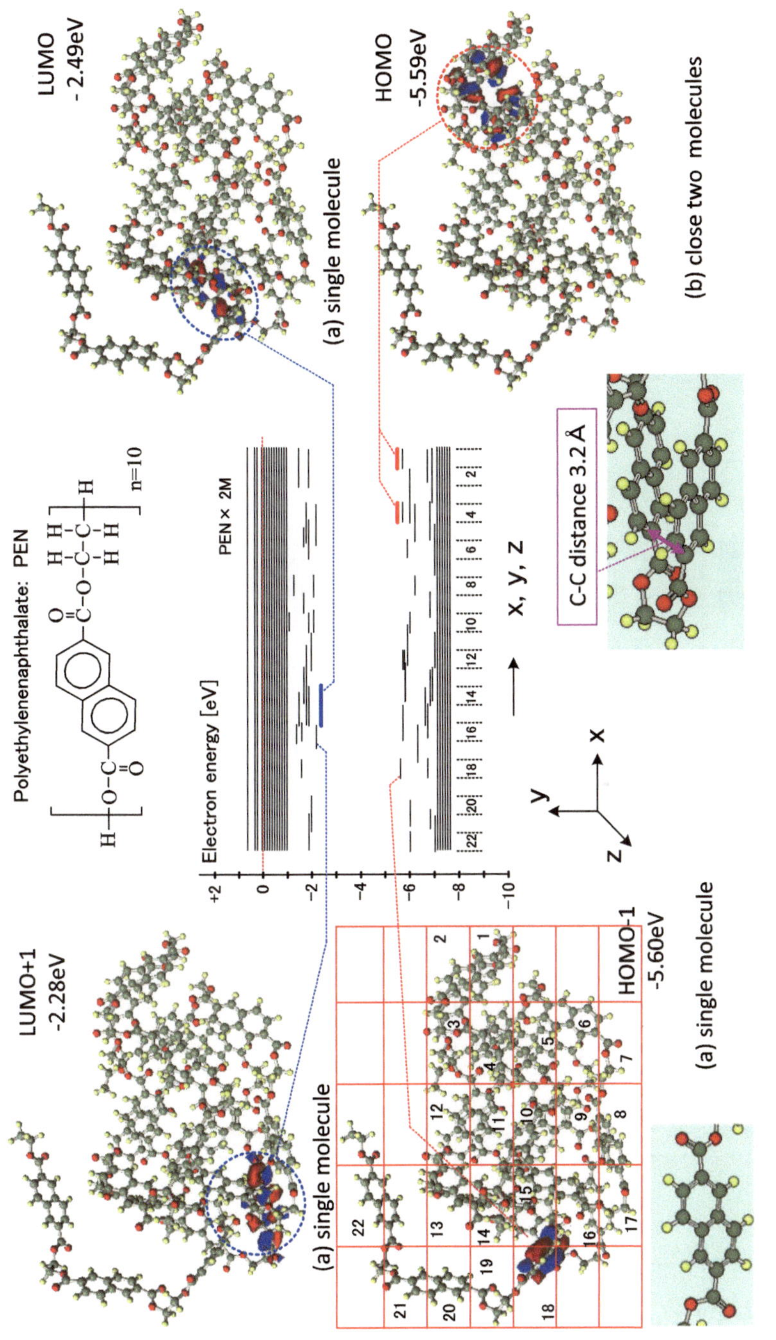

Fig. 13.17 Observation of molecular orbital and distribution of trap site of PEN

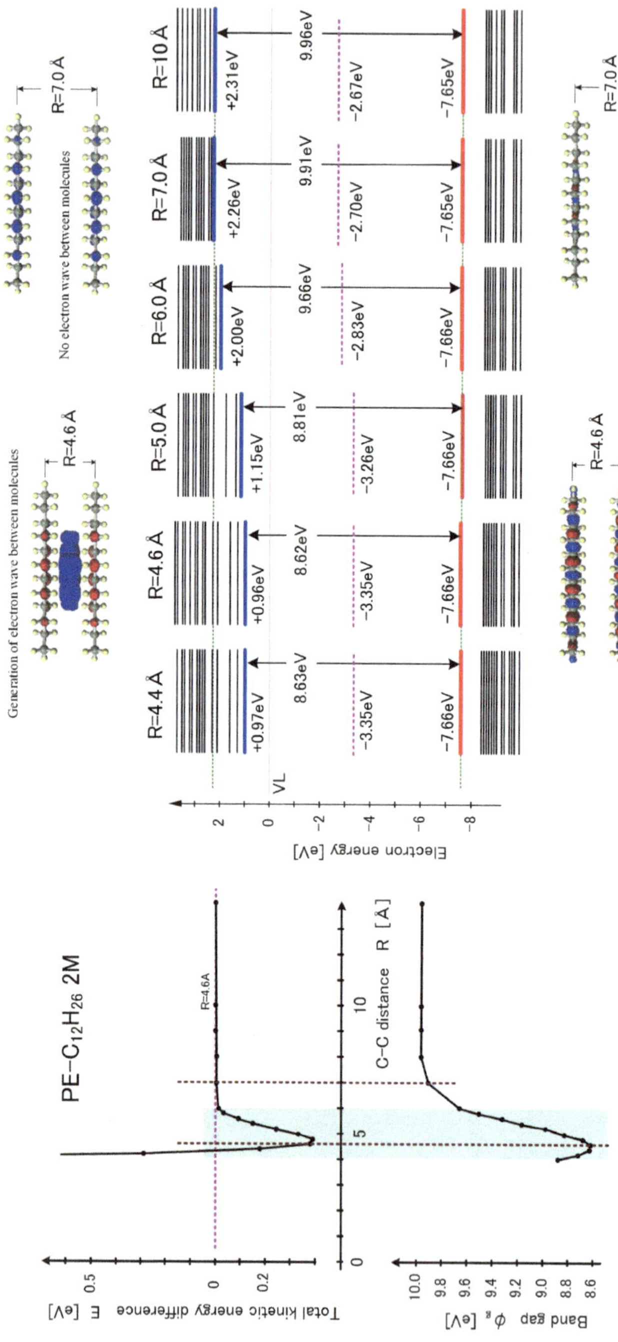

Fig. 13.18 Observation of changes in LUMO and HOMO levels and molecular orbitals during parallel approaching process of 2 PE molecules

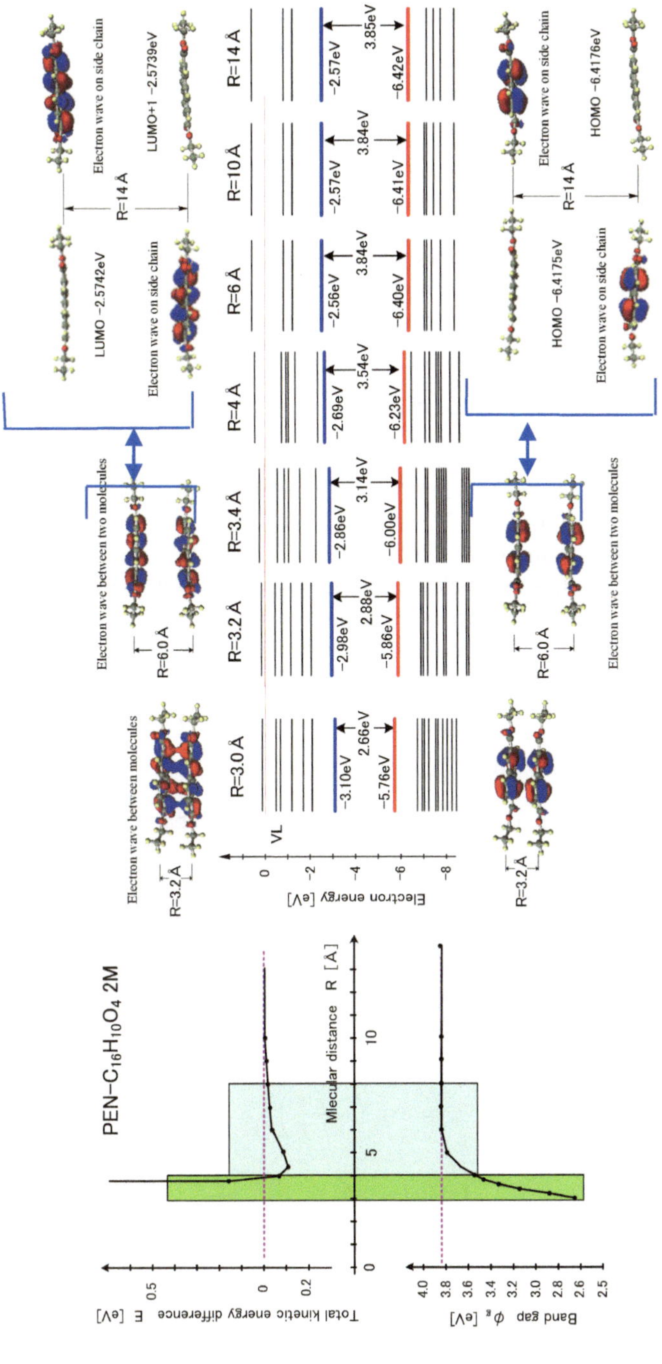

Fig. 13.19 Observation of changes in LUMO and HOMO levels and molecular orbitals during approaching parallel process of 2 PEN molecules

Since the LUMO level of PE and the HOMO level of PEN can change locally, that place becomes a trap site. Observation of changes in LUMO and HOMO levels and molecular orbitals can be analysed during parallel approach of 2 PE and 2 PEN molecules.

13.5 Hopping Transfer of Charge

The formation of trap sites due to the chemical structure is described in Sect. 13.3 above. Here, we describe a transfer model in which charges are hopping between the trap sites. The PET is selected as an example to explain the hopping movement of electrons and holes.

(1) **Hopping Transfer of Charge in PET**

In the part (1) of Sect. 13.2, we explain that the energy levels of PET are spatially localized and the hopping sites of electrons and holes are formed. Figure 13.20 shows the energy level distribution of the PET. The LUMO/HOMO level in the terephthalic acid part ⑮ is narrow, but that of the ⑤ ethylene part is wide. Therefore, ⑮ terephthalic acid part becomes a trap site for electrons and holes, and ⑤ ethylene part becomes a conduction level. Since the trap sites in the distributed energy level diagram are complex, a simplified hopping model is drawn in Fig. 13.20. The effective depth of the electron trap is $\phi_{\mathrm{te}} = 1.9$ eV, and its hopping probability is θ_{HOP}. The effective hole trap depth is $\phi_{\mathrm{th}} = 0.79$ eV, and the hopping probability is θ_{HOP}.

(2) **Effective Probability and Effective Trap Depth**

Consider the electron trap levels in the ⑮ terephthalic acid part in Fig. 13.20, there are many trap levels. Therefore, the carrier migration needs to overcome each energy level difference ϕ_{tn}. The respective probabilities θ_{n} are given by Eq. (13.1). This probability θ_{n} occurs independently, and the final effective probability θ_{HOP} is given by Eq. (13.2). The effective probability of electrons θ_{HOP} is shown by Eq. (13.3), and the effective probability of holes is given by Eq. (13.4). From these relationships, the effective trap depth ϕ_{t} is determined by the largest energy level difference. (See section 13.5 in Takada Text)

$$\theta_n = \exp\left(-\frac{\varphi_{tn}}{kT}\right) \tag{13.1}$$

$$\frac{1}{\theta_{HOP}} = \frac{1}{\theta_1} + \frac{1}{\theta_2} + \frac{1}{\theta_3} + \ldots = \sum_{n=1}^{N} \frac{1}{\theta_n} \tag{13.2}$$

$$\theta_{HOP}(electron) = \exp\left(-\frac{\varphi_{te}}{kT}\right) \tag{13.3}$$

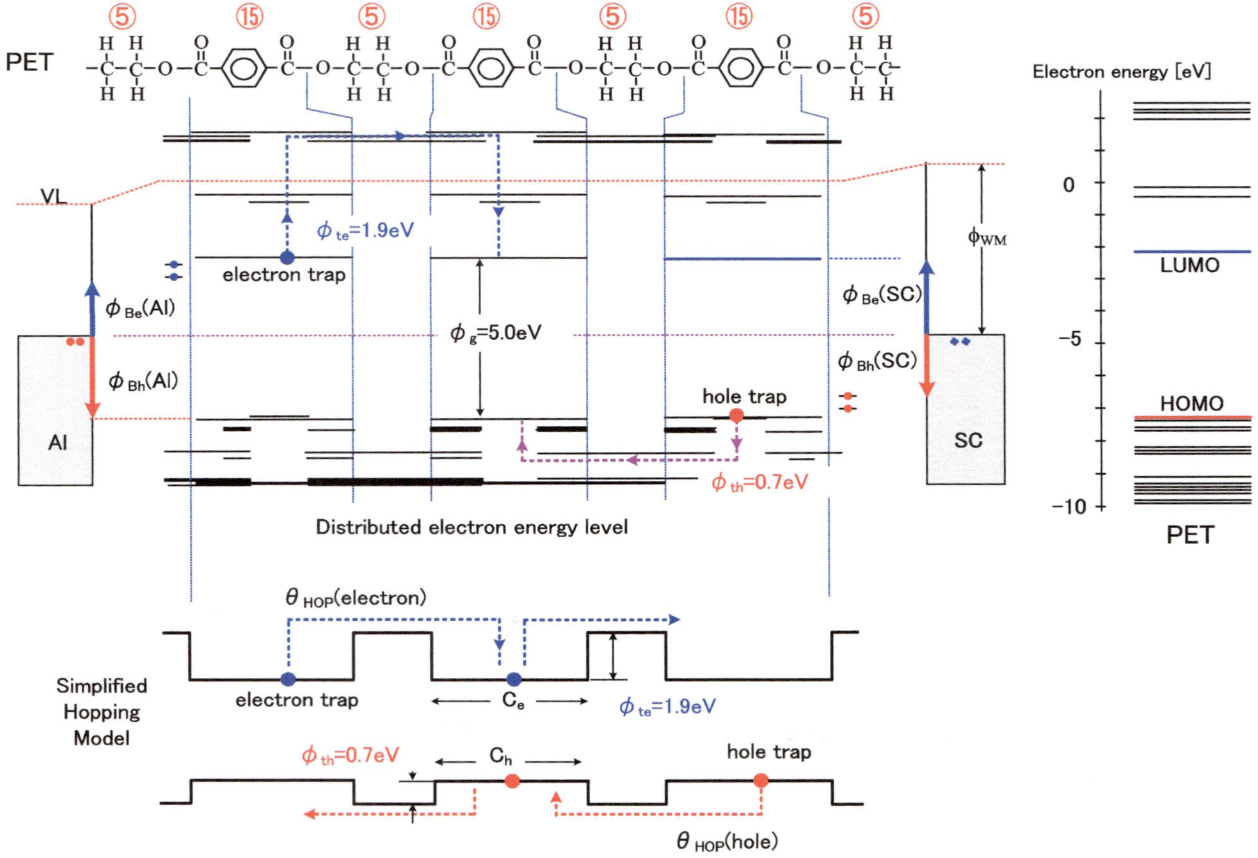

Fig. 13.20 A trap site for the distributed energy level diagram of PET with a simplified hopping model

$$\theta_{HOP}(hole) = \exp\left(-\frac{\varphi_{th}}{kT}\right) \tag{13.4}$$

(3) **Hopping Movement Model**

When an electric field E [V/m] is applied to the dielectric sample, a current J [A/m^2] flows through it shown in the left of Fig. 13.21. This is because the electric charge (electron) hops between the traps with a depth of ϕ_t [eV] and a current flows as described in the part (1) (also center of Fig. 13.21). On the right side of Fig. 13.21, the hopping movement of electrons is plotted with the position z as the vertical axis and time T as the horizontal axis.

If there is no trap ($\phi_t = 0$), the electron can move to the counter electrode (anode) at time T_c as drawn by the red dotted line. When there is a trap (ϕ_t), an electron can hop to an adjacent trap during the time (hopping time) τ_c shown by the red solid line, and it remains in the new trap during the time (trapping time) τ_t shown by the blue solid line. This hopping process is repeated until the charge moves to the counter electrode. When the trap is shallow ($\phi_t =$ small), the electron can move to the counter electrode (anode) in a short time, while the movement time is long when it is deep (Fig. 13.21, right). In the figure, c is set as the trap interval.

(4) **Hopping Probability**

Therefore, the hopping probability θ is defined by the ratio of τ_c and (τ_c+τ_t) in Eq. (13.5). The probability is equal to the Maxwell-Boltman distribution function shown by the final term of Eq. (13.5), since the electron overcomes the barrier (ϕ_t) due to the thermal energy (kT). Therefore, the probability of hopping electrons and holes is given by Eqs. (13.3) and (13.4), where ϕ_{te} [eV] and ϕ_{th} [eV] are the effective trap depths of electrons and holes.

$$\theta_{hop} = \frac{\tau_c}{(\tau_c + \tau_t)} = \exp\left(-\frac{\varphi_t}{kT}\right) \tag{13.5}$$

$$\tau_t + \tau_c = \tau_c \exp\left(+\frac{\varphi_t}{kT}\right) \tag{13.6}$$

$$\tau_t = \tau_c \exp\left(+\frac{\varphi_t}{kT}\right) \tag{13.7}$$

(5) **Graph of Trap Depth and Trapping Time**

Eq. (13.5) in the previous part (4) is transformed into Eq. (13.6). Here, the hopping time is $\tau_c = 8.7 \times 10^{-15}$ s, which is an extremely short time, as described in the part (6) below. On the other hand, the trapping time of space charge is in the range of τ_t=10^{-3} s to 1 year (=86,400 s). Therefore, since τ_c<<τ_t holds, Eq. (13.6) can be approximated to the Eq. (13.7). Figure 13.22 shows the relationship between the trap depth ϕ_t [eV] as the ordinate and the trap time τ_t [s] as the logarithmic scale, which is calculated by Eq. (13.7).

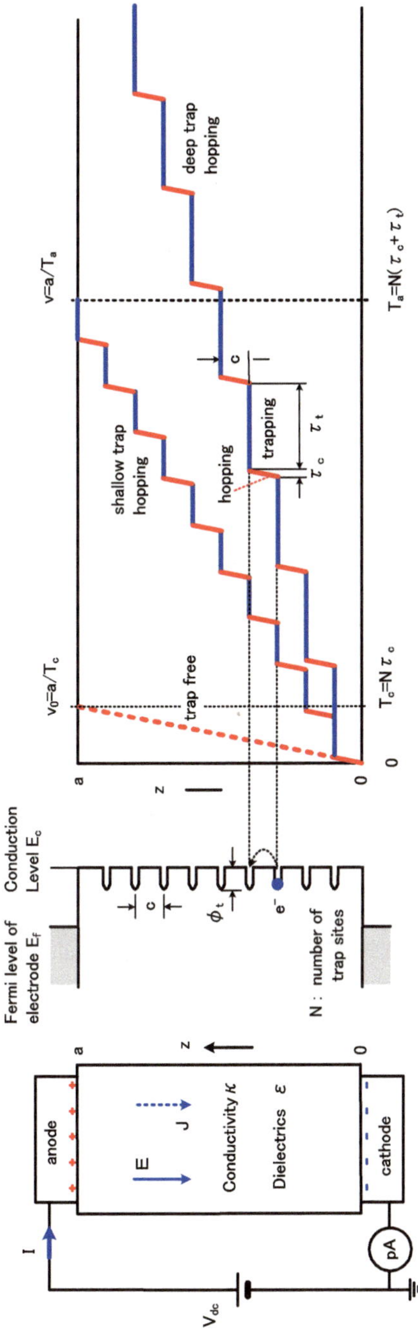

Fig. 13.21 A model in which electron carrier hops between traps with a trap depth of ϕ_t [eV]

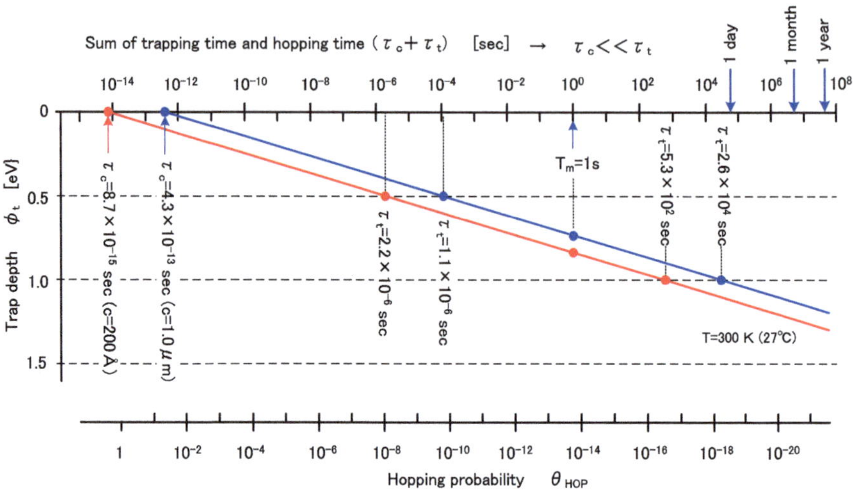

Fig. 13.22 Relationship between trap depth and trapping time

For example, at $T = 27°\mathrm{C}$, the time during which an electron or hole stays at a trap depth of $\phi_t = 1.0$ eV is $\tau_t = 1200$ s. At $\phi_t = 0.5$ eV, the trapping time is $\tau_t \approx 5\times 10^{-6}$ s, which is extremely short. Further, Eq. (13.7) can be calculated as a straight line in the semi-logarithmic graph when the temperature is 27 °C, 57 °C, and 87 °C. At $\phi_t = 1.0$ eV, the trapping time is $\tau_t = 1200$ s at $T = 27$ °C, $\tau_t = 72$ s at $T = 57$ °C and $\tau_t = 1.4$ s at $T = 87$ °C. It means the trapping time is shorter due to temperature rise.

(6) Estimation of Hopping Time τ_c

The hopping time $\tau_c = 8.7 \times 10^{-15}$ s is used in the previous part (5). Here, the value of this hopping time is calculated.

The hopping of the electron wave in the polyethylene molecule and the hopping time τ_c are shown in Fig. 13.24. This polyethylene molecule has a trap site in the ketone group and has a spacing *c*. Electron waves can hop between the trap interval *c* within time τ_c.

Electrons have particle nature and wave properties. The momentum *p* of the electron in Eq. (13.8) and the energy *E* in Eq. (13.9) connects the two properties. Where, *m* is the mass of the electron, v_e is the velocity of the electron, λ is the wavelength, and *h* is Planck's constant. Estimate the value of τ_c from the electron hopping distance *c* in Eq. (13.11).

$$p = m\ v_e = \frac{h}{\lambda} \tag{13.8}$$

$$E = \frac{1}{2} m\ v_e^2 = h\ \nu \tag{13.9}$$

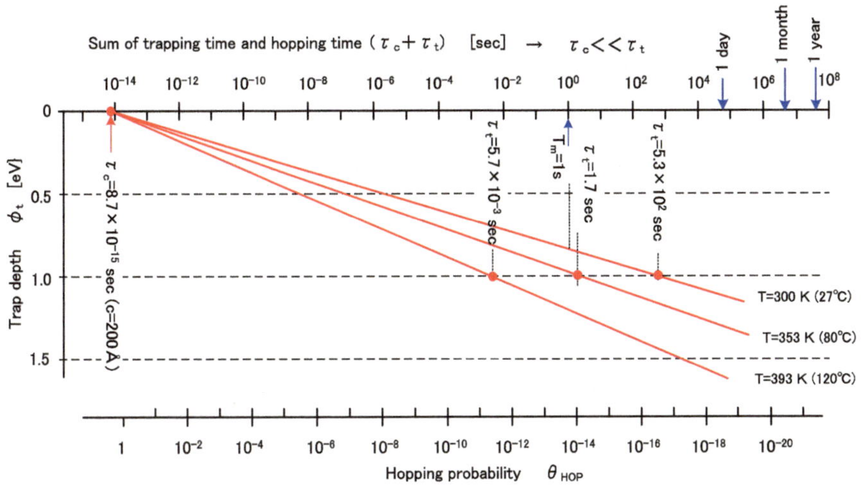

Fig. 13.23 Relationship between trap depth ϕ_t [eV] and trapping time τ_t [s] with its temperature dependence

As shown in Fig. 13.24, the relation between wavelength λ and carbon atom δ is $\lambda = 2\delta$. An electron with the wavelength λ can move in the periodic field of the electric potential between carbon atoms. If the vibration frequency at this time is ν_0, this period τ_0 (=$1/\nu_0$) can be derived from Eqs. (13.8) and (13.9) to Eq. (13.10).

Next, we estimate the time τ_c during which electrons move between the trap interval c. Here, by assuming the relationship between the period τ_0 in Eq. (13.10) and the wavelength λ(=2δ), the hopping time τ_c in Eq. (13.11) is obtained.

$$\tau_0\left(=\frac{1}{\nu_0}\right)=\frac{2\,m}{h}\,\lambda^2=\frac{8\,m}{h}\,\delta^2 \tag{13.10}$$

$$\frac{c}{\tau_c}=\frac{2\lambda}{\tau_0}=v_e \tag{13.11}$$

For example, since the carbon atom spacing of polyethylene is $\delta = 1.54$ Å, the wave period of Eq. (13.10) is $\tau_0 = 2.61 \times 10^{-16}$ s, and the vibration frequency is $\nu_0 = 3.84\times10^{15}$ Hz. As shown in Fig. 13.24, the electron of linear polyethylene molecule is estimated to be a one-dimensional standing wave and Eq. (13.11) is applied approximately. If there is no trap ($\phi_t = 0$), the trapping time becomes $\tau_t = 0$, as shown in Fig. 13.23. When the straight line of semi-logarithmic graph crosses the logarithmic axis of $\phi_t = 0$, the intercept is the hopping time τ_c. If the hopping distance is $c = 200$ Å within polyethylene, the hopping time is $\tau_c = 8.7 \times 10^{-15}$ s. If $c = 20000$ Å (= 2.0 μm), which is two digits long, then $\tau_c = 8.7 \times 10^{-13}$ s as shown in Fig. 13.22.

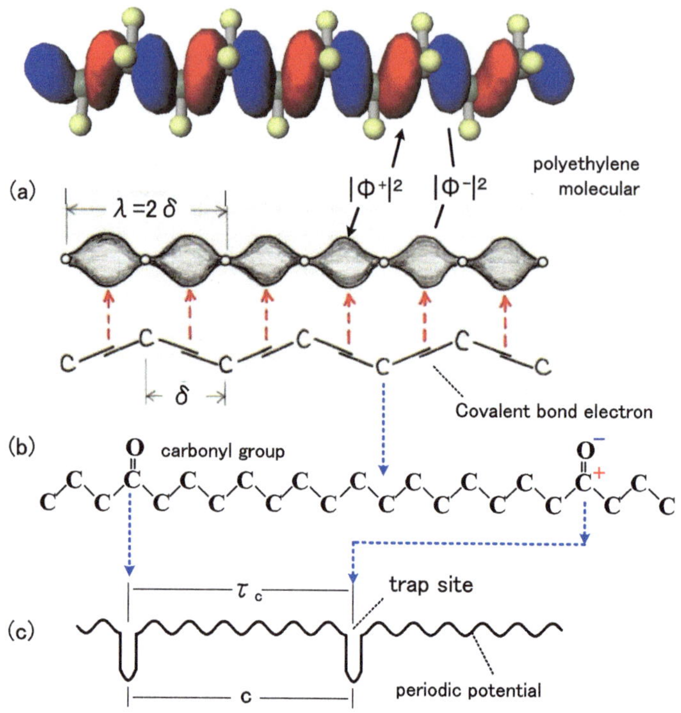

Fig. 13.24 Electron waves in polyethylene molecule with hopping time τ_c

13.6 Dominant Current Law

(1) Charge Injection Current and Excitation Current in Band Gap

Band gap ϕ_g [eV] is described in Sects. 12.5 and 13.4. Injection barrier ϕ_B [eV] of electrode charge is described in Sect. 12.7. Physical and chemical charge trap ϕ_t [eV] is described in Sect. 13.1. The electric current J [A/m^2] flowing through the dielectric under the electric field E [V/m] is governed by the values of ϕ_g [eV], ϕ_B [eV], and ϕ_t [eV]. The model of the current law drawn in Fig. 13.25 is explained.

First, consider J_1 [A/m^2] in Eq. (13.12). This J_1 [A/m^2] becomes a current in which the charge is injected overcoming the injection barrier ϕ_B [eV] and the electrode charge hops between traps ϕ_t [eV] (Fig. 13.25a). Here, θ_{INJ} is the probability of overcoming the injection barrier ϕ_B [eV] in Eq. (13.14), and θ_{HOP} is the hopping probability between traps ϕ_t [eV]. In addition, e [C] is the elementary charge, n_0 [particles/m^3] is the carrier density, and μ_0 [m^2/Vs] is the carrier mobility. Next, consider J_2 [A/m^2] in Eq. (13.13). This J_2 [A/m^2] becomes a current in which electrons and holes are excited in the band gap ϕ_g [eV] and hop between traps ϕ_g [eV] (Fig. 13.25a). Here, θ_{GAP} is the excitation probability between the band gaps ϕ [eV] in Eq. (13.16).

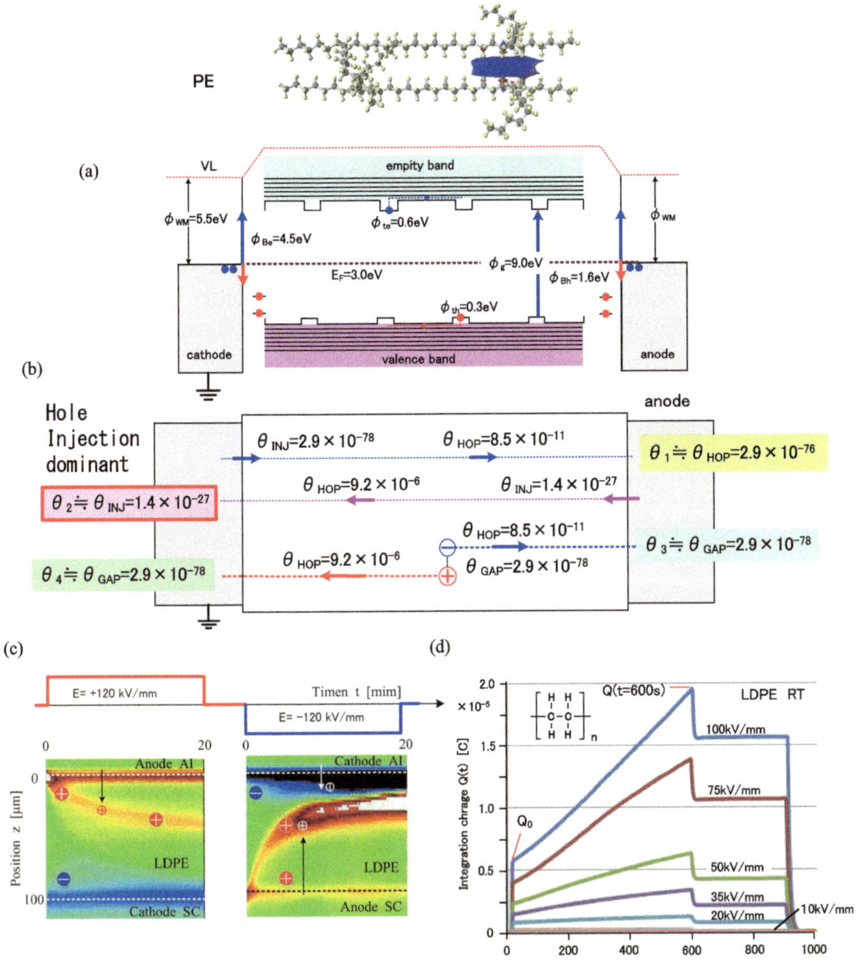

Fig. 13.25 Model of PE current law with measured PEA and Q(t) data

$$J_1 = en\mu E = en_0\theta_{INJ} \times \mu_0\theta_{HOP} \times E \tag{13.12}$$

$$J_2 = e\,n\,\mu\,E = e\,n_0\,\theta_{\mathrm{GAP}} \times \mu_0\theta_{HOP} \times E \tag{13.13}$$

$$\theta_{INJ} = \exp\left(-\frac{\varphi_B}{k\,T}\right) \tag{13.14}$$

$$\theta_{HOP} = \exp\left(-\frac{\varphi_t}{k\,T}\right) \tag{13.15}$$

$$\theta_{GAP} = \exp\left(-\frac{0.5\,\varphi_g}{k\,T}\right) \tag{13.16}$$

(2) Probability Dependent Phenomenon

Focus on J_1 [A/m^2] in Eq. (13.12), it is the product of the charge injection probability $\theta_{\rm INJ}$ and the hopping probability $\theta_{\rm HOP}$. This is a dependent phenomenon of the probability $\theta_{\rm INJ}$ that determines the number of carriers, and the probability $\theta_{\rm HOP}$ determines the carrier hopping and move. Here, we evaluate which of $\theta_{\rm INJ}$ and $\theta_{\rm HOP}$ is more strongly regulated. Eq. (13.17) is a strongly-regulated probability equation. θ_1 is for electron carriers and θ_2 is for hole carriers.

Next, pay attention to J_2 [A/m^2] in Eq. (13.13). It is the product of the charge generation probability $\theta_{\rm GAP}$ in band gap and the hopping probability $\theta_{\rm HOP}$. Eq. (13.18) is a strongly-regulated probability equation. θ_3 is for electron carriers and θ_4 is for holes.

$$\frac{1}{\theta_1} = \frac{1}{\theta_{INJ}} + \frac{1}{\theta_{HOP}} \quad \frac{1}{\theta_2} = \frac{1}{\theta_{INJ}} + \frac{1}{\theta_{HOP}} \tag{13.17}$$

$$\frac{1}{\theta_3} = \frac{1}{\theta_{GAP}} + \frac{1}{\theta_{HOP}} \quad \frac{1}{\theta_4} = \frac{1}{\theta_{GAP}} + \frac{1}{\theta_{HOP}} \tag{13.18}$$

(3) In PE, Hole Injection from Anode is Dominant

Which of the PE currents in Eqs. (13.12) and (13.13) mainly flows in Fig. 13.25? It is determined according to which of the electrons and holes in Eqs. (13.17) and (13.18) mainly flows. Then, the results of PEA measurement and Q(t) measurement can be compared.

Figure 13.25a, b shows the calculated values of $\phi_{\rm g}$[eV], $\phi_{\rm B}$ [eV], $\phi_{\rm te}$ [eV] and $\phi_{\rm th}$ [eV] of PE by DFT calculation. By substituting $\phi_{\rm g}$[eV], $\phi_{\rm B}$ [eV], $\phi_{\rm te}$ [eV] and $\phi_{\rm th}$ [eV] obtained by DFT calculation into Eqs. (13.14) to (13.16), the probability $\theta_{\rm INJ}$, $\theta_{\rm HOP}$, $\theta_{\rm GAP}$ results can be calculated, as shown in Fig. 13.25. Further, the probabilities calculated in Eqs. (13.17) and (13.18) are obtained, and the values obtained by θ_1, θ_2, θ_3 and θ_4 are marked in Fig. 13.25b. The largest value among θ_1, θ_2, θ_3 and θ_4 is θ_2 (hole) $= \theta_{\rm INJ} = 1.4 \times 10^{-27}$.

From this result, it can be explained that in PE, the charge accumulation and the leakage current depends on whether the holes with θ_2 (hole) $= \theta_{\rm INJ}$ are easily injected into the anode. In the PEA (Fig. 13.25c) measurement result, positive charge is injected from the anode, and the charge accumulation can also be confirmed from the Q(t) (Fig. 13.25d) result. Therefore, since the evaluation of the dominant current law is in agreement with the measurement result, the current law model is considered to be correct.

(4) In PET, Electron Hopping Movement is Dominant

A similar study is conducted for the PET current in Fig. 13.26. As in the contents of part (3), $\phi_{\rm g}$[eV], $\phi_{\rm B}$ [eV], $\phi_{\rm te}$ [eV] and $\phi_{\rm th}$ [eV] of PET are calculated by DFT calculation, and the values obtained from θ_1, θ_2, θ_3 and θ_4 are calculated as shown in Fig. 13.26. The largest value among them is θ_1 (electron) $= \theta_{\rm HOP} = 1.3 \times 10^{-32}$.

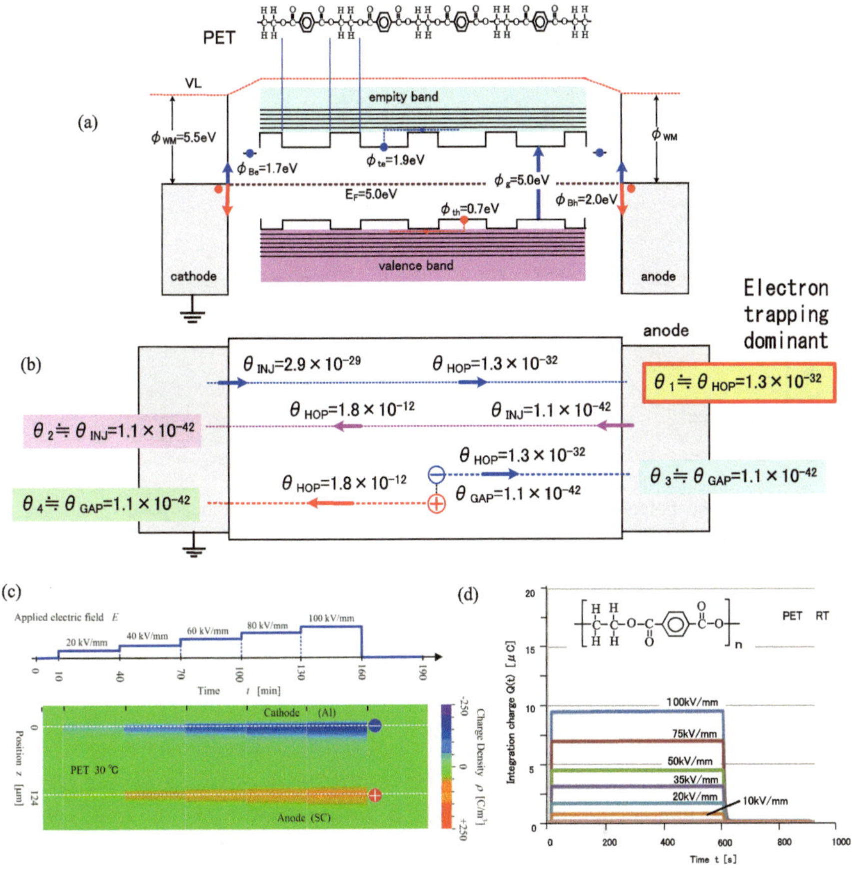

Fig. 13.26 Model of PET current law with measured PEA and Q(t) data

From this result, it can be explained that charge accumulation and leakage current in PET depends on whether the injected electrons easily hop and move. In the PEA (Fig. 13.26c) result, positive charge is injected from the anode, and the charge accumulation can also be confirmed from the Q(t) (Fig. 13.26d) result. The negative charge injected from the cathode with $\theta_1(\text{electron}) = \theta_{\text{HOP}}$ is slightly accumulated in the vicinity of the cathode, and the Q(t) (Fig. 13.26d) result also shows that the charge is accumulated. The measurement results are the same.

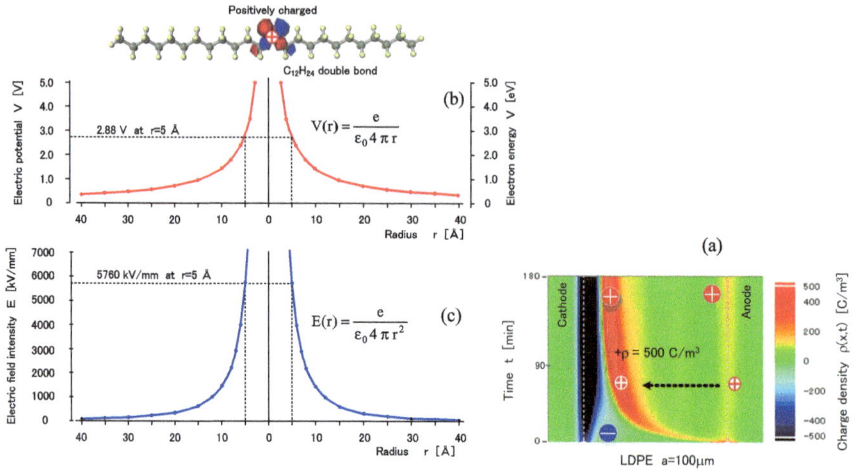

Fig. 13.27 Electric potential distribution and field distribution of trapped positive charges in $C_{12}H_{24}$ (double bond)

13.7 Scope of Application of Classical Theory and Quantum Theory

(1) Coulomb Potential of Point Charge

In the previous section 13.6, we introduce the results that positive charges are injected from the anode and accumulated in LDPE. It is shown again in Fig. 13.27a. If the charge accumulates, the energy level distribution in the LDPE molecule should be disturbed by the Coulomb potential. Therefore, according to the classical theory, the potential distribution of Eq. (13.19) and the electric field distribution of Eq. (13.20) of the positive point charge (+e = 1.6×10^{-19} C) in the trap of $C_{12}H_{24}$ (double bond) are shown in Fig. 13.27b, c.

$$V(r) = \frac{q}{\varepsilon_0 \varepsilon_r 4\pi r} \tag{13.19}$$

$$E(r) = \frac{q}{\varepsilon_0 \varepsilon_r 4\pi r^2} \tag{13.20}$$

According to this result, the potential at r=5 Å from the point charge becomes $V(r = 5\text{ Å}) = 2.88$ V, which is 1/3 of the LDPE band gap of $\phi_g = 9.0$ eV. Furthermore, the electric field strength is $E(r = 5\text{ Å}) = 5760$ kV/mm, which is 10 times more than the breakdown electric field of LDPE. However, no insulation breakdown occurs. This is because classical theory is used and in fact the use range of quantum theory should be considered.

(2) Deformation of Energy Level Distribution Due to Electrification

Figure 13.28 shows the distribution of electron energy levels in the $C_{32}H_{64}$ (double bond) molecule for (a) positive charge, (b) neutral charge, and (c) negative charge, calculated by DFT in the molecular axis direction. (a) The positively charged energy level distribution is "valley-shaped" centered on the positive charge (⊕), and (c) negatively charged is "mountain-shaped" centered on the negative charge (⊖). Furthermore, (a) the positively charged (⊕) LUMO level (+3.66 eV) and HOMO level (−1.96 eV) are higher than (b) the neutral LUMO level (+0.88 eV) and HOMO level (−6.36 eV).) Also, they are lower than the negatively charged LUMO level (−3.88 eV) and HOMO level (−8.78 eV) (⊖).

In this energy level distribution diagram, the potential distribution of (a) positively charged (red dotted line) and (c) negatively charged (blue dotted line) cases are drawn with Eq. (13.19). A comparison between the two figures shows that the energy level distribution and the potential distribution of the point charge have similar tendencies. Therefore, the energy level distribution of 1S orbit of carbon C is plotted. This plot directly shows the potential distribution in the $C_{32}H_{64}$ molecule, and the energy level of the 1S orbital shows the potential only at the location of carbon C. The potential distribution of the neutral 1S orbit is a constant value, while the positively charged distribution has a "valley" distribution.

(3) The Boundaries of Application of Classical Theory and Quantum Theory

Here, the potential distribution of the 1S orbit of carbon C calculated by quantum theory (short horizontal bar in the figure) is compared with the results by Eq. (13.19) of classical theory (red dotted line and blue dotted line).

(a) **In the case of positive charge**:
Both the potential distributions match when the distance is more than $r = 7.7$ Å from the point charge, but they do not match inside the boundary range of $r =$ 7.7 Å.

(b) **In the case of negative charge**:
The boundary is $r = 14$ Å. That is, it is shown that Eq. (13.19) in classical theory for the point charge cannot be applied inside this boundary range, wihle the electron wave model of quantum theory can be applied.

Therefore, the potential and the electric field do not extremely increase near the point charge. Electrons are not point charges, but they are treated by a wave function with a spread. Therefore, it has a constant finite potential in the vicinity of the charge.

The boundary between classical theory and quantum theory is roughly equal to the radius of $r = 7.7$ Å from the point charge in the case of positive charge and the radius of r=14 Å in the case of negative charge.

(4) Energy Level Shift Due to Electrification

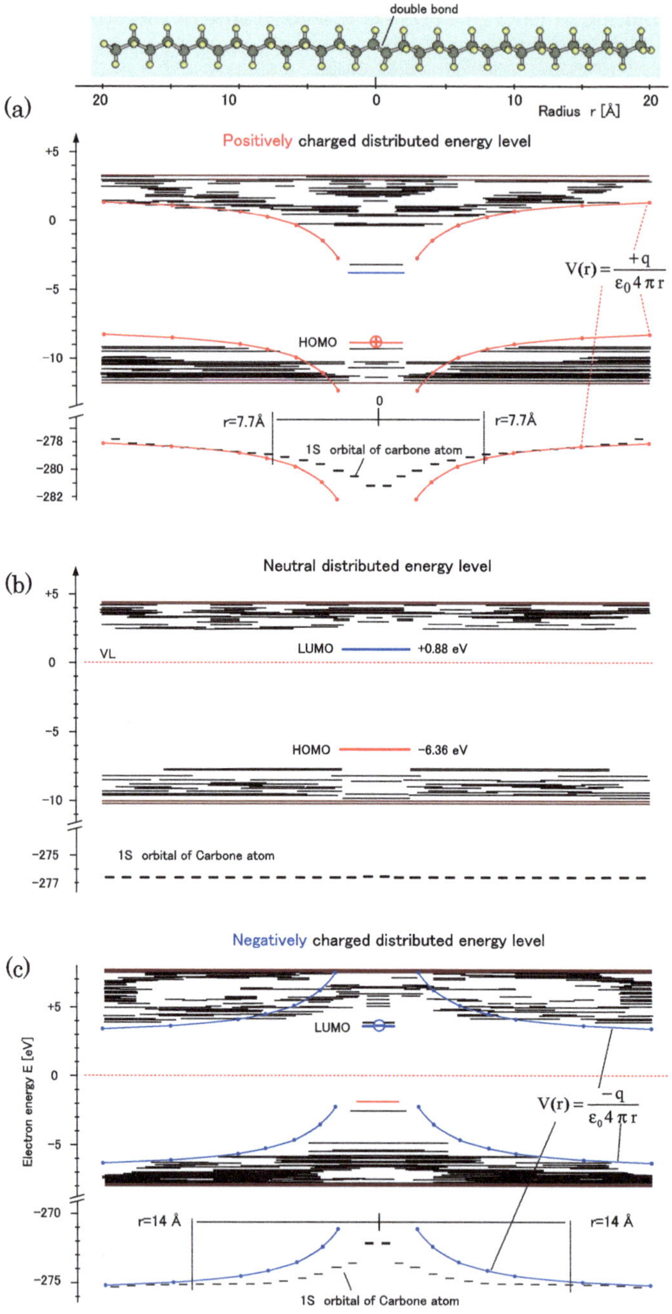

Fig. 13.28 $C_{32}H_{64}$ (double bond) molecule's electron energy levels. **a** Positively charged. **b** Neutrally charged. **c** Negatively charged

In the above part (3), the electron energy levels of (a) positively charged, (b) neutrally charged and (c) negatively charged cases of $C_{32}H_{64}$ (double bond) molecule are analyzed. Here, the energy levels of the three cases are compared (Fig. 13.29). As shown in the figure, the positively charged Fermi level of $E_F(\text{Pos}) = -6.33$ eV is lower than the neutral one of $E_F(\text{Neu}) = -2.74$ eV and the shift is −3.59 eV. Moreover, the negatively charged Fermi level of $E_F(\text{Neg}) = +0.85$ eV is higher than the neutral one and the shift is +3.59 eV. In the case (a), since one electron is deficient in the positively charged case, the energy level of the $C_{32}H_{64}$ molecule drops. On the contrary, in the case (c), the molecule acquires one electron, and thus the energy level o f he $C_{32}H_{64}$ molecule rises.

(5) Spread of Electrostatic Potential Distribution Due to Charging Process

Furthermore, we observe the electrostatic potential distribution $V(x, y, z)$ of the $C_{32}H_{64}$ molecule, as shown in the bottom of Fig. 13.29. In the neutral state, hydrogen H is localized with the positive potential (yellow-orange) and carbon C is localized

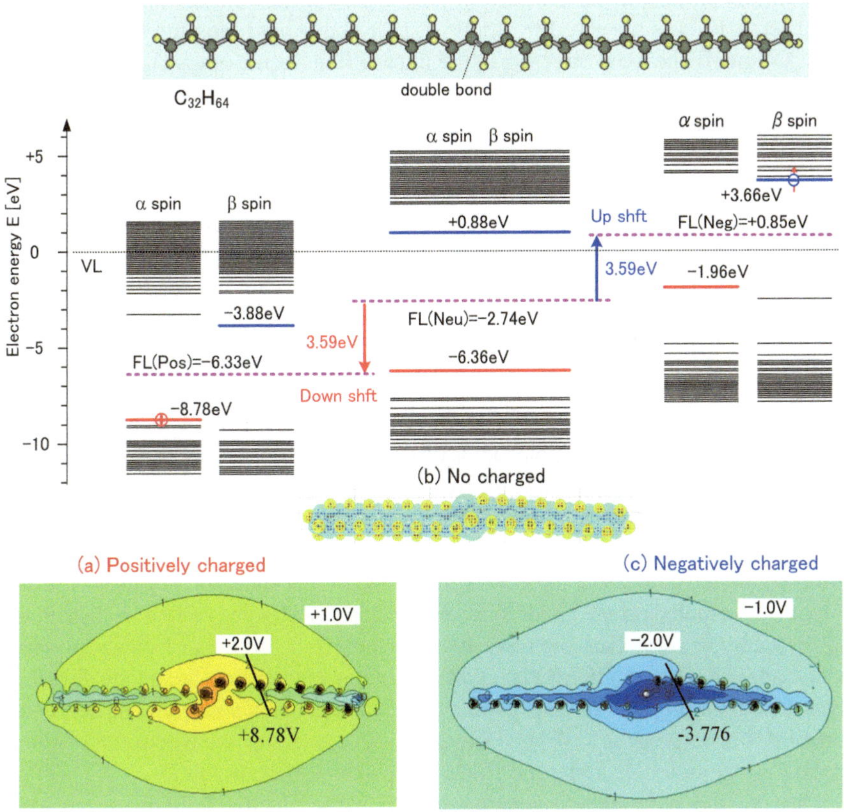

Fig. 13.29 Comparison of **a** positively charged, **b** neutrally charged and **c** negatively charged electron energy levels of $C_{32}H_{64}$ (double bond) molecules

with the negative potential (light blue). In contrast, the positively charged and negatively charged cases show widely distributed potentials along the $C_{32}H_{64}$ molecular axis around the trapped charges. The spreading potential of the trapped negative charges decays from -3.66 V (potential of trapped electron = $(-1) \times (+3.66)$) to -2.00 V and -1.00 V, and that of the trapped positive charges decays from $+8.78$ V (potential of trapped hole = $(-1) \times (-8.78)$) to $+2.00$ V and $+1.00$ V.

13.8 Electric Field Application and Band Gap

In Sect. 13.7, it is described that the energy level changes due to positive and negative charges. This is because the energy level of the electron wave changes due to the local potential distribution $V(x, y, z)$. Here, it is confirmed by DFT calculation that the energy level can also change due to the electric field applied from the outside.

(1) Apparently Narrow Band Gap

The energy level is observed by applying an electric field ranging from $E = 0$ to 3080 kV/mm in the axial direction of the linear PE molecule of $C_{20}H_{42}$ (Fig. 13.31). When $E = 0$, it is as wide as $\phi_g = 10.04$ eV, but at $E = 1030$ kV/mm it is $\phi_g = 8.0$ eV, and when $E = 3080$ kV/mm, it is $\phi_g = 4.52$ eV. Does the band gap really become narrow? To answer this question, the relationship between each energy level and the corresponding molecular orbitals are summarized and drawn in Fig. 13.30. Figure 13.30a shows the energy levels and molecular orbitals at $E = 0$. Figure 13.30b corresponds to $E = 3080$ kV/mm. From Fig. 13.30b, it is found that the band gap is not narrowed, but the molecular orbital is biased by the external electric field.

(2) External Electric Field Application and Inclination of Intramolecular Potential

Here, focus on the case of $E = 3080$ kV/mm in Fig. 13.30b, and the inclination of the potential distribution in the molecule is observed. When an external electric field of 3080 kV/mm is applied, the molecular orbitals of HOMO 81H and 80H in the valence band are deviated to the cathode side. The molecular orbitals such as 22H, 21H and 2H are deviated to the anode side. In the 2D potential distribution calculated by the Milliken atomic charges of the $C_{20}H_{42}$ molecule in this state, the negative potential (blue) is distributed on the anode side and the positive charge (yellow) is distributed on the cathode side. This indicates that the molecular orbitals of electron waves which are deeper than HOMO level are more concentrated on the anode side than those near HOMO level. As a result, the $C_{20}H_{42}$ molecule is polarized and the dipole moment is as large as $\mu = 4.186$ Debye. This is a result of the electron movement in the $C_{20}H_{42}$ molecule under the effect of the external electric field.

Observe the potential distribution in which the electric field in this $C_{20}H_{42}$ molecule is canceled out. The result of the potential distribution of the 1S orbital of 20 carbons C along the molecular axis is shown by the short horizontal bar at the

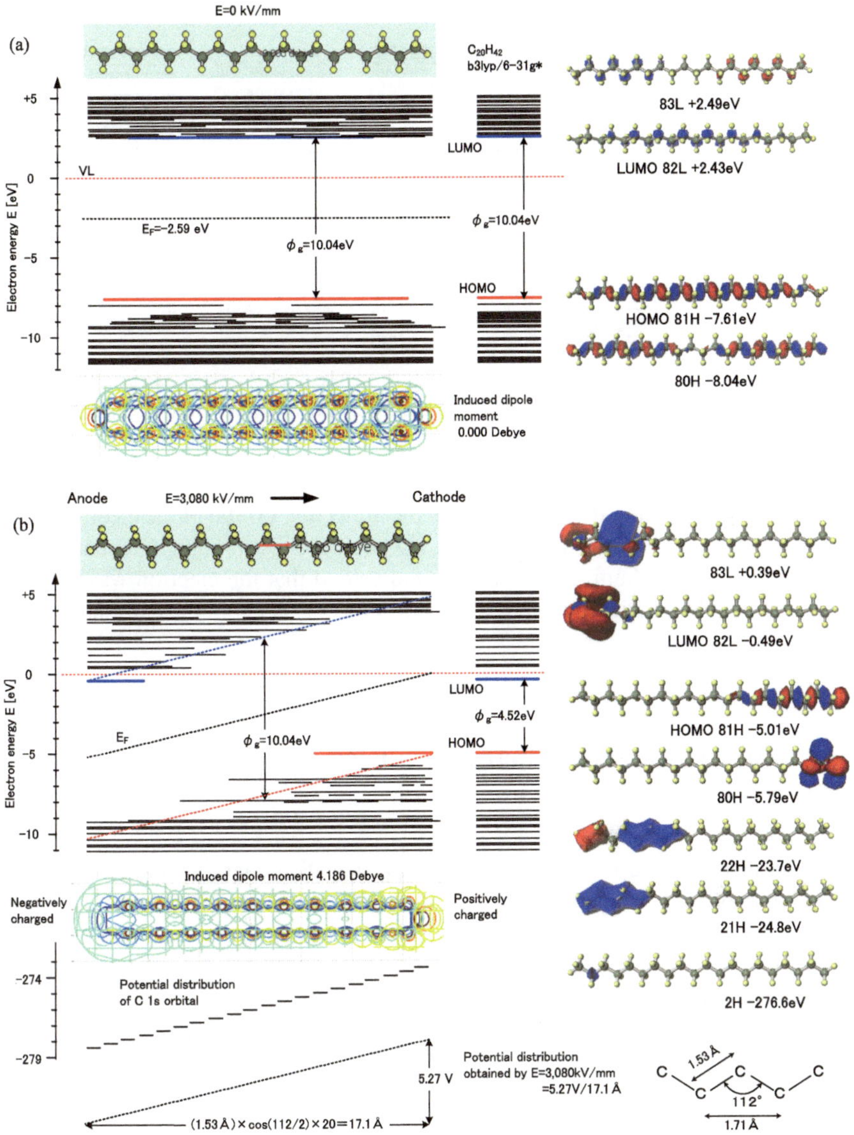

Fig. 13.30 Comparison of distribution energy levels of linear $C_{20}H_{42}$ PE molecule under applied electric fields of 0 and 3080 kV/mm

bottom of Fig. 13.30b. Furthermore, the potential difference in the $C_{20}H_{42}$ molecule due to the external electric field is calculated as 5.27 V by the product of the applied electric field of $E = 3080$ kV/mm and the molecular length of 17.1 Å. The slope of this potential and the slope of the 1S orbit are exactly the same.

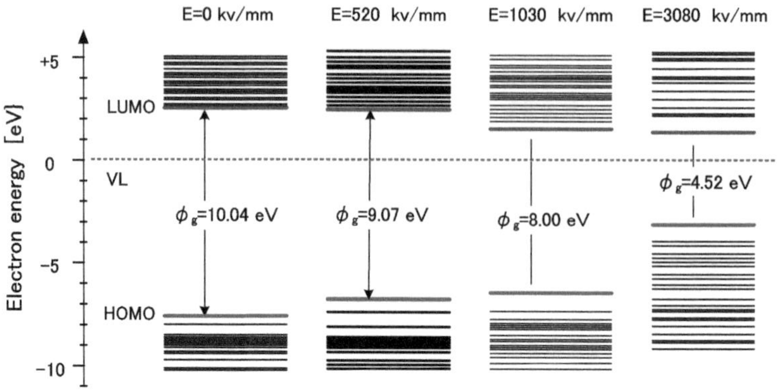

Fig. 13.31 Energy levels of linear $C_{20}H_{42}$ PE molecule under applied electric fields from 0 to 3080 kV/mm

When $E = 0$ kV/mm, there is no bias in the molecular orbital of the electron wave in Fig. 13.30a, and there is no bias in the 2D potential distribution. Compared with the case of $E > 0$ kV/mm, it can be well understood that the electron waves in the $C_{20}H_{42}$ molecule are biased toward the anode side to cancel out the external electric field.

13.9 Polarization and Charge Transfer Under Electric Field Application

When an electric field is applied, the electron (negative charges) wave function is shifted to the anode side. As a result, the molecule is polarized. Furthermore, under the electric field, holes (positive charges) move to the cathode side, and electrons (negative charges) move to the anode side. This movement becomes an electric current. The movement also depends on the trap depth of holes and electrons. This phenomenon is specifically observed in this section by DFT calculation.

(1) **Polarization of Electron Wave Under Electric Field**

The induced polarization is confirmed by observing the movement of molecular orbitals when an electric field is applied in the axial direction of the linear $C_{35}H_{72}$ molecule. First, we focus on the molecular orbitals of valence band 133H, 132H, 35H, 34H and empty band 134L without electric field. The molecular orbital is axially symmetrical.

Next, the molecular orbitals are observed under positive and negative electric field applications ($E = \pm 514$ kV/mm). The molecular orbital of the empty band 134L moves to the anode side, as shown in the center of Fig. 13.32. However, the molecular orbitals of 133H and 132H levels as the orbits of electrons move to the cathode side. It becomes a contradiction that negative charges move to the cathode side. On the

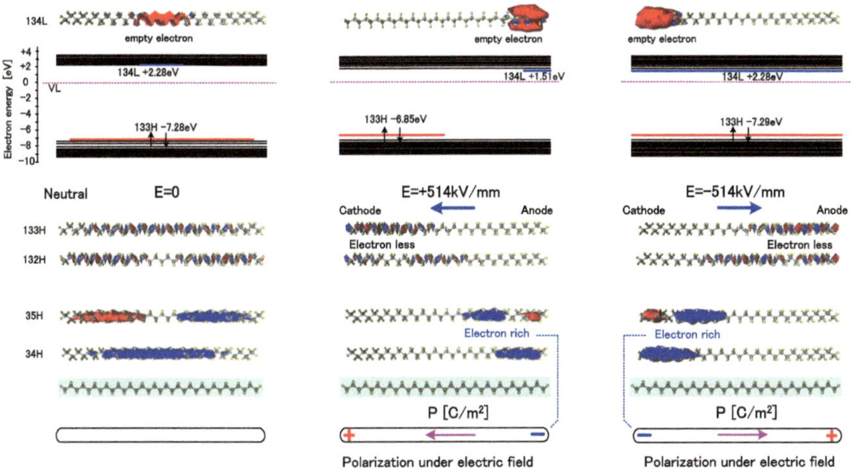

Fig. 13.32 Movement of molecular orbitals and induced polarization of a linear $C_{35}H_{72}$ molecule in an applied electric field

other hand, the molecular orbitals of 35H and 34H levels as the orbits of electrons move to the anode side, which is consistent with the theory. Thus, when observing the molecular orbitals of electrons in the full valence band, there are two kinds including low-level molecular orbitals (sufficient electrons) and high-level molecular orbitals (a little electrons), and the number of former is more. Therefore, the negative electrons in the molecular orbital of the entire $C_{35}H_{72}$ linear molecule move to the anode side. The negative polarity P [C/m^2] occurs on the anode side and the positive polarity is on the cathode side as shown in the center of Fig. 13.32. In the case of the negative electric field application of $E = -514$ kV/mm, the positive polarity P [C/m^2] occurs on the cathode side and the negative polarity is on the anode side, as shown in the right of Fig. 13.32.

(2) Electrostatic Potential Distribution and Electric Dipole Moment

In Sect. 13.8, we have already observed the polarization and electric dipole moment μ [Cm] of the $C_{20}H_{42}$ molecule by applying an ultrahigh electric field of $E = 3080$ kV/mm to the linear PE molecule of $C_{20}H_{42}$ in Fig. 13.30. Here, we observe the electric dipole moment μ [Cm] and the electrostatic potential distribution when an electric field of $E = 520$ to 3080 kV/mm is applied to the linear PE molecule of $C_{12}H_{26}$ in Fig. 13.33. A negative charge (⊖) on the anode side and a positive charge (⊕) on the cathode side are induced. As a result, the $C_{12}H_{26}$ molecule has a dipole moment μ [Cm], and the relationship of the dipole moment and applied electric field is plotted in the graph. It can be seen that μ is proportional to E.

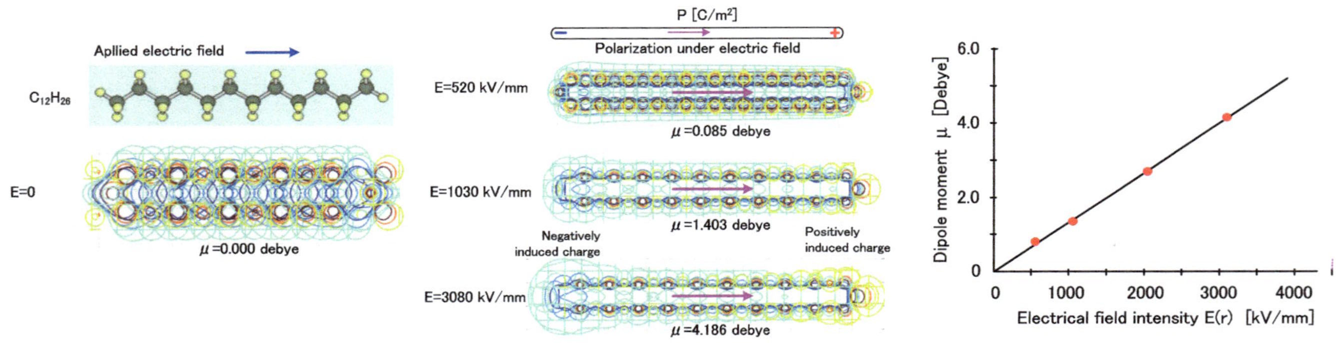

Fig. 13.33 Polarization and electric dipole moment of a linear $C_{12}H_{26}$ PE molecule by applying an electric field

(3) Charge Transfer in Shallow Traps

The hydroxyl group as the shallow trap is created in the $C_{32}H_{65}O$ molecule, and it is positively and negatively charged, as shown in Fig. 13.34. Then, it is analyzed how much electric field is applied to release the trapped charges and make them move. An electron trap ($\phi_{te} = 0.60$ eV) and a hole trap ($\phi_{th} = 0.75$ eV) can be confirmed at the location of the hydroxyl group from the energy level diagram when no electric field is applied.

Positive charge: When the molecule is positively charged by excluding the β electron at HOMO level, the β electron at HOMO level does not exist in the trap, and all the molecular orbitals extend among the molecule. Holes (absence of β electrons) spread to the HOMO level. Even if $E = 103$ kV/mm is applied, holes (absence of β electrons) still do not move. α-electron is deviated to the anode side due to polarization. When a higher electric field of $E = 514$ kV/mm is applied, holes (positive charge, absence of β electrons) are released from the trap and move to the cathode side. α-electrons (negative charge) are still deviated to the anode side due to polarization.

Negative charge: The α electrons at LUMO level are trapped in the hydroxyl group. By applying $E = 103$ kV/mm and 514 kV/mm, β electrons are released from the trap and move to the anode side. Since the electron trap ($\phi_{te} = 0.60$ eV) is shallower than the hole trap ($\phi_{th} = 0.75$ eV), electrons are easy to be released from the trap at the low electric field of $E = 103$ kV/mm.

(4) Charge Transfer in Deep Traps

The $C_{32}H_{64}$ molecule is created with double bonds as deep traps, and it is positively and negatively charged (Fig. 13.35). An electron trap ($\phi_{te} = 1.78$ eV) and a hole trap ($\phi_{th} = 1.40$ eV) can be confirmed at the location of the double bond from the energy level diagram when no electric field is applied.

Then, it is observed how much electric field is applied to release the trapped charges and make them move. At the electric field of $E = 514$ kV/mm, trapped electrons cannot be released, but the migration of trapped holes is observed. At a higher electric field of $E = 2570$ kV/mm, electrons and holes can be released and move. It is confirmed that a higher electric field is required to release the trapped electrons and holes charged in the $C_{32}H_{64}$ molecule with the double bond as deep traps, as compared with the $C_{32}H_{65}O$ molecule with the hydroxyl group as shallow traps in the above part (3).

(5) Charge Transfer Between Two Series of Molecules

In the previous parts (3) and (4), it is observed that the electrons/holes release from the trap and move inside the molecule. Here, a molecule ($C_{16}H_{33}O$) with a trap and a molecule ($C_{16}H_{34}$) without traps are taken as an example in a calculation system (Fig. 13.36). The two molecules are arranged linearly. When this two-molecule system is positively and negatively charged, the charge is always trapped on the $C_{16}H_{33}O$ molecule side. When an electric field is applied to this two-molecule system,

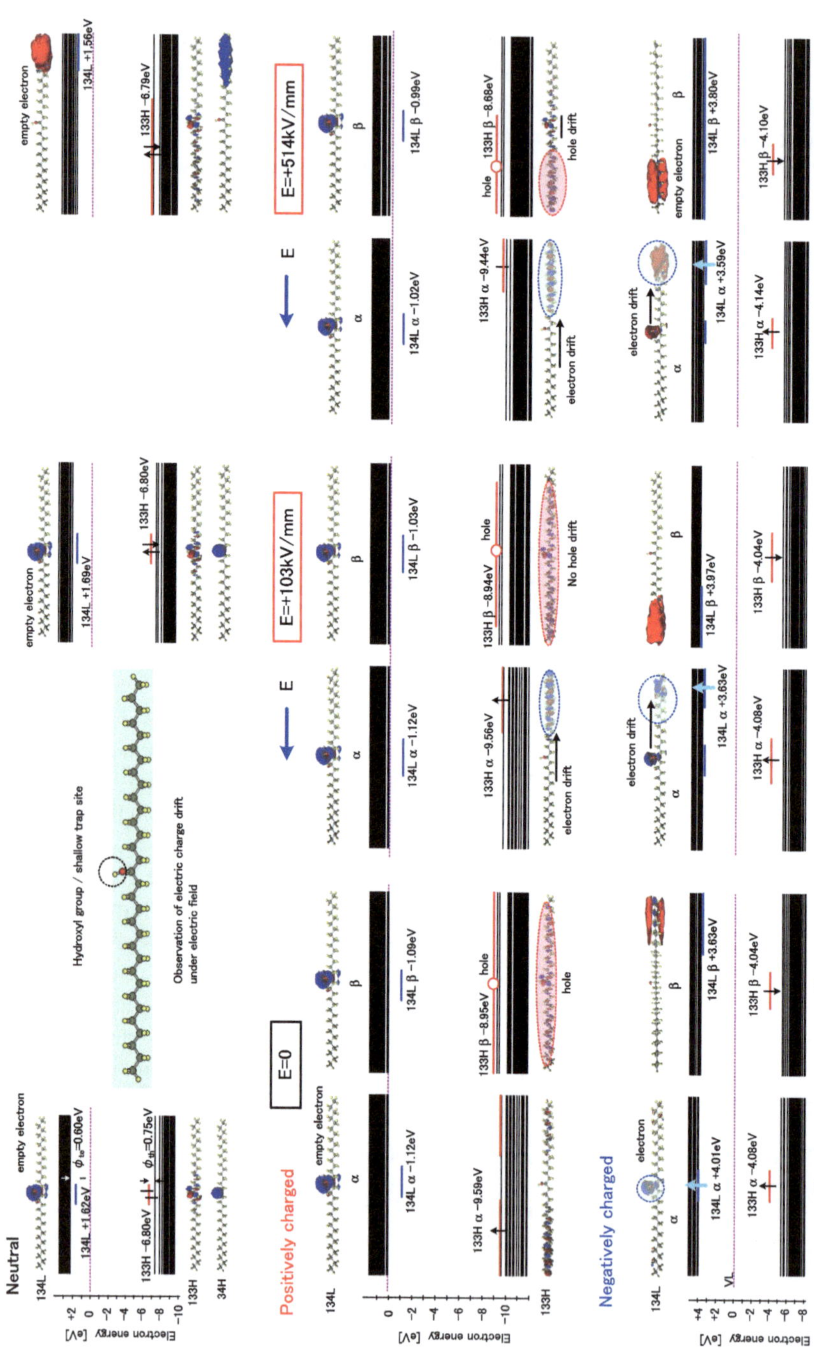

Fig. 13.34 Observation of detrapping and movement of trapped charges by applying an electric field to $C_{32}H_{65}O$ molecule

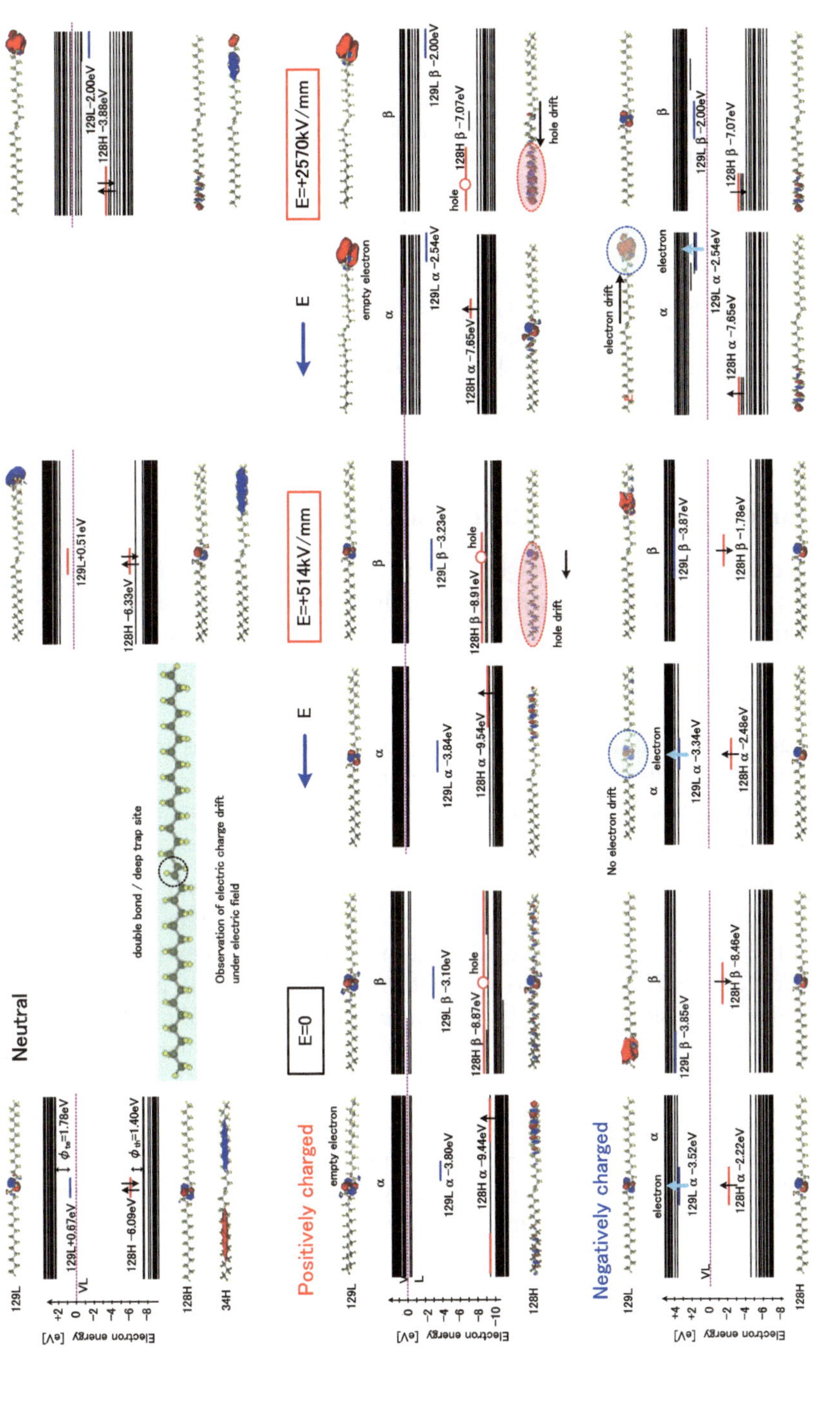

Fig. 13.35 Observation of detrapping and movement of trap charges by applying an electric field to $C_{32}H_{64}$ molecule

the charge of the $C_{16}H_{33}O$ molecule is released from the trap and move to the adjacent $C_{16}H_{34}$ molecule.

Positive charge: When $E = 0$ and the $C_{16}H_{33}O$ molecule is positively charged, β electrons at the HOMO level are absent and holes are accumulated there. When an electric field of $E = +514$ kV/mm is applied, holes move intermolecularly to the adjacent $C_{16}H_{34}$ molecule. In contrast, when a reverse electric field of $E = -514$ kV/mm is applied, holes remain within the same molecule. In other words, there are no adjacent molecules, and thus holes remain unchanged.

Negative charge: When $E = 0$ and the $C_{16}H_{33}O$ molecule is negatively charged, α electrons are trapped in the hydroxyl group at the LUMO level. When an electric field of $E = +514$ kV/mm is applied there, α electrons are released from the trap and move to the anode side. The α electrons stop at the end of the molecule because there are no adjacent molecules. On the other hand, when a reverse electric field of $E = -514$ kV/mm is applied, α electrons are released from the trap and move to the adjacent $C_{16}H_{34}$.

(6) Charge Transfer Between Three Parallel Molecules

In the previous part (5), the charge transfer between two series of molecules is described. Here, the charge transfer between three parallel molecules is analyzed. An electric field is applied at right angles to three parallel molecules. The $C_{16}H_{32}$ molecule with a double bond as deep traps is placed in the center, and two $C_{16}H_{34}$ molecules are placed in parallel on both sides.

No charge (Fig. 13.37): In the neutral state where $C_{16}H_{32}$ molecules are not charged, α and β electrons at the HOMO level remain in the deep traps of the double bond even when an ultrahigh positive or negative electric field of $E = 2570$ kV/mm is applied.

Positive charge (Fig. 13.38 upper): β electrons at the HOMO level of the $C_{16}H_{32}$ molecule with a double bond are absent and holes are remained there. At the same time, both the adjacent $C_{16}H_{34}$ molecules also lack β electrons. When a positive electric field of $E = 2570$ kV/mm is applied there, the distribution of holes moves to the cathode (lower) side. Conversely, when a negative electric field of $E = -2570$ kV/mm is applied, the distribution of holes moves to the cathode (upper) side.

Negative charge (Fig. 13.38 lower): When an electron is externally introduced to the $C_{16}H_{32}$ molecule with a double bond, α electron is charged to the LUMO level. When a positive electric field of $E = 2570$ kV/mm is applied there, the distribution of α electrons moves to the anode (upper) side. Conversely, when a negative electric field of $E = -2570$ kV/mm is applied, the distribution of α electrons moves to the anode (lower) side. Therefore, when a sufficiently high electric field is applied, the trapped charge is released and the intermolecular transfer of charges can be confirmed. Actually, since the intermolecular transfer of charges is evaluated by probability (see Sect. 13.6), it is not necessary to apply a high electric field ($E =$

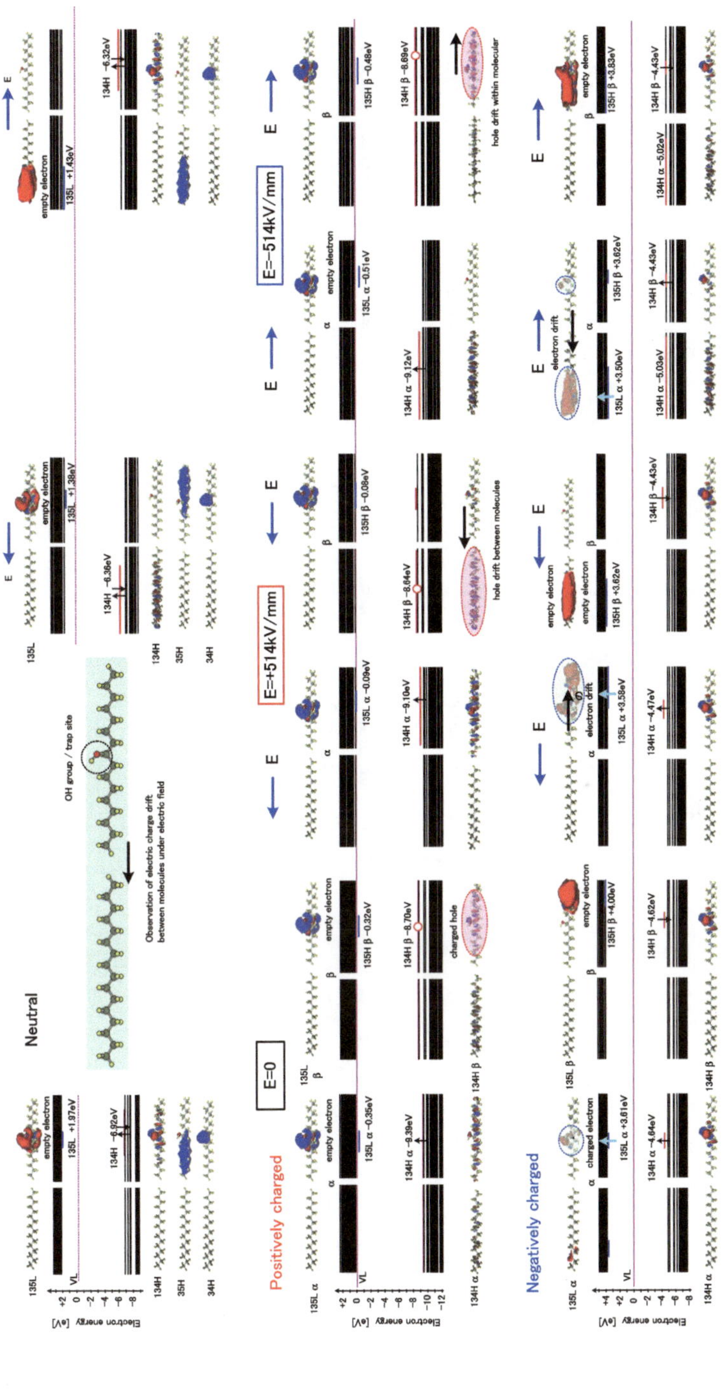

Fig. 13.36 Observation of charge transfer from series intermolecular molecule ($C_{16}H_{33}O$) to molecule ($C_{16}H_{34}$)

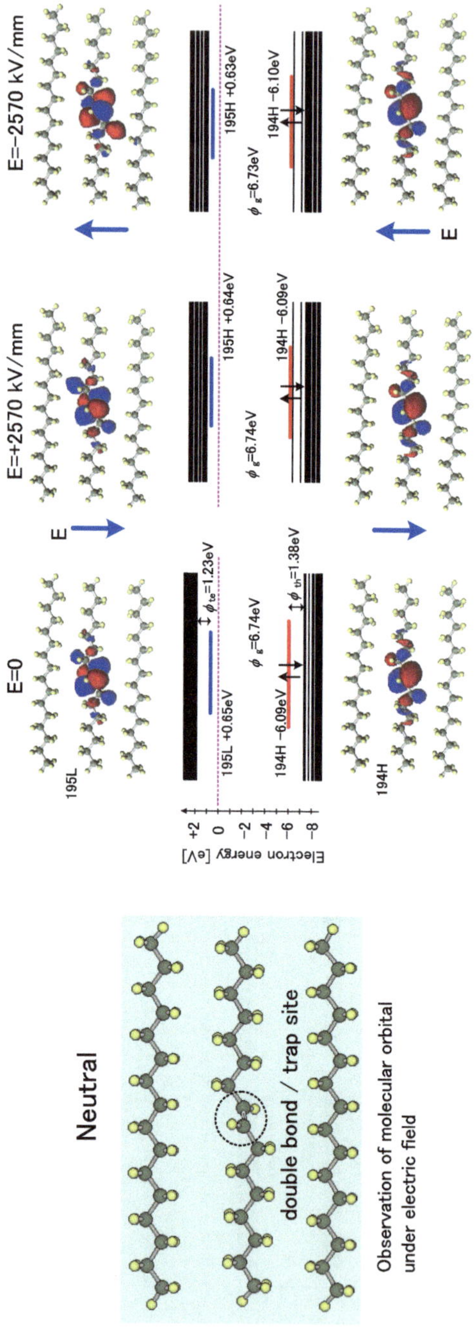

Fig. 13.37 No charge transfer phenomenon between 3 parallel and neutral molecules

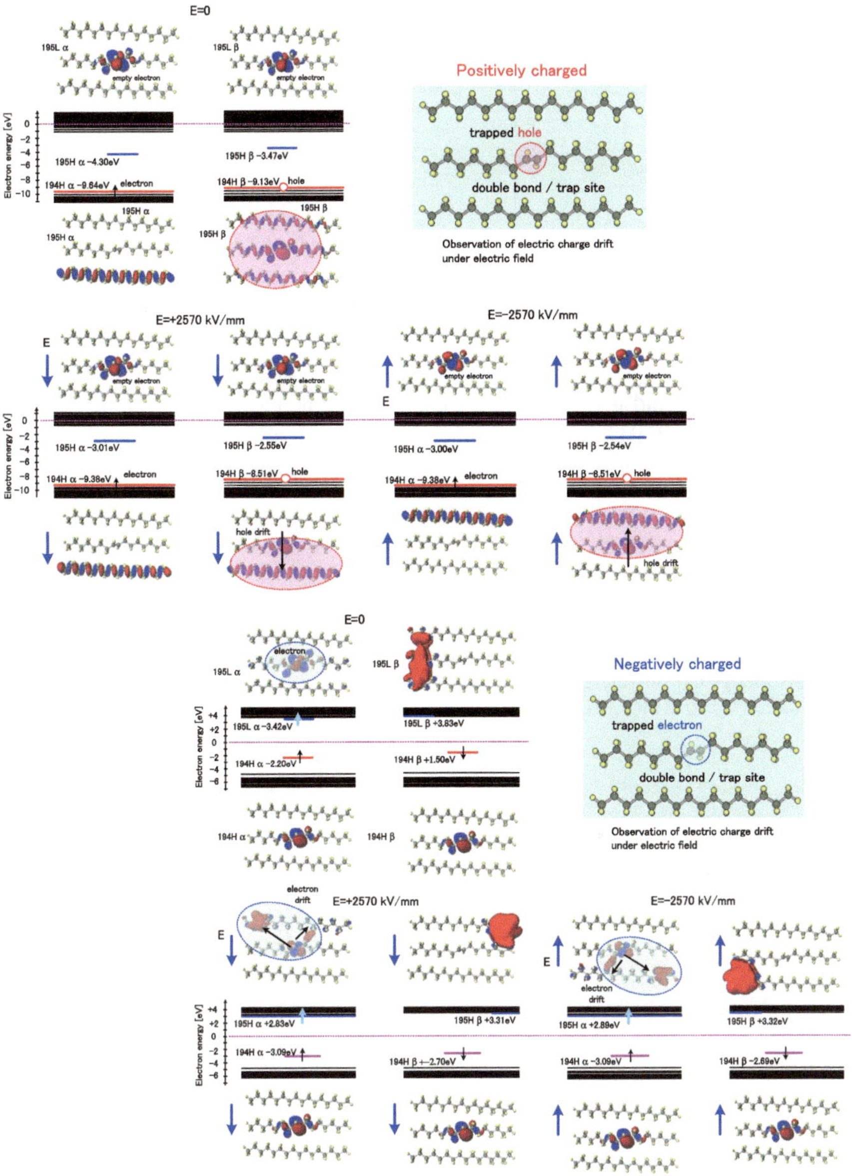

Fig. 13.38 Observation of charge transfer between 3 parallel and charged molecules

2570 kV/mm), and we can think that the intermolecular transfer of charges in a lower electric field can also occur with a probability.

13.10 Summary

In Sect. 13.5, charge hopping transfer is explained based on the probability that charge carriers are released from the trap depth. In parts (3) and (4) of Sect. 13.9, we observe by DFT simulation how charge carriers are released from shallow traps and deep traps under the application of a high electric field. In parts (5) and (6), the state in which the trapped charge carriers migrate to the adjacent molecule under the application of a high electric field is observed by DFT simulation. There remains a problem of comparing these DFT simulation results with the experimental observation results of space charge transfer in an actual polymer insulating material.

Reference

1. W. Wang, T. Takada, Y. Tanaka, S. Li, Trap-controlled charge decay and quantum chemical analysis of charge transfer and trapping in XLPE, *IEEE Transactions on Dielectrics and Electrical Insulation*, 24(5), 3144–3153 (2017).

Chapter 14
Analysis Examples by Quantum Chemical Calculation

14.1 Analysis of Charge Accumulation in PE

(1) Charge Accumulation Characteristics by Q(t) Method

Figure 14.1a, b show the Q(t) data when a DC electric field ranging from E = 10 kV/mm to 100 kV/mm is applied to a low-density polyethylene (LDPE) sample with a thickness of 100 μm at the temperatures of 20 °C and 80 °C, respectively.

Focus on the Q(t) data at the temperature of 20 °C in Fig. 14.1a. At a low electric field of 10 kV/mm, the initial value Q_0 and the value $Q(t = 600\ \mathrm{s})$ after 600 s is equal, and no charge accumulation is observed in the sample. At 20 kV/mm, $Q_0 \geqq Q(t = 600\ \mathrm{s})$ and charge accumulation starts. At an electric field of 50 kV/mm or more, $Q_0 > Q(t = 600\ \mathrm{s})$, and the leakage current becomes dominant in addition to the charge accumulation. Furthermore, at the temperature of 80 °C in Fig. 14.1b, charge accumulation starts at a low electric field of 1 kV/mm. When the electric field is 2 kV/mm or more, $Q_0 > Q(t = 600\ \mathrm{s})$, the leakage current is dominant, and the insulating property is lost.

(2) Charge Ratio $Q(t)/Q_0$

The measurement results of Fig. 14.1a, b are summarized in the relationship between the charge ratio $Q(t = 600\ \mathrm{s})/Q_0$ and the applied electric field (Fig. 14.1c). When the charge ratio is $Q(t = 600\ \mathrm{s})/Q_0 = 1$, it can be determined that the charge exists only on the electrode surface and there is no charge accumulation inside the sample.

When the charge ratio is in the range of $1 < Q(t = 600\ \mathrm{s})/Q_0 < 2$, it can be determined that the charge is accumulated in the sample. Furthermore, when the charge ratio is $2 < Q(t = 600\ \mathrm{s})/Q_0$, it can be determined that the leakage current is dominant and the material loses the insulating property.

For more information on the Q(t) measurement method and charge ratio $Q(t)/Q_0$, see Sect. 3.1, Chap. 3, Part A "Fundamentals and Applications of Q(t) method".

T. Takada et al., *Electric Charge Accumulation in Dielectrics: Measurement and Analysis*,
https://doi.org/10.1007/978-981-19-6156-4_14

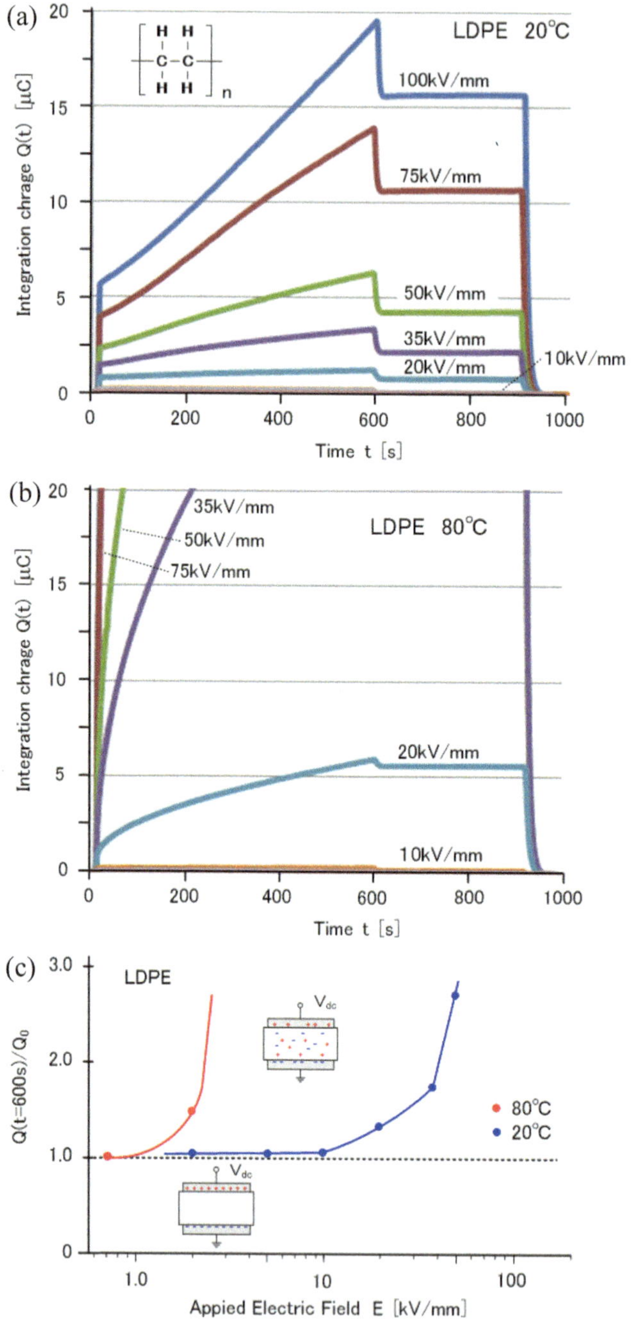

Fig. 14.1 Q(t) results of LDPE sample

(3) Space Charge Distribution Characteristics by PEA Method

Figure 14.2 shows the space charge distribution $\rho(x, t)$ [C/m^3] by the PEA measurement when a DC electric field of $E = 120$ kV/mm is applied to a LDPE sample with a thickness of 100 μm at the temperature of 20 °C. For details of PEA measurement, refer to the Sect. 8.1, Chap. 8, Part B "Fundamentals and Application of Pulsed Electro-Acoustic Method".

Figure 14.2b shows the space charge distribution $\rho(x, t)$ [C/m^3], and Fig. 14.2c shows the internal electric field distribution $E(x, t)$ [V/m]. Figure 14.2a shows the $\rho(x, t)$ distribution in color. Immediately after the application of a DC electric field of 380 kV/mm, a positive charge (red) of $\rho(x, t)$ is injected from the anode and moves toward the cathode. After 20 mins, the electric field between the accumulated positive charges and the cathode reaches a high value of 550 kV/mm, and the LDPE sample shows a dielectric breakdown.

After the breakdown, the remained positive charge $\rho(x, t)$ moves toward the negative image charge on both ground electrodes and disappears after 40 s. The electric field distribution $E(x, t)$ in the sample shows positive and negative polarities toward the ground electrodes, but it also disappears after 40 s.

PEA measurement can observe the time dependence of space charge distribution $\rho(x, t)$ and internal electric field distribution $E(x, t)$, and then we are able to imagine the charge $\rho(x, t)$ and electric field $E(x, t)$ in dielectric insulating materials. It is a good tool to observe the charge behavior spatially and temporally.

(4) Analysis of Charge Injection by Quantum Chemical Calculation

As described in the part (3) above, the result that positive charge is always injected into LDPE from the anode and moves to the cathode is obtained. Then, the question arises: "Why is positive charge injected into LDPE from the anode". Therefore, we investigate this question by Quantum Chemical Calculation/DFT [1, 2].

When the LDPE and the electrode come into contact with each other, an electric double layer of positive charges on the LDPE surface and negative charges on the electrode surface is formed (Fig. 14.3e). Therefore, the positive charges exist on the LDPE surface only when contacting it with the electrode. When a DC voltage is applied there, the positive charges on the LDPE surface can move toward the cathode. The energy level diagrams of the LDPE and the electrode are calculated by DFT calculation to explain the formation of the electric double layer.

Figure 14.3 shows the energy level diagram and the measurement results of $\rho(x, t)$ when the LDPE sample is sandwiched between SC electrode (semiconductor) and Al electrode (metal) with an application of DC voltage. Figure 14.3a shows the positive voltage (+120 kV/mm) applied to the SC electrode (semiconductor), and Fig. 14.3b shows the negative voltage (−120 kV/mm) applied at $\rho(x, t)$. From the measurement result, it has been confirmed that positive charges ($+\rho(x, t)$) are always injected from the anode. Figure 14.3c depicts the energy level diagram when the electrodes (SC and Al) and LDPE are separated and independent. The work function of Al electrode is $\phi_{WM}(\text{Al}) = 4.25$ eV, and that of SC electrode is $\phi_{WM}(\text{SC}) = 5.55$ eV. For LDPE, the HOMO level is −8.88 eV, the LUMO level is +1.88 eV, and the Fermi level is

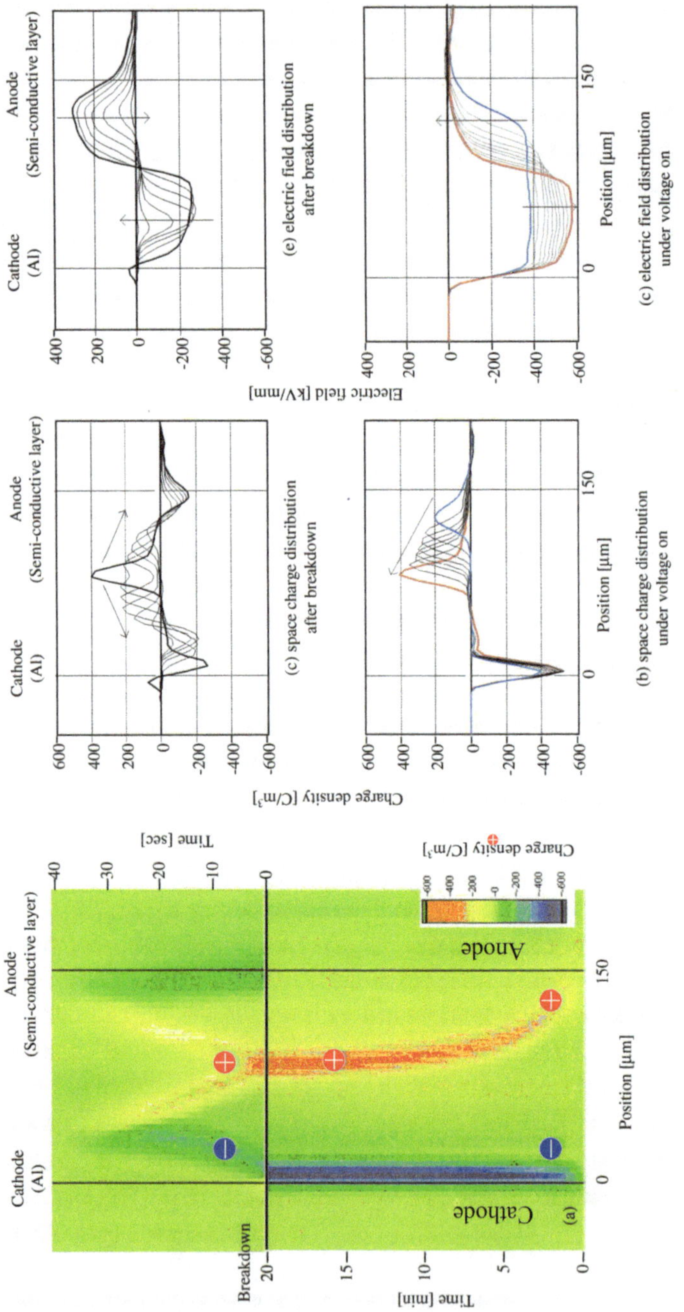

Fig. 14.2 PEA results of LDPE sample

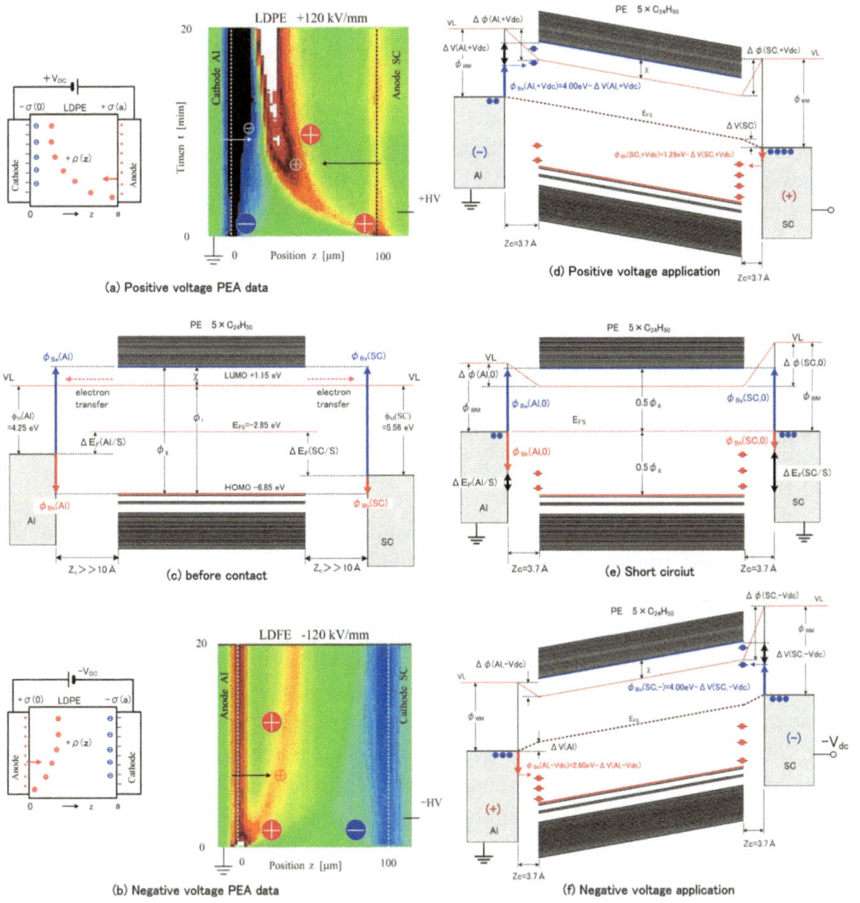

Fig. 14.3 The energy level diagram and measurement results of the LDPE sample

$E_{FS} = -2.85$ eV calculated by Quantum Chemical Calculation. In the calculation, LDPE is a molecular model in which five linear $C_{24}H_{50}$ molecules are arranged in parallel.

Figure 14.3e is the energy level diagram before the application of electric field (closed circuit) when the electrodes (SC and Al) are in contact with LDPE. The contact means that the electrode and LDPE are close to each other until $Z_c = 3.7$ Å. In this state, the Fermi levels of the electrode and LDPE match. In the process of matching the Fermi level, the electrons are transferred from the LDPE with a high Fermi level ($E_{FS} = -2.85$ eV) toward the low electrodes ($\phi_{WM}(Al) = 4.25$ eV and $\phi_{WM}(SC) = 5.55$ eV). As a result, the LDPE surface is charged with positive charges (holes) and the electrode surface is charged with negative ones (electrons), and an electric double layer is formed. For details of electric double layer, see Sect. 12.7.

Figure 14.3d is the energy level diagram when a positive voltage is applied to the SC electrode and Fig. 14.3f corresponds to an applied negative voltage. Since the electron energy is drawn positively (upward), the shape of the positive electrode is downward and that of the negative one is upward. Next, consider the charge movement. In the case of Fig. 14.3d in which a positive voltage is applied, positive charges (holes) on the LDPE surface charged on the SC electrode side move in the sample toward the cathode. The $+\rho(x, t)$ in Fig. 14.3a also moves from the positive electrode (SC) to the negative electrode (Al) under the positive voltage application, which is consistent with the model in Fig. 14.3d. Similarly, in Fig. 14.3f, even when a negative voltage is applied to the SC electrode, the positive charges (holes) charged on the LDPE surface of the Al electrode side move in the sample toward the cathode (SC). The $+\rho(x, t)$ in Fig. 14.3b also moves from the positive electrode to the negative electrode under negative voltage application, which is consistent with the model in Fig. 14.3f.

(5) **Charge Transfer Analysis**

In the previous part (4), charge injection from electrodes is examined. This part (5) considers the movement of carriers. By simulating the characteristics of low-density polyethylene, the LDPE molecule has a structure with two $C = 10$ side chains in a linear PE molecule with $C = 32$ carbons. The spatial distribution of energy level of this LDPE molecule and the distribution of molecular orbitals of HOMO level and LUMO level are obtained from DFT calculation, as shown in Fig. 14.4.

The molecular orbital of LUMO level in Fig. 14.4 exists at a localized location, and the location becomes the charge trap site. Since such a trap site exists above the LUMO level, the trap site is represented by a horizontal bar line. The electron trap depth is spatially distributed in the range of $\phi_{te}(x, y, z) = 0.1$–0.6 eV. Similarly, below the HOMO level, the hole trap depth is also spatially distributed within the range of $\phi_{th}(x, y, z) = 0.1$–0.6 eV. See Sect. 13.1 for details of molecular orbitals and charge trap sites. In the LDPE molecule, both electron and hole carriers can move easily.

Now, let us look at the formation of the electric double layer described in the part (4) for the amorphous LDPE molecule with a side chain structure. Figure 14.5a shows the energy levels when the amorphous LDPE molecule and the electrode are sufficiently separated. Figure 14.5b shows the levels when the LDPE molecule is contact with the electrode. As in the case of Fig. 14.3, an electric double layer is formed at the interface, and the surface of the LDPE molecule is charged with positive charges (holes). As described in the current law of Sect. 13.5, in the LDPE molecule, both electrons and holes can easily move, but the current is regulated by charge injection from the electrode described in the part (4).

From the PEA data of PE in Fig. 14.2, it is found that when a high electric field of 100 kV/mm is applied at room temperature (20 °C), positive charges are injected from the anode and move in the sample toward the cathode. The injection mechanism of electric charge is analyzed by the DFT calculation of electric double layer in Figs. 14.3 and 14.5. Furthermore, the trap depth of charge transfer is calculated by DFT calculation, and it is possible to analyze that positive charges can move within

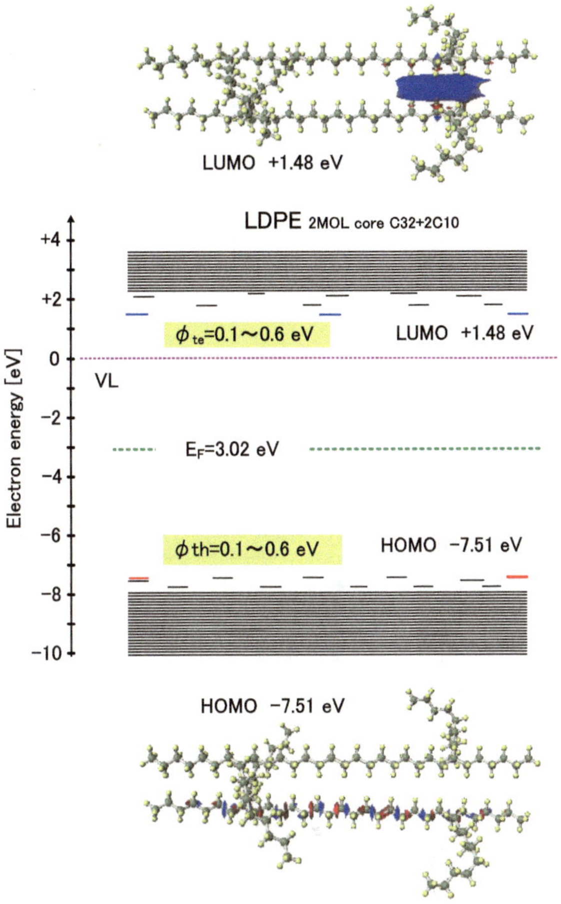

Fig. 14.4 The energy level of LDPE molecule and molecular orbitals

the measurement time in Fig. 14.4. Both the Q(t) data in Fig. 14.1 and PEA data in Fig. 14.2 have the same charge characteristics, and they are in agreement with the DFT calculation analysis.

Summary: It is clear that the charge accumulation studies can be analyzed from Q(t) measurements, PEA measurements and Quantum Chemical/DFT calculations.

14.2 Analysis of Charge Accumulation in PET

(1) Charge Accumulation Characteristics by Q(t) Method

Figure 14.6 shows the Q(t) data when a DC electric field ranging from $E = 10$ kV/mm to 100 kV/mm is applied to a Polyethylene-terephthalate (PET) sample with the thickness of 124 μm under the temperature of 20 °C and 80 °C. In Fig. 14.6a, the

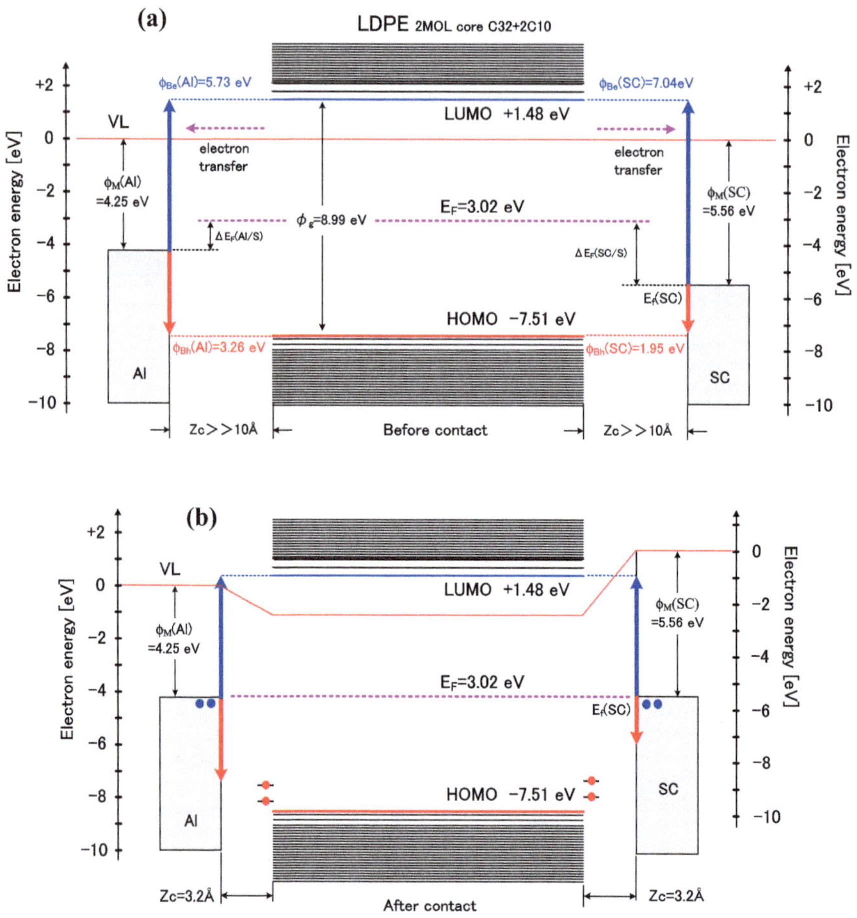

Fig. 14.5 The energy levels under different contact ways of LDPE sample and electrodes

initial charge value Q_0 and the value $Q(t = 600\ \text{s})$ after 600 s shows $Q_0 = Q(t = 600\ \text{s})$ over the whole electric field range at the temperature of 20 °C. No charge accumulation is observed. At a temperature of 80 °C in Fig. 14.6b, $Q_0 \geqq Q(t = 600\ \text{s})$ is found from the electric field of 35 kV/mm, and some charge accumulation appears.

(2) Charge Ratio $Q(t)/Q_0$

Figure 14.6c shows the results of 14.6a and b arranged into the relationship between the charge ratio $Q(t = 600\ \text{s})/Q_0$ and the applied electric field.

When the charge ratio is $Q(t = 600\ \text{s})/Q_0 = 1$, it can be determined that the charge exists only on the electrode and there is no charge accumulation in the sample. Further, when the charge ratio is in the range of $1 < Q(t = 600\ \text{s})/Q_0 < 2$, it can be determined that the charge is accumulated in the sample. Therefore, in the PET sample, at the

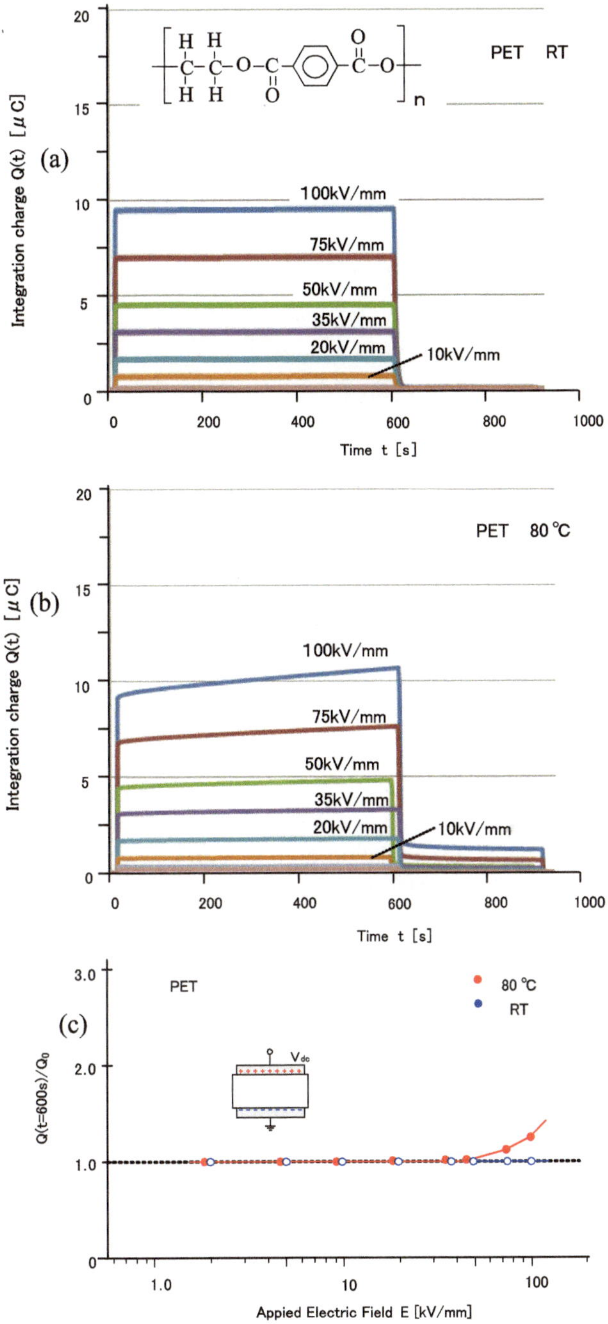

Fig. 14.6 Q(t) results of PET sample

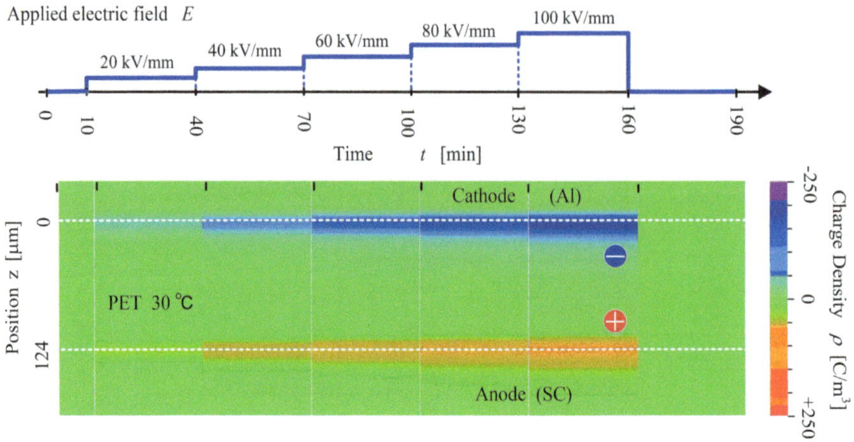

Fig. 14.7 PEA results of PET sample

temperature of 80 °C and the electric field of 100 kV/mm, the charge accumulation of about $Q(t = 600\text{ s})/Q_0 = 1.1$ appears.

(3) Space Charge Distribution Characteristics by PEA Method

Figure 14.7 shows a color display of the space charge distribution $\rho(x, t)$ [C/m^3] when a DC electric field of $E = 20$ to 100 kV/mm is applied to a PET sample with the thickness of 124 μm at a temperature of 30 °C. No significant charge accumulation is observed even when an electric field of 100 kV/mm is applied. The accumulation of some negative charges is observed. It can be said that PET is an excellent insulating material with less charge accumulation even at high temperature and high electric fields, compared with PE introduced in Sect. 14.1.

The PEA method at the temperature of 80 °C is currently difficult to measure the correct charge distribution due to the lack of the detection element of pressure waves and attenuation of pressure waves at high temperature.

(4) Analysis of Charge Injection by DFT Calculation

The PEA data in Fig. 14.7 are obtained under the system of negative Al electrode/PET/positive SC electrode. At a high electric field of 80 kV/mm, negative charge accumulation is observed from the negative Al electrode. In order to consider this negative charge accumulation, the energy level diagram is calculated by DFT. The upper part of Fig. 14.8 is the energy level diagram before the contact between the PET and the electrode, and the lower part is the diagram after the contact. After the contact, an electric double layer is formed at the interface between the electrode and PET. The PET surface on the negative electrode side is charged with negative charges (electrons), and the surface on the positive electrode side is charged with positive charges (holes). Indeed, the electron injection barrier of $\phi_{\mathrm{Be}}(\mathrm{Al}) = 1.72$ eV of the negative electrode is lower than the hole injection barrier of $\phi_{\mathrm{Bh}}(\mathrm{SC}) = 1.97$ eV of

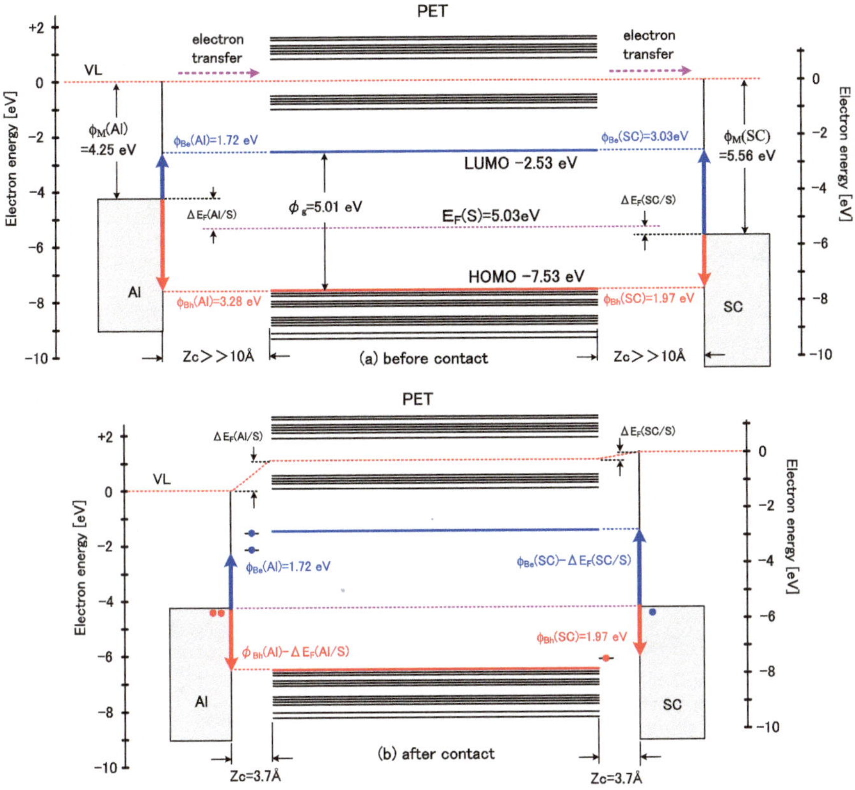

Fig. 14.8 The energy levels under different contact ways of PET sample and electrodes

the positive electrode. Further, the charge accumulation requires the consideration of both the injection barrier and the trap depth for charge transfer hopping. Next, consider the trap depth.

(5) Analysis of Charge Transfer by Quantum Chemical Calculation

Here, the trap depth of charge transfer is examined. Figure 14.9 shows the DFT calculation results of the PET energy levels. PET is a polymer molecule with ethylene and terephthalic acid. The intervals of LUMO and HOMO in the ethylene part are wide, and that of the terephthalic acid part is narrow. As a result, the LUMO level of terephthalic acid becomes an electron trap site and the HOMO level becomes a hole trap site. The electron trap depth ϕ_{te} is deeper than the hole trap depth ϕ_{th}. Therefore, the movement of electrons becomes more difficult than that of holes. The explanation of negative charge accumulation in Fig. 14.7 becomes difficult. This is a phenomenon observed in a high electric field with a high temperature. Therefore, the details of this negative charge accumulation are explained by the current law in the part (4) of Sect. 13.6.

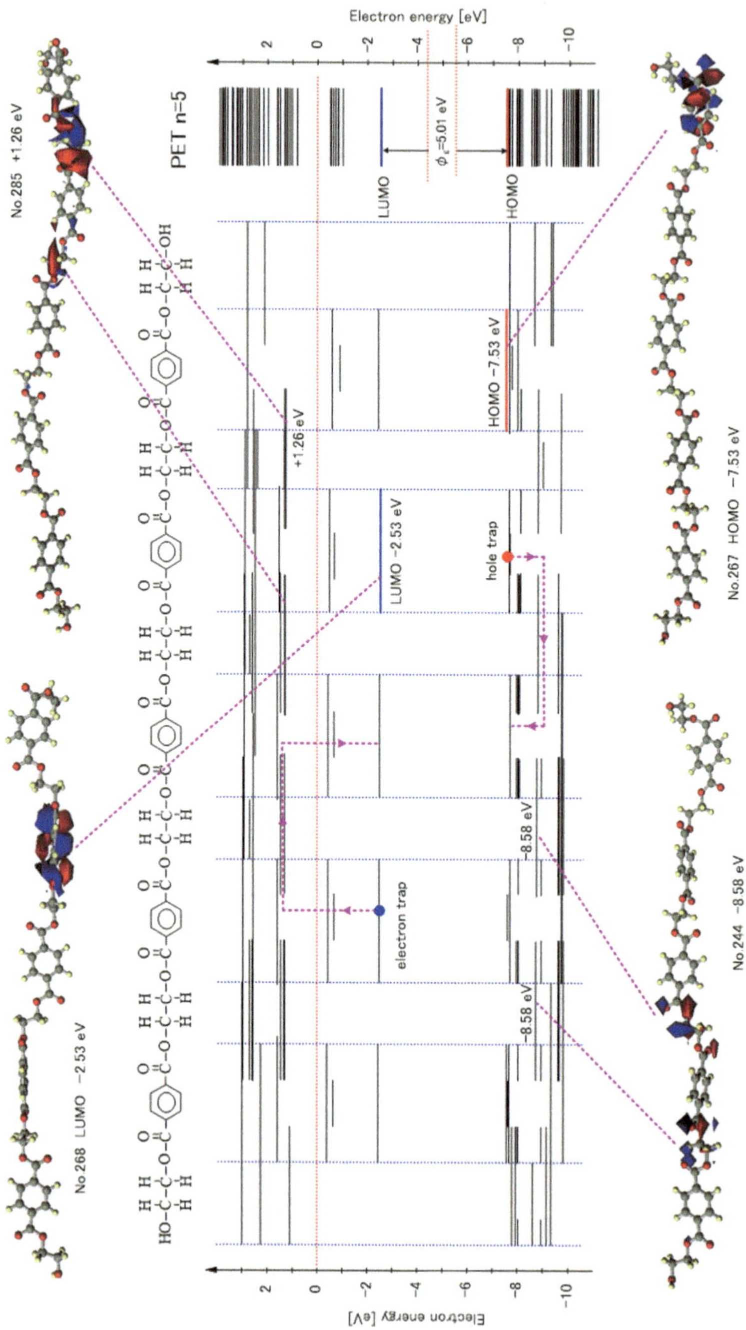

Fig. 14.9 The DFT calculation results of the PET energy levels

Future research issues: It is necessary to obtain the PEA data at higher temperatures (difficult to measure by PEA) and higher electric fields, which causes a significant difference in the accumulation of positive and negative charges. Therefore, it is desirable to carefully observe the Q(t) measurement under higher electric fields and higher temperatures and to analyze the data with a significant difference in the charge accumulation by the Quantum Chemical Calculation.

14.3 Analysis of PI Charge Accumulation

(1) Charge Accumulation Characteristics by Q(t) Method

Figure 14.10a, b shows the Q(t) data when a DC electric field ranging from $E =$ 2.5 kV/mm to 100 kV/mm is applied to a Polyimide (PI/Kapton) sample with the thickness of 126 mm at the temperatures of 20 °C and 80 °C.

Focus on the Q(t) data at the temperature of 20 °C in Fig. 14.10a. At a low electric field of 10 kV/mm, the initial value Q_0 and the value of $Q(t = 600$ s) after 600 s shows $Q_0 = Q(t = 600$ s), and no charge accumulation is observed in the sample. At 100 kV/mm, $Q(t = 600$ s)$\geqq Q_0$ and charge accumulation starts. Furthermore, at the temperature of 80 °C in Fig. 14.10b, no charge accumulation is observed at 20 kV/mm or less, but at 50 kV/mm or more, $Q(t = 600$ s) $\geqq$ Q_0 and charge accumulation is observed. The PI sample has better charge accumulation characteristics than the LDPE sample introduced in Sect. 12.1.

(2) Charge Ratio $Q(t)/Q_0$

Figure 14.10c shows the results of Fig. 14.10a, b arranged into the relationship between the charge ratio $Q(t = 600$ s)/Q_0 and the applied electric field. When the charge ratio is $Q(t = 600$ s)/$Q_0 = 1$, it can be determined that the charge exists only on the electrode and there is no charge accumulation in the sample. When the charge ratio is in the range of $1 < Q(t = 600$ s)/$Q_0 < 2$, it can be determined that the charge is accumulated in the sample.

(3) Space Charge Distribution Characteristics by PEA Method

The space charge distribution $\rho(x, t)$ [C/m^3] by PEA measurement is shown in Fig. 14.11 [2], in which a DC electric field ranging from $E = 30$ kV/mm to 120 kV/mm is applied to a PI sample with the thickness of 126 μm at a temperature of 80 °C.

When a DC electric field of 60 kV/mm and 80 kV/mm is applied, positive charges (red) are injected from the anode and negative charges (blue) are injected from the cathode. They remain near the electrodes. That is, a homo charge accumulation is formed. Like LDPE in Sect. 14.1, they do not move to the inside of the sample. Furthermore, when a high electric field of 120 kV/mm is applied, electron-hole pairs are generated from the inside of the sample, neutralizing the already formed homo-charges. The electrons (negative charges) move toward the anode side and holes (positive charges) are on the cathode side, that is, they form a hetero-charge. After

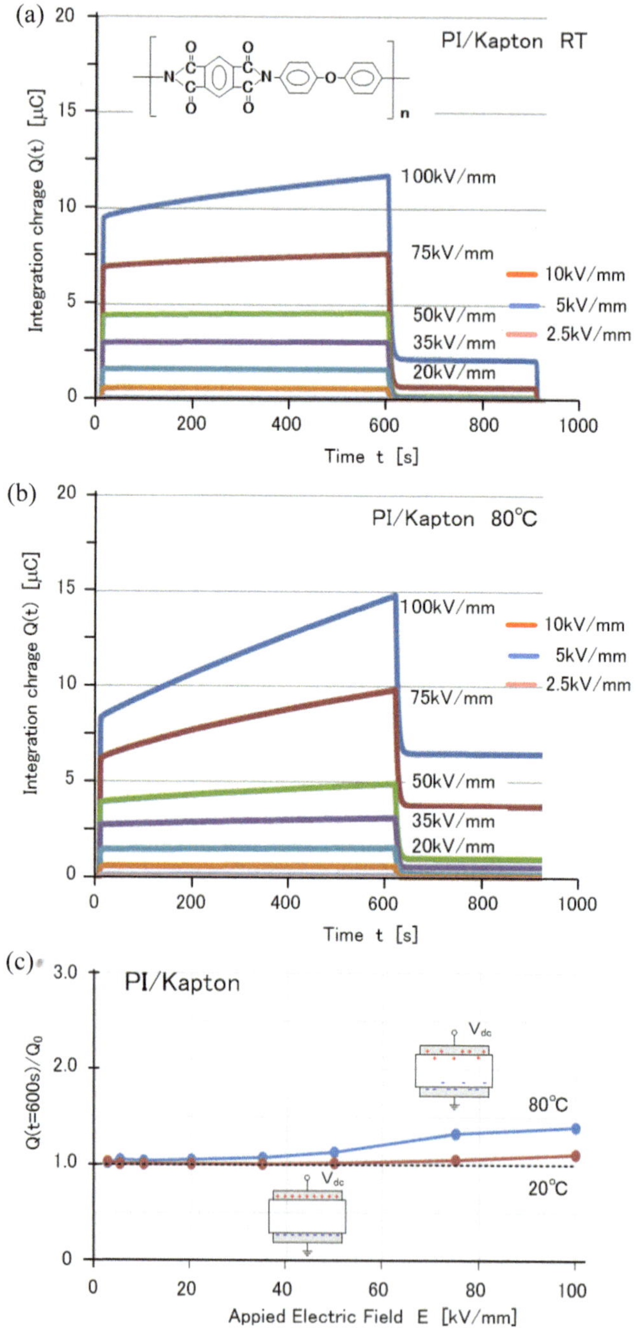

Fig. 14.10 Q(t) results of PI sample

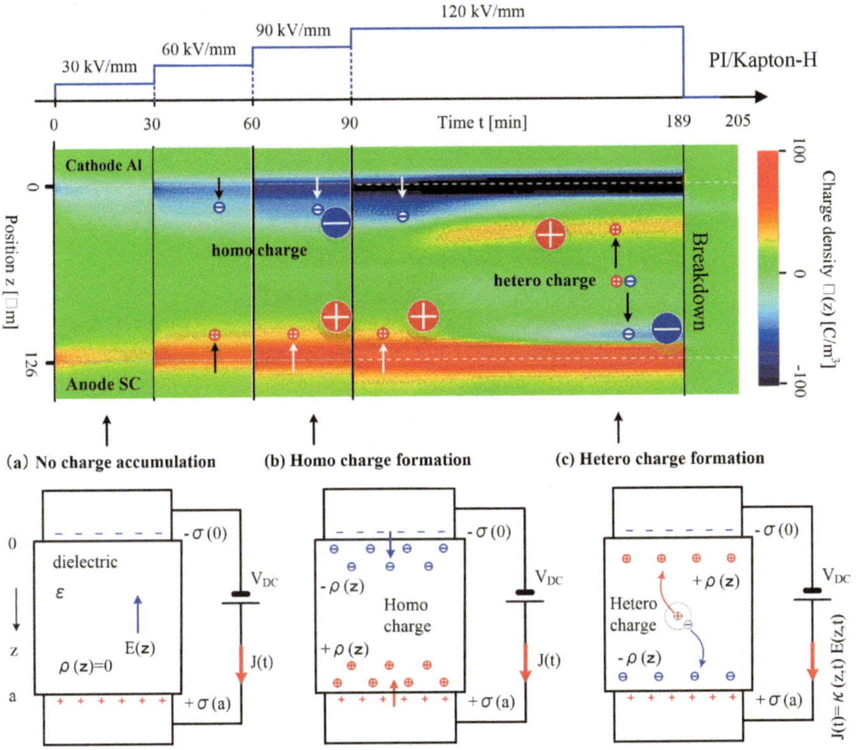

Fig. 14.11 PEA results of PI sample

99 min (the elapsed whole time is 189 min) after applying a high electric field of 120 kV/mm, the dielectric breakdown happens.

It is observed from the PEA data that positive hetero charge moves toward the cathode. After the time of 189 min, the electric field between the accumulated positive charges and the cathode reaches the breakdown electric field, and the PI sample is then broken down. After the breakdown, the remained positive charge $\rho(x, t)$ moves toward the negative image charge on both ground electrodes and disappears. For the electric field distribution $E(x, t)$ in the sample, a positive electric field $+E(x, t)$ and a negative electric field—$E(x, t)$ are generated toward the ground electrodes, but they also disappear finally.

(4) Analysis of Charge Generation and Transfer by DFT Calculation

The $\rho(x, t)$ [C/m^3] data of PI in Fig. 14.11 is obtained under the system of negative Al electrode/PI/ positive SC electrode. At a high electric field of 60 kV/mm, negative charges are injected from the negative Al electrode and positive charges are injected from the positive electrode. In order to consider the injection from the negative and positive electrodes, the energy level diagram is calculated by DFT.

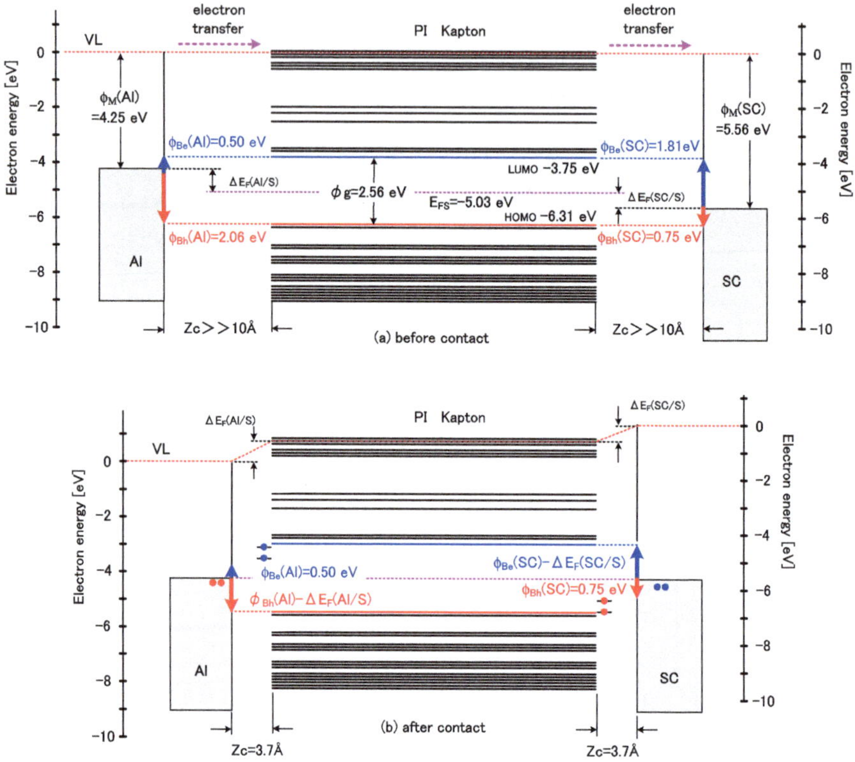

Fig. 14.12 The energy levels under different contact ways of PI sample and electrodes

The upper part of Fig. 14.12 is the energy level diagram before the contact between the PI and the electrode, and the lower part of Fig. 14.12 is the diagram after the contact. After the contact, negative charges exist on the PI surface of the negative Al electrode side (Fig. 14.1). Moreover, the electron injection barrier of the negative Al electrode is as low as $\phi_{\mathrm{Be}}(\mathrm{Al}) = 0.50$ eV. Then, under a high electric field, the negative charges on the PI surface start moving toward the anode. Positive charges already exist on the PI surface on the opposite SC electrode side. The hole injection barrier of positive SC electrode (Fig. 14.1) is as low as $\phi_{\mathrm{Be}}(\mathrm{SC}) = 0.75$ eV. Similarly, the positive charge on the PI surface starts to move toward the cathode under a high electric field.

Next, consider the hopping transfer of negative charges and positive charges. Figure 14.13 shows the result of the DFT calculation of the trap depth of the charge transfer. The trap depth of negative electrons is $\phi_{\mathrm{te}} = 1.26$ eV, and the depth of positive holes is $\phi_{\mathrm{th}} = 0.72$ eV. By comparing the electron injection barrier of $\phi_{\mathrm{Be}}(\mathrm{Al}) = 0.50$ eV and trap depth of $\phi_{\mathrm{te}} = 1.26$ V, electrons are regulated by hopping movement. In addition, from the comparison of hole injection barrier of $\phi_{\mathrm{Be}}(\mathrm{Al}) = 0.75$ eV and trap depth of $\phi_{\mathrm{th}} = 0.72$ eV, holes are regulated by both injection and hopping movement.

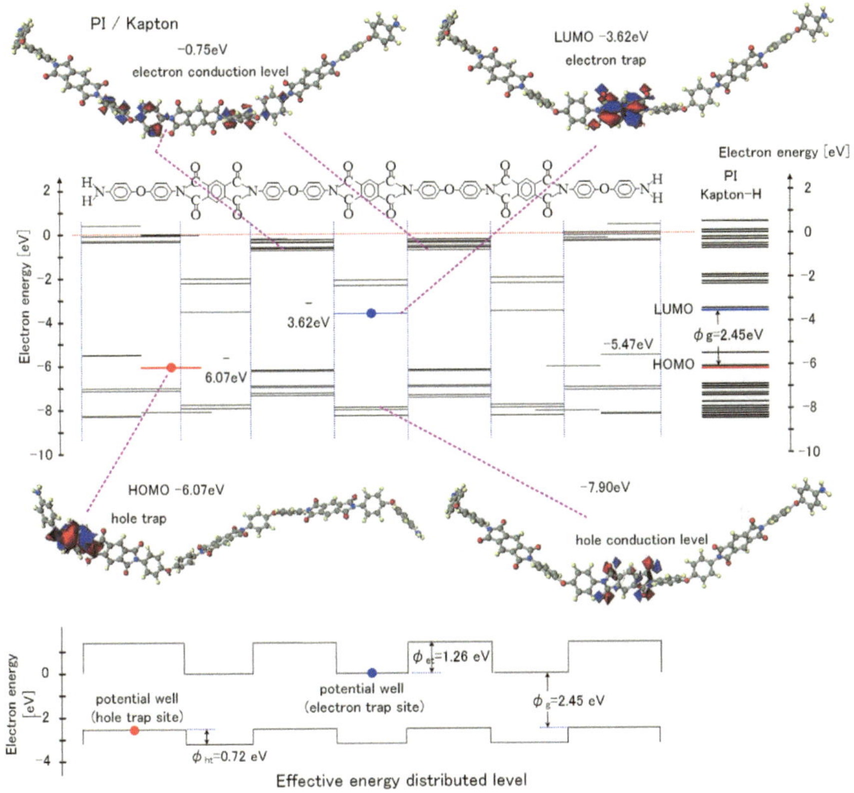

Fig. 14.13 The results of the DFT calculation of PI sample

(5) Visible Light Irradiation and Charge Generation

PI/Kapton-H film is dark brown and semi-transparent. Long wavelength (Red) can be transmitted, but short wavelength (Violet) is opaque. Opacity is the loss of light energy due to light absorption and electron–hole pair generation in Kapton-H.

<See Takada Text Vol. III, Part 4, Chap. 21>

PEA measurement under light irradiation (Fig. 14.14): PI/Kapton-H film (band gap: $\phi_g = 2.45$ eV) is guided by visible light (Red: ~2.0 eV; Voilet: ~3.1 eV) with an optical fiber [3]. Observe the space charge formation process within 10 mins by this PEA method under a DC electric field of 30 kV/mm. In this case, Al material is used for the ground electrode and ITO film is used for the high voltage electrode.

Charge accumulation data of Kapton-H under visible light irradiation:

Irradiation with a bright light wavelength ($h\nu = 2.46$ eV) (the upper part of Fig. 14.14): The electron-hole pairs are uniformly generated in the bulk. Negative charges (electrons) on the anode side and positive charges (holes) on the cathode side are observed. However, it is easy to inject charges from the ITO electrode and

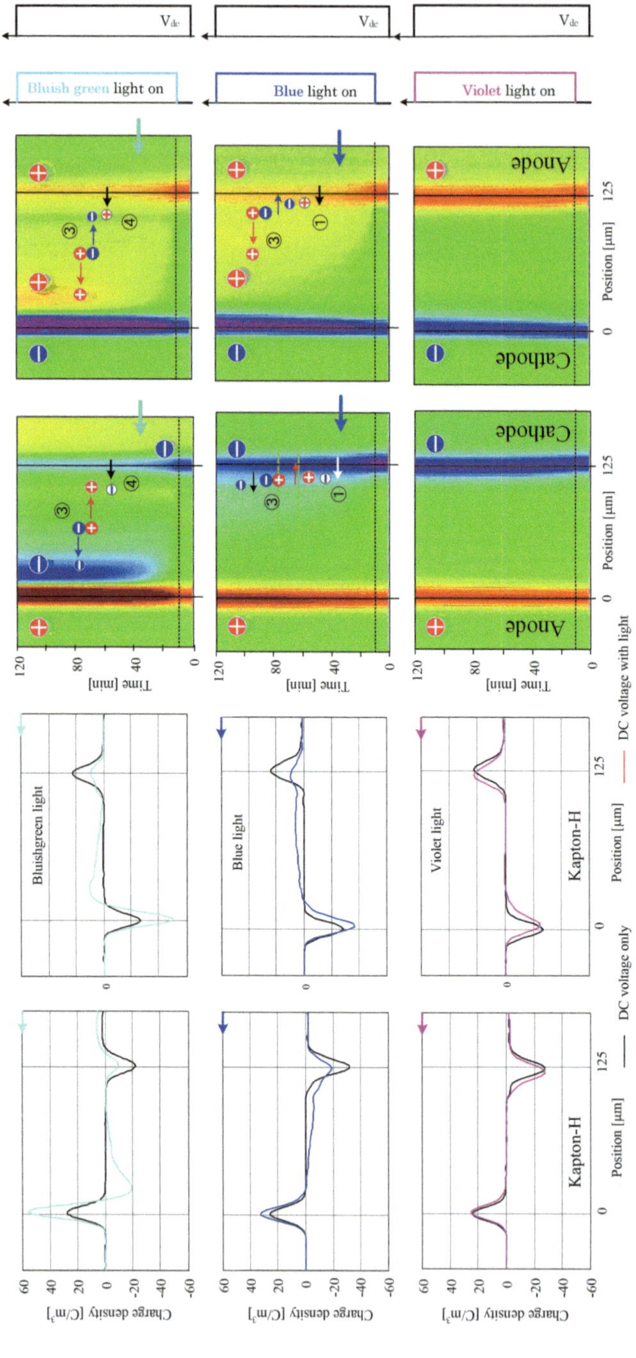

Fig. 14.14 Charge accumulation data of Kapton-H under visible light irradiation

they are easily neutralized with the transferred charges. Since the energy of Bulish light ($hv = 2.46$ eV) is about the same as the band gap of Kapton-H ($\phi_g = 2.28$ eV), electron-hole pair generation occurs.

Irradiation with Blue light wavelength ($hv = 2.64$ eV) (the middle part of Fig. 14.14): The absorption of short wavelength light becomes remarkable, and this light absorption occurs near the sample surface. The electron-hole pairs are generated near the surface. More electron-hole pair generation occurs on the surface of the irradiation side than that of the irradiation of Bulish light.

Irradiation with Violet light wavelength ($hv = 3.10$ eV) (the lower part of Fig. 14.14): The absorption of shorter wavelength light becomes more significant, and this light cannot reach the inside of the sample. Therefore, since the electron-hole pair generation occurs near the ITO electrode, no charge accumulation is observed.

Summary: It is clear that the charge accumulation studies can be analyzed from Q(t) measurements, PEA measurements and Quantum Chemical/DFT calculations.

14.4 Charge Accumulation Characteristics of PEN

(1) Charge Accumulation Characteristics by Q(t) Method

Figure 14.15a, b show the Q(t) data when a DC electric field ranging from $E =$ 2.5 kV/mm to 100 kV/mm is applied to a Polyethylen-naphtharate (PEN) sample with the thickness of 128 μm under the temperature of 20 °C and 80 °C. Both the Q(t) data at temperatures of 20 °C and 80 °C show $Q_0 = Q(t = 600\text{ s})$, and no charge accumulation is observed in the sample. Similarly, the charge ratio in Fig. 14.15c also shows $Q(t = 600\text{ s})/Q_0 = 1$, and no charge is accumulated in the sample.

Compared with LDPE in Sect. 14.1 and PI/Kapton in Sect. 14.3, we can see that PEN is a dielectric with excellent properties.

(2) Space Charge Distribution Characteristics by PEA Method

Figure 14.16 shows the $\rho(x, t)$ data of PEA at a temperature of 30 °C and a DC electric field ranging from 20 to 100 kV/mm applied to PEN with the thickness of 128 μm. Only positive charge (red) and negative charge (blue) are observed on the electrodes and no charge accumulation (green) is observed in the sample. This result shows that no charge is accumulated in the sample, as in the Q(t) data in the part (1).

(3) Challenges of PEA Measurement

There is a desire to obtain PEA data at higher temperatures. However, at this stage, there is no piezoelectric element that can detect nanosecond pressure waves at high temperature, and it is difficult to measure $\rho(x, t)$ correctly because the viscoelastic property of the sample changes at high temperature. In contrast, the Q(t) measurement can record the charge accumulation measurement at high temperatures (up to 200 °C). Therefore, it is recommended to acquire PEA data at low temperatures and Q(t) data over the entire temperature range.

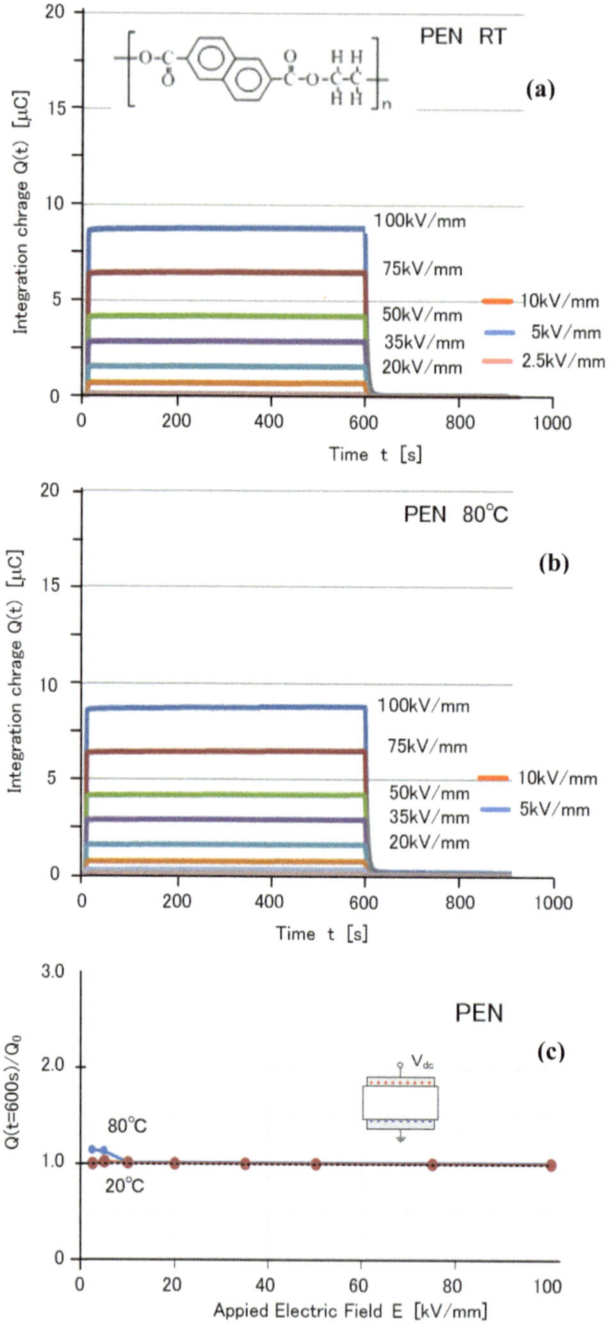

Fig. 14.15 Q(t) results of PEN sample

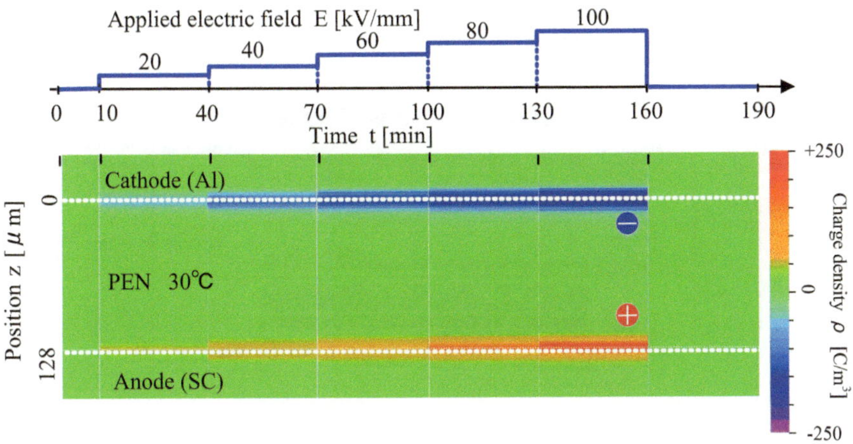

Fig. 14.16 PEA results of PEN sample

(4) **Contact Charging Between PEN and Electrode**

No charge accumulation is observed in the PEN sample under high temperature (80 °C) and high electric field (100 kV/mm) application. The reason is considered from the properties of the energy level of PEN. Figure 14.17 shows the energy levels of PEN used for the DFT calculation of Quantum Chemical Calculation.

Figure 14.17 is a model in which an electric double layer is formed by the contact between the PEN and an electrode. Since the Fermi level of PEN is $E_F(S) = 4.74$ eV and the work function of the Al electrode is $\phi_M(Al) = 4.25$ eV, it shows $_M(Al) > E_F(S)$. Therefore, at the time of contact, electrons move from the Al electrode to PEN, and the PEN surface is negatively charged (electrons). On the other hand, the work function of the SC electrode is $\phi_M(SC) = 5.56$ eV, so $\phi_M(Al) < E_F(S)$. Therefore, at the time of contact, electrons move from the PEN to the SC electrode, and the PEN surface is positively charged (holes). Therefore, the contact between the PEN and the electrode forms an electric double layer in which electrons (negatively charged) are present at the PEN surface on the Al electrode side and holes (positively charged) are present at the PEN surface on the SC electrode side.

(5) **Hopping Movement in PEN**

Next, we discuss the trap depth at which contact-charged electrons and holes on the PEN surface hop through the sample. The energy levels obtained by the DFT calculation are shown in Fig. 14.18a. The LUMO level is −2.93 eV, the HOMO level is −6.54 eV, and the band gap $\phi_g = 3.61$ eV. From the relationship between each energy level and molecular orbital (electron wave), the energy level distribution for each binding molecule is shown in Fig. 14.18b. Also, the effective trap level distribution is summarized in Fig. 14.18c. As a result, the effective electron trap depth is about $\phi_{te} = 1.26$ eV, and the effective hole trap depth is about $\phi_{th} = 0.72$ eV. Here, the law of charge accumulation is considered. The electron injection barrier

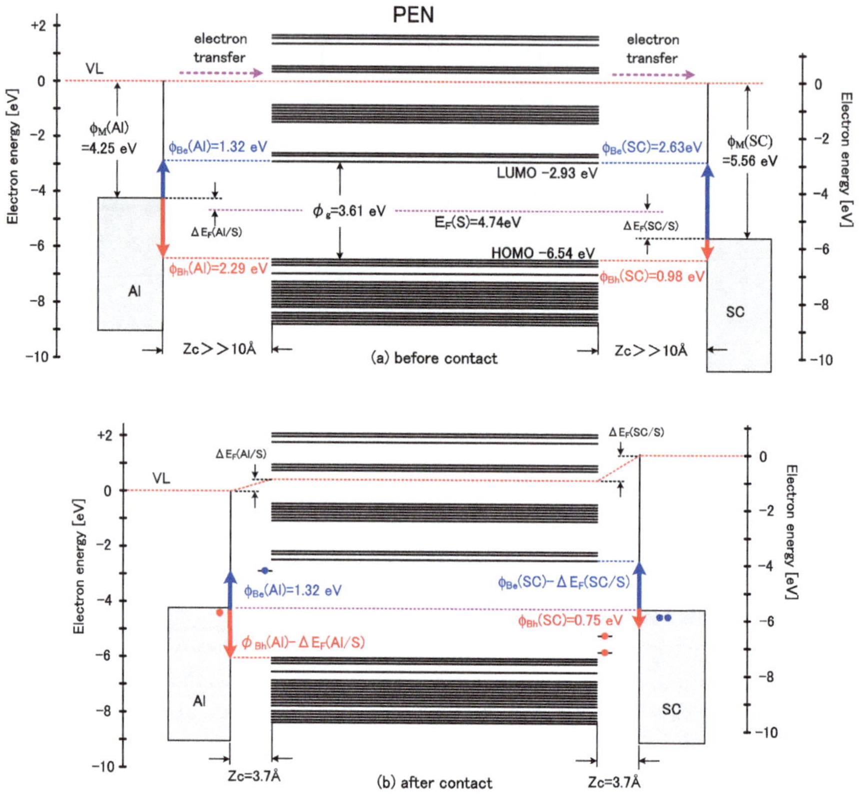

Fig. 14.17 The energy levels under different contact ways of PEN sample and electrodes

from the Al cathode is $\phi_{\mathrm{Be}}(\mathrm{Al}) = 1.32$ eV, and the effective electron trap depth is about $\phi_{\mathrm{te}} = 1.26$ eV. Thus, negative charge accumulation (electrons) is regulated by the injection barrier. On the other hand, the hole injection barrier is $\phi_{\mathrm{Bh}}(\mathrm{Al}) = 0.75$ eV and the effective trap depth is about $\phi_{\mathrm{th}} = 0.72$ eV from the positive SC electrode. Thus, the injection barrier is slightly larger, which regulates the positive charge accumulation. In this way, the charge accumulation of PEN can be evaluated, and it can be determined that it is a material in which charge accumulation is difficult.

Summary: It is clear that the charge accumulation studies can be analyzed from Q(t) measurements, PEA measurements and Quantum Chemical/DFT calculations.

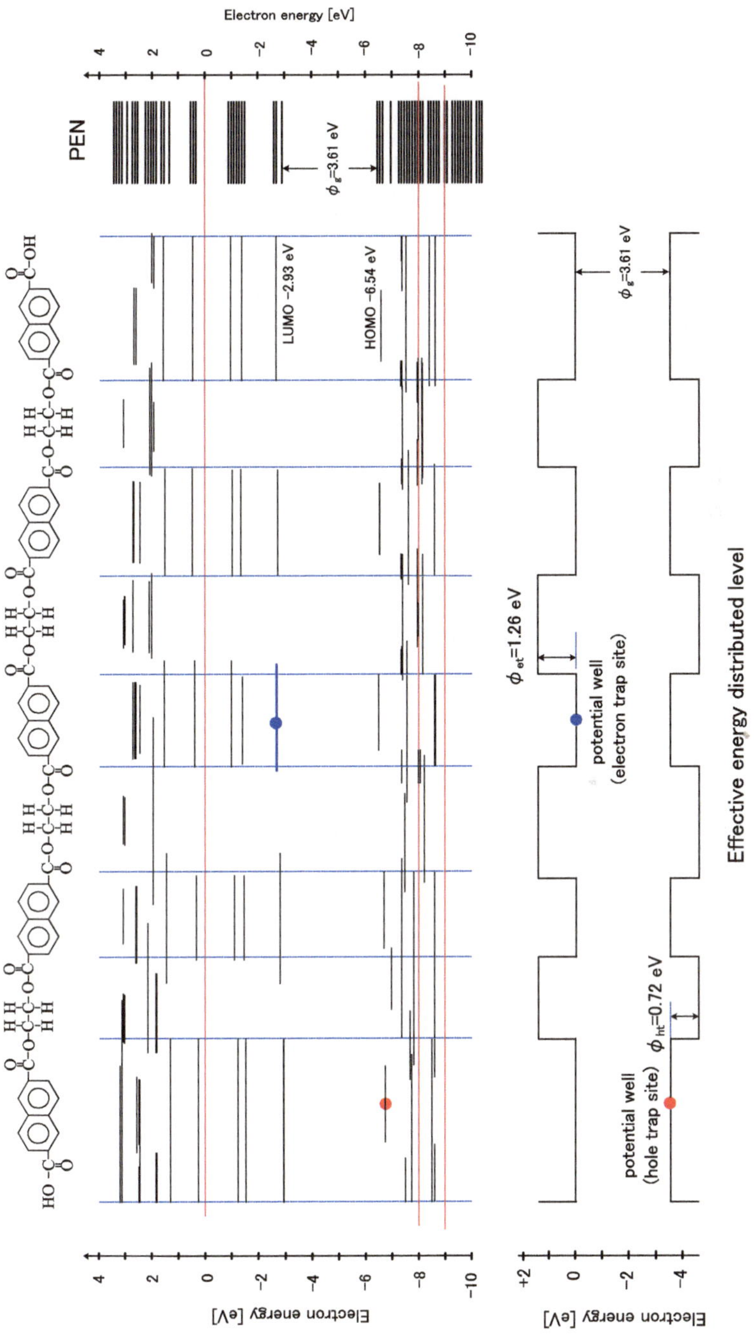

Fig. 14.18 The energy levels of PEN sample obtained by the DFT calculation

14.5 Induced Trap Site of MgO and Fullerene Under Applied Electric Fields

(1) Effect of MgO Particles on Suppressing Charge Accumulation

In the process of developing a XLPE cable for DC high voltages, the charge storage characteristics are significantly improved by adding MgO particles into XLPE. Figure 14.19 shows the measurement results of the space charge distribution $\rho(x, t)$ [C/m^3] obtained by the PEA method. Figure 14.19a shows the measurement results of $\rho(x, t)$ for single LDPE without MgO particles, and Fig. 14.19b represents the LDPE+MgO case. The color display in the upper part of Fig. 14.19 shows the result of $\rho(x, t)$ at $E = 150$ kV/mm, and the lower part shows the result of $\rho(x, t)$ and the electric field distribution $E(x, t)$ at $E = 200$ kV/mm.

LDPE (Fig. 14.19a): Positive charge is injected from the anode and moves toward the cathode. When $\rho(x, t)$ is accumulated, the internal electric field $E(x, t)$ is also distorted. The positive charge distribution is a characteristic of PE polymer, and the details are described in Sect. 12.1.

LDPE+MgO (Fig. 14.19b): The LDPE mixed with MgO particles shows $\rho(x, t) = 0$ and $E(x, t)$ is also uniform. By mixing the MgO particles, a DC insulating material without charge accumulation is developed.

(2) Electrostatic Potential Due to Field Induction Spheres with Fine Particles

Why can LDPE mixed with MgO particles suppress charge accumulation? The relative permittivity $\varepsilon_r = 9.8$ of MgO is larger than the $\varepsilon_r = 2.3$ of the LDPE. Furthermore, the MgO particles placed under a high electric field are polarized by electric field induction, and an electrostatic potential well is formed in the LDPE. We propose a model in which this potential well traps charge carriers [4]. (See Takada text Vol. 2 Chap. 10, Appendix 2).

Figure 14.20a depicts a model in which a spherical dielectric with a dielectric constant ε_2 (radius a) i s placed in a matrix dielectric with a dielectric constant ε_1. The spherical dielectric to which an electric field E_0 is applied is polarized. The spherical dielectric has more orientation polarization than that of the base dielectric. Then, polarization charges ($+\sigma_{p2}$, $-\sigma_{p2}$) are induced on the surface of the dielectric sphere, and polarization charges ($-\sigma_{p1}$, $+\sigma_{p1}$) are induced on the outer surface. As a result, there is a polarization charge difference ($+\Delta\sigma_p$, $-\Delta\sigma_p$) on the surface of the dielectric sphere.

Figure 14.20b: the polarization in the dielectric sphere is represented by positive point charge + q and negative point charge −q, and the two point charges determine the bipolar moment $m = q\delta$ in the interval δ.

Figure 14.20c: the electric potential distribution $V(r, \phi, \theta = \pi)$ created by this dipole moment $m = q\delta$ is derived as shown in Eq. (14.1).

Figure 14.20d: it shows the calculation result of the potential distribution $V(r, \phi, \theta = \pi)$ of Eq. (14.1) on the x-y plane of $\theta = \pi$. The vertical axis represents the potential distribution $V(r, \phi, \theta = \pi)$ in the range of ±2.5 V. The positive potential

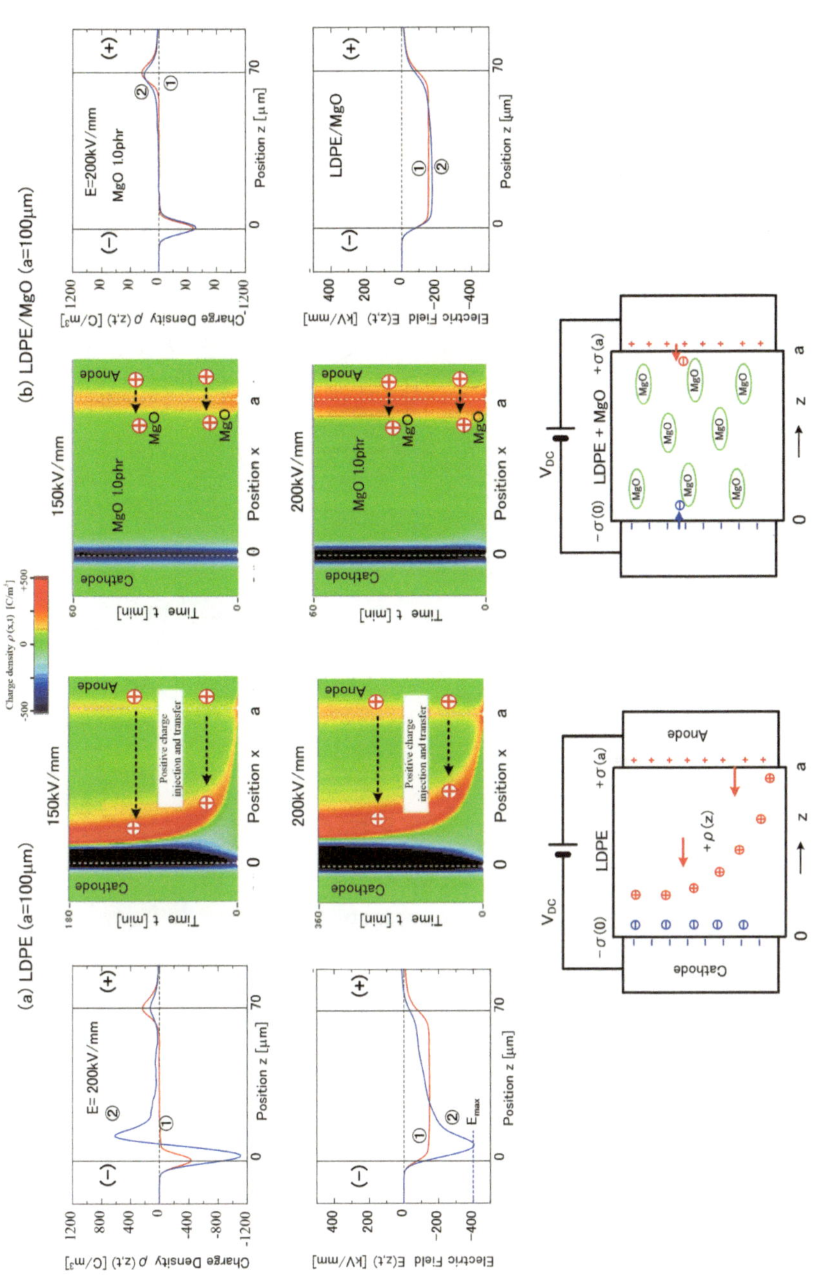

Fig. 14.19 PEA results of XLPE samples under the effect of MgO particles

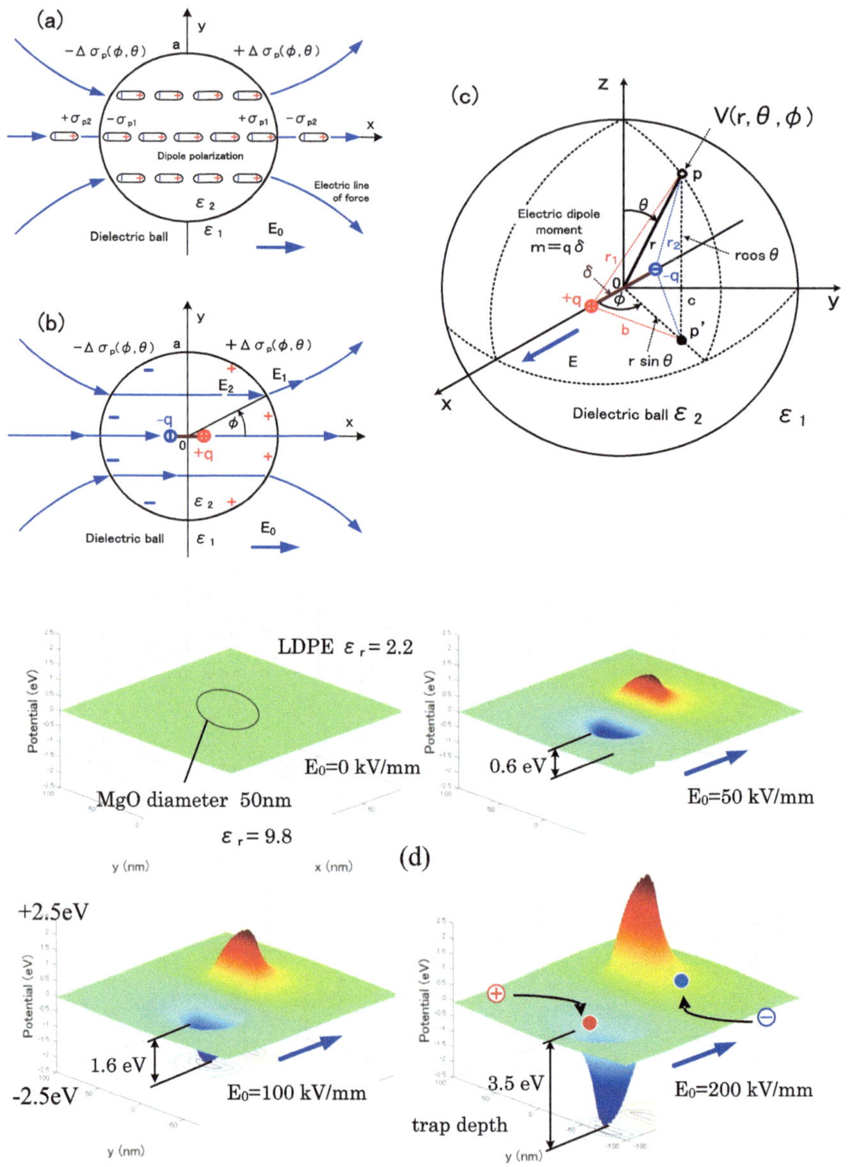

Fig. 14.20 A model for analyzing the effect of MgO particles

distribution is displayed in warm colors and the negative one is displayed in cold colors. Focus on Eq. (14.1), $V(r, \phi, \theta = \pi)$ depends on the applied electric field E_0, the size of MgO particles (radius: a), and the dielectric constant ratio ($\varepsilon_2/\varepsilon_1$). Firstly, the dependence of the electric field E_0 is analyzed.

$$V_p(r, \varphi, \theta = \pi) = aE_0\left[1 - \frac{3}{2 + (\varepsilon_2/\varepsilon_1)}\right] \times \frac{1}{4\pi}\int_{\varphi'=-\pi}^{+\pi}\int_{\theta'=0}^{+\pi}\frac{\sin^2\theta'\, d\theta' \ \cos\varphi'\, d\varphi'}{\sqrt{1 + (r/a)^2 - 2(r/a)\ \sin\theta' \cos(\varphi - \varphi')}} \tag{14.1}$$

$$\sigma_p(a, \varphi = 0, \theta = 0) = \varepsilon_1\, E_0\left[1 - \frac{3}{2 + (\varepsilon_2/\,\varepsilon_1)}\right] \tag{14.2}$$

Figure 14.20d shows the calculation results of $V(r, \phi, \theta = \pi)$ at $E_0 = 0, 50, 100, 200$ kV/mm. The ellipse in the figure depicts the cross section of MgO particles ($\varepsilon_r = 9.8$). A negative potential well is formed on the anode side of the applied voltage and serves as a trap for positive charges (hole carriers). A positive potential well is formed on the cathode side and negative charges (electron carriers) are trapped. When $E_0 = 100$ kV/mm is applied, the depth of the potential well becomes 1.5 eV. With this trap depth, the trapping time is extremely long, which is about one month. This means that the trapped charges cannot move. (See Takada text Vol. 2 Chap. 13, Sect. 13.4).

(3) **Dependence of Electrostatic Potential on Permittivity Ratio**

Next, the dependence of the permittivity ratio ($\varepsilon_2/\varepsilon_1$) in Eq. (14.1) is analyzed. In Fig. 14.21, a spherical dielectric ε_2 is placed in the matrix ε_1 (relative permittivity of $\varepsilon_r = 2$). The potential distribution $V(r > a, \phi = 0, \theta = \pi)$ on the x-axis is shown when the relative permittivity of the sphere is changed to $\varepsilon_r = 2, 4, 12, \infty$ (metal). At this time, the induced charge density $\sigma_p(a, \phi = 0, \theta = \pi)$ on the spherical dielectric surface is given by Eq. (14.2), and its potential is $V(a$ when $r = a$ in Eq. (14.1), $\phi = 0, \theta = \pi)$ and drawn in the upper part of Fig. 14.21. The lower part of Fig. 14.21 shows the potential distribution (red line) based on the induced charge density $\sigma_p(a, \phi = 0, \theta = \pi)$ without the external potential distribution. (See Takada text Vol. 2, Chap. 10, Appendix 2).

Summary: Assume that the MgO particles are polarized due to electric field induction by applying a high electric field and an electrostatic potential well is formed in LDPE. This potential well is analyzed. As a result, we obtain the result that the potential well depth is as high as 1.5 eV when $E_0 = 100$ kV/mm is applied. Basically, the mechanism of charge injection inhibition of LDPE mixed with MgO particles can be explained.

(4) **Energy Level of MgO/PE Interface**

The previous part introduces an analysis of permittivity and electric field induction using the classical theory.

Here, we introduce the analysis of the energy levels at the interface between MgO and PE by Quantum Chemical Calculation using quantum theory (Fig. 14.22) [5]. This calculation uses the Hybrid Functional HSE06 based on the Vienna Ab initio Simulation Package (VASP) software of the density functional theory (DFT).

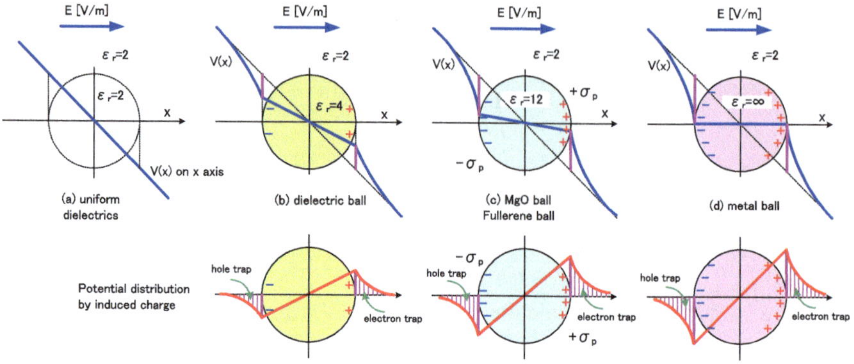

Fig. 14.21 The electrostatic potential due to permittivity ratio

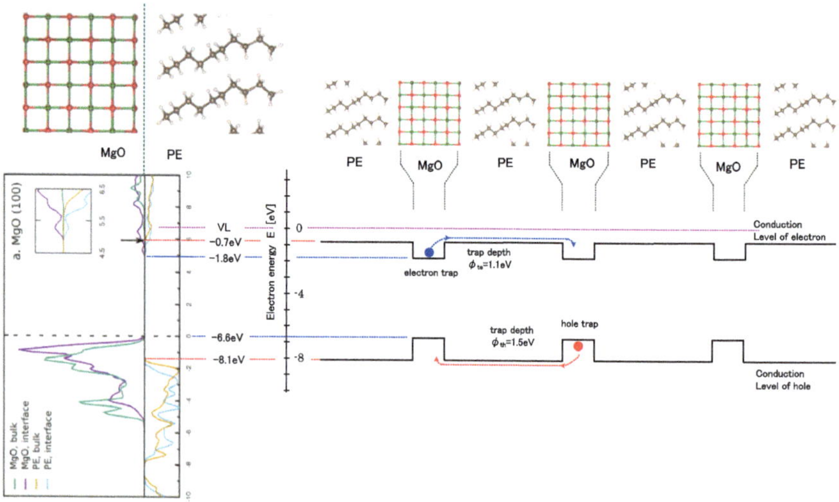

Fig. 14.22 The DFT calculation results without chemical bond at the close interface between PE and MgO

Figure 14.22 shows the DFT calculation results when there is no chemical bond at the close interface between PE and MgO. The LUMO level of PE is −0.7 eV and the HOMO level is −8.2 eV. The LUMO level of MgO is −1.8 eV and the HOMO level is −6.5 eV. Based on these values, when the PE and MgO are mixed, the LUMO level and the HOMO level are schematically arranged in Fig. 14.22, where the formation of electron and hole traps is also depicted. The LUMO and HOMO levels of MgO are close to each other, and those of PE are far from each other. As a result, the LUMO level of MgO becomes the electron trap site and the HOMO level becomes the hole trap site. The electron trap depth is expected to be $\phi_{te} = (1.8–0.7) = 1.1$ eV, and the hole trap depth is expected to be $\phi_{th} = (8.2–6.5) = 1.7$ eV. Actually, there are

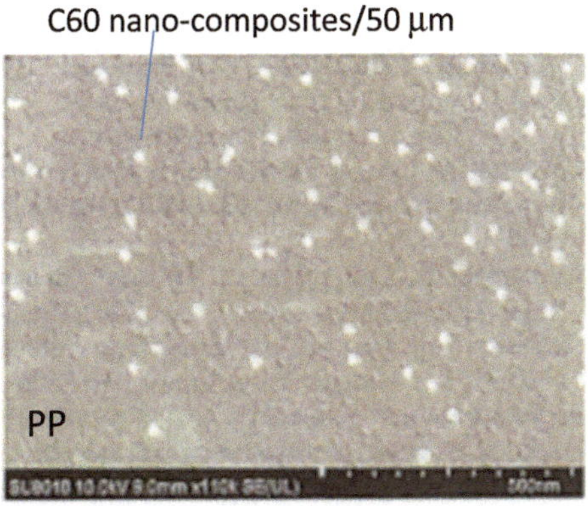

Fig. 14.23 The sphere shape of Fullerene/C60

several levels between the LUMO level of MgO and the LUMO level of PE, which is not shown in the figure. Therefore, the effective electron trap depth is shallower than $\phi_{te} = 1.1$ eV. The effective hole trap depth is also shallower than $\phi_{th} = 1.7$ eV. For details of this effective trap depth, see Takada text Vol. 2, Chap. 13, Appendix 4.

(5) **PEA Measurement Data of Fullerene/C60 for Charge Accumulation Prevention** [6]

As shown in Fig. 14.25, Fullerene/C60 is a sphere shape with a diameter of 7.125 Å which is bonded only by 60 carbons C. It is a molecule with a structure of 5 rings or 6 rings and bonded by π electrons and σ electrons. We create a test sample by making clusters using these C60 and adding these particles with the diameter about 50 μm into PP (Polypropylene). Figure 14.23 is a SEM photograph of the PP/C60 sample. PP/C60 samples with the C60 particles of 0.0, 0.1, 0.5 and 1.0 wt% are prepared. Figure 14.24 shows the PEA data of space charge distribution $\rho(x, t)$ [C/m^3] when an electric field of 60 kV/mm is applied to each PP/C60 sample. In the case of 0.0 wt%, negative charge is injected from the cathode and accumulated after 600 s and 1800s. Negative charge injection is suppressed by increasing the C60 mixing part to 0.1, 0.5, 1.0 wt%, and no charge accumulation is observed at 1.0 wt%. As described above, the mixing of C60 has an effect of preventing charge accumulation.

(6) **Induced Potential by Electric Field of Fullerene/C60**

The LDPE mixed with MgO particles prevents the charge accumulation due to the deep trap effect on induced potential well by the electric field in part (1) and (2). The same induced potential effect is also investigated here for the C60 case. In Sect. 12.1, we investigate the depth of the potential well. Since the relative permittivity of C60 particles is unknown now, an electric field is applied to C60 particles using Quantum Chemical Calculation here. The depth of the well is examined from the electrostatic potential.

Figure 14.25a shows the structure of C60 molecule with the diameter of 7.125 Å. When an electric field of $E = 5140$ kV/mm is applied, the electron waves are biased

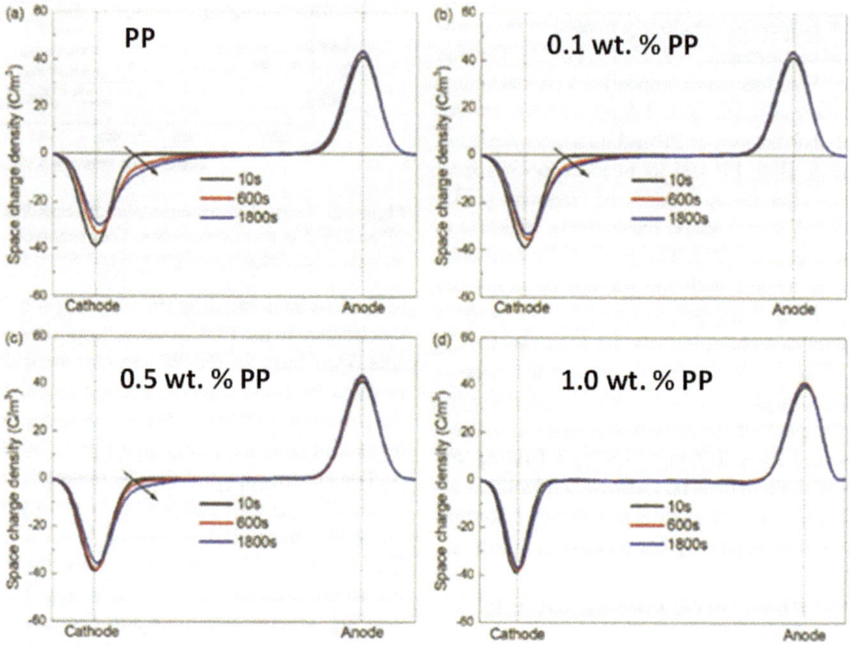

Fig. 14.24 PEA results of PP/C60 samples

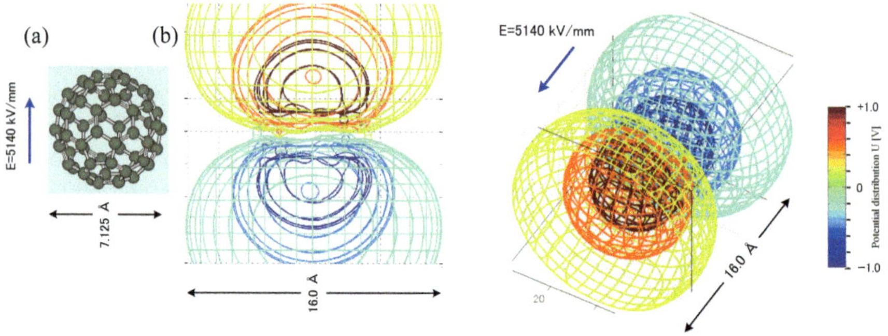

Fig. 14.25 The potential distribution of C60 molecule

by the electric field. Negative charges (electrons) are induced on the anode side and positive charges (holes) are induced on the cathode side, and C60 molecules are polarized. Figure 14.25b shows the two-dimensional distribution potential $V(x, y)$ of the cross section of the C60 molecule, and Fig. 14.25c shows the three-dimensional distribution $V(x, y, z)$. The positive potential distribution is displayed in warm colors and the negative part is displayed in cold colors. The potential distribution is within ±1.0 V. Here, the applied electric field is extremely large with the value of $E =$

5140 kV/mm. When the diameter of C60 particle is as large as 50 μm, it is shown that a trap with a depth of ±1.0 V can be formed by the applied electric field of $E = 0.72$ kV/mm using the following similar rule. The blocking effect of charge accumulation in PP mixed with C60 can also be explained. The similar side of electric field application is $E = (7.125$ Å/50 mm) × 5140 kV/mm = 0.00014 × 5140 kV/mm = 0.72 kV/mm.

(7) **Charge Trap Due to Difference in Energy Level Distributions Between C60 and PP**

Polymer materials such as PET and PEN are polymerized polymers of monomers with different energy levels. It is described in Sect. 14.3 that the energy level difference between the monomers forms a trap. Therefore, the energy levels of C60 and PP are analyzed, and it is discussed whether charge trap sites are formed there. Figure 14.26a shows the energy levels obtained from the respective DFT calculation of C60 and PP. The Fermi level of C60 is $E_F = -4.78$ eV, and that of PP is $E_F = -2.73$ eV. The Fermi levels of them match with each other in the process of contact. Figure 14.26b) shows the energy levels of the mixed C60 and PP sample between the Al electrodes. The HOMO and LUMO levels of C60 exist in the band gap of PP. The HOMO level of C60 becomes a hole trap and the LUMO level of C60 becomes an electron trap. The positive charge (red circle) and negative charge (blue circle) of the electric double layer are drawn at the interface between C60 and PP in the figure. The formation of hole and electron traps can also be confirmed from this figure.

Research subject: In Fig. 14.26, the energy levels of the C60 and PP molecules are calculated independently, and then the trap sites in which the energy levels of both are analyzed by drawing circles. The measurement result in Fig. 14.24 corresponds to a sample in which PP molecules and C60 with the 1.0 wt% are mixed. Therefore, it is proposed to make a mixed sample by molecular dynamics and study the trap depth of the energy level difference by DFT calculation.

An example of analyzing the charge accumulation characteristics of MgO/PE mixed system and C60/PP mixed system by DFT calculation is introduced above. Furthermore, the Graphene/Epoxy mixed system is also simply shown, as shown in Fig. 14.27 [7]

14.6 Analysis of Positive and Negative Charge Accumulation in ETFE

(1) **Measurement Result of Charge Distribution in ETFE Under an Electron Beam Irradiation**

Figure 14.28 shows the measurement results of the space charge distribution when a DC electric field of 100 kV/mm is applied to ETFE (Ethylene/tetrafluoroethylene) irradiated with an electron beam for 10 mins (acceleration energy: 60 keV, irradiation

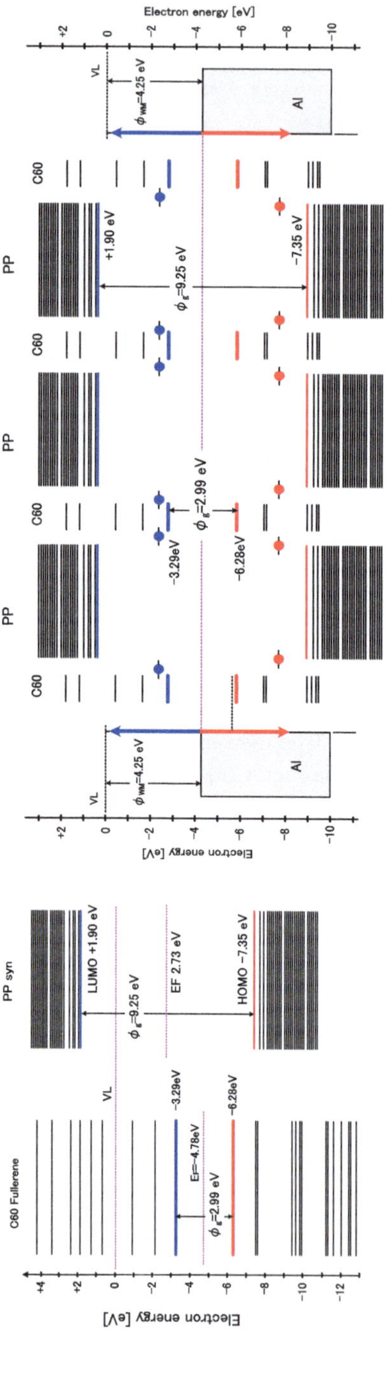

Fig. 14.26 The energy levels of the C60 and PP samples

Fig. 14.27 The independently calculated energy levels of the C60 and PP molecules

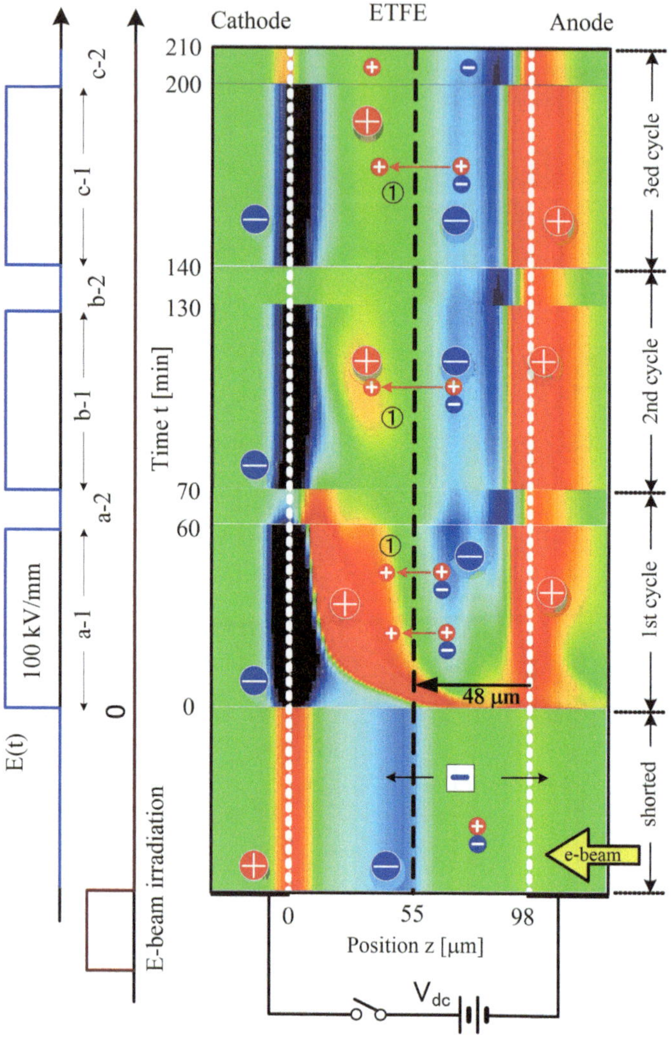

Fig. 14.28 PEA results of ETFE sample irradiated with an electron beam

current density: 80 nA/cm^2). Both the electrodes contacted with ETFE are short-circuited and stored for 24 h after electron beam irradiation.

Before voltage application: The electron penetration depth of 60 keV energy shows that negative charges (electrons) are accumulated at a position deeper than 43 μm (= 98 − 55 μm) from the irradiated surface [8].

Applying a DC electric field of 100 kV/mm for 60 mins: In the range of 55 μm < z < 98 μm, electron–hole pairs remain due to the high-energy electron irradiation. The remained hole carriers start moving toward cathode. As a result, the holes neutralize the accumulated electrons (negative charges in 0 < z < 55 μm), and the positive charges (holes) are accumulated in the non-irradiated region of $0 < z < 55$ μm. Meanwhile, negative charges (electrons) remain in the range of 55 μm $< z <$ 98 μm where positive charges (holes) have passed.

By the subsequent second and third voltage application, the positive charges (holes) disappear while the negative charges (electrons) still remain. To summarize this characteristic, it can be imagined that positive charges (holes) can move by hopping in shallow traps, but negative charges (electrons) are trapped in deep traps and difficult to move.

(2) Calculated Energy Level o f ETFE

ETFE is a copolymer composed of ethylene ($C_2H_2 = C_2H_2$) and tetrafluoro-ethylene ($C_2F_2 = C_2F_2$). Briefly, it is a polymer material of copoly-merization in which polyethylene (PE) and Teflon (TF) are alternately arranged.

Therefore, we design the small ETFE molecule in Fig. 14.29 by the alternating copolymerization of Teflon $(C_2F_2\text{-}C_2F_2)_3$-3 pairs and polyethylene $(C_2H_2\text{-}C_2H_2)_3$-2 pairs. The energy level of this small ETFE molecule is calculated by the DFT method, and the result is shown in the figure.

The LUMO level in the PE part is high, but that in the TF part is low. As a result, the LUMO level in the low TF part becomes the electron trap site. On the other hand, since the HOMO levels of the PE and TF parts are almost the same, hole trap sites do not appear clearly. It is clear from Fig. 14.29 that ETFE is a polymer material

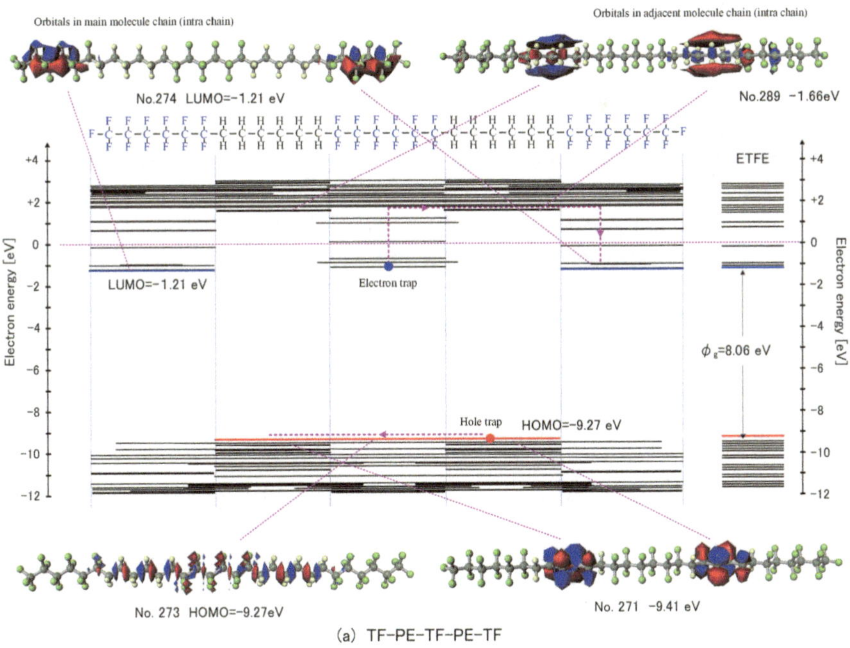

Fig. 14.29 The energy level of ETFE material by the DFT

that forms deep electron traps and shallow hole traps, which is the same as the result obtained in irradiated ETFE (Fig. 14.28) showing that positive charges (holes) can move by hopping in shallow traps and negative charges (electrons) are trapped in deep traps.

In this way, the trap depth of the energy level in the Quantum Chemical Calculation/DFT calculation can be studied, and the characteristics of the electron beam irradiation to ETFE can also be explained clearly.

(3) Electrostatic Potential Distribution in ETFE

It is estimated from the energy level diagram in Fig. 14.29 that negative charges (electrons) are trapped in the low LUMO level of the TF part. Therefore, the ETFE molecule is negatively charged (electrons) and positively charged (holes) in the following DFT calculation, and the relationship between the electrostatic potential distribution and the corresponding position in the ETFE molecule is observed, as shown in Fig. 14.30. Figure 14.30a shows the DFT calculation results of the electrostatic potential distribution of ETFE molecule without charge. Figure 14.30b is the result with positive charge, and Fig. 14.31c is the result with negative charge.

Uncharged ETFE molecule in Fig. 14.30a: Carbon C in TF part is positively charged, and fluorine F is biased negatively (Fig. 14.31a-1). Since the electronegativity of carbon C is 2.5 and that of fluorine F is 4.0, the electrons of carbon C are attracted to the fluorine F side. Thus, carbon C is positively charged, and fluorine F is negatively charged. Conversely, carbon C in PE part (electronegativity 2.5) is negatively charged, and hydrogen H (electronegativity 2.1) is positively charged. Therefore, the two-dimensional electrostatic potential distribution of the ETFE molecule is shown in Fig. 14.31a-3 and the three-dimensional distribution is shown in Fig. 14.31a-4. Obviously, according to the magnitude of electronegativity of each atom, it can be observed that they are positively and negatively charged. In the three-dimensional electrostatic potential distribution, the positive potential of hydrogen H in the PE part and the negative potential of fluorine F in the TFE (tetrafluoroethylene) part protrude outside due to the charging polarity of the atom in side chain.

Positive charge and negative charge cases of ETFE molecule: Fig. 14.30b shows the electrostatic potential distribution when ETFE molecule is negatively charged (electrons) and Fig. 14.30c shows the positively charged case (holes), respectively. The positive potential distribution in the upper part of Fig. 14.30b distributes over the entire molecular chain. On the other hand, the negative potential distribution in the upper part of Fig. 14.31c is concentrated around the TF molecule. This is because the electron traps in the TF molecule are deep.

(4) Energy Levels of Positive Charge and Negative Charge Cases

Figure 14.31 is the energy level diagrams for which the center is neutral, the left is positively charged, and the right is negatively charged. Since the positively charged ETFE molecule lacks one electron, the energy of the whole molecule is lower than that of the uncharged one. Conversely, the negatively charged ETFE molecule has one electron in excess, and thus the overall energy of the molecule is higher than that

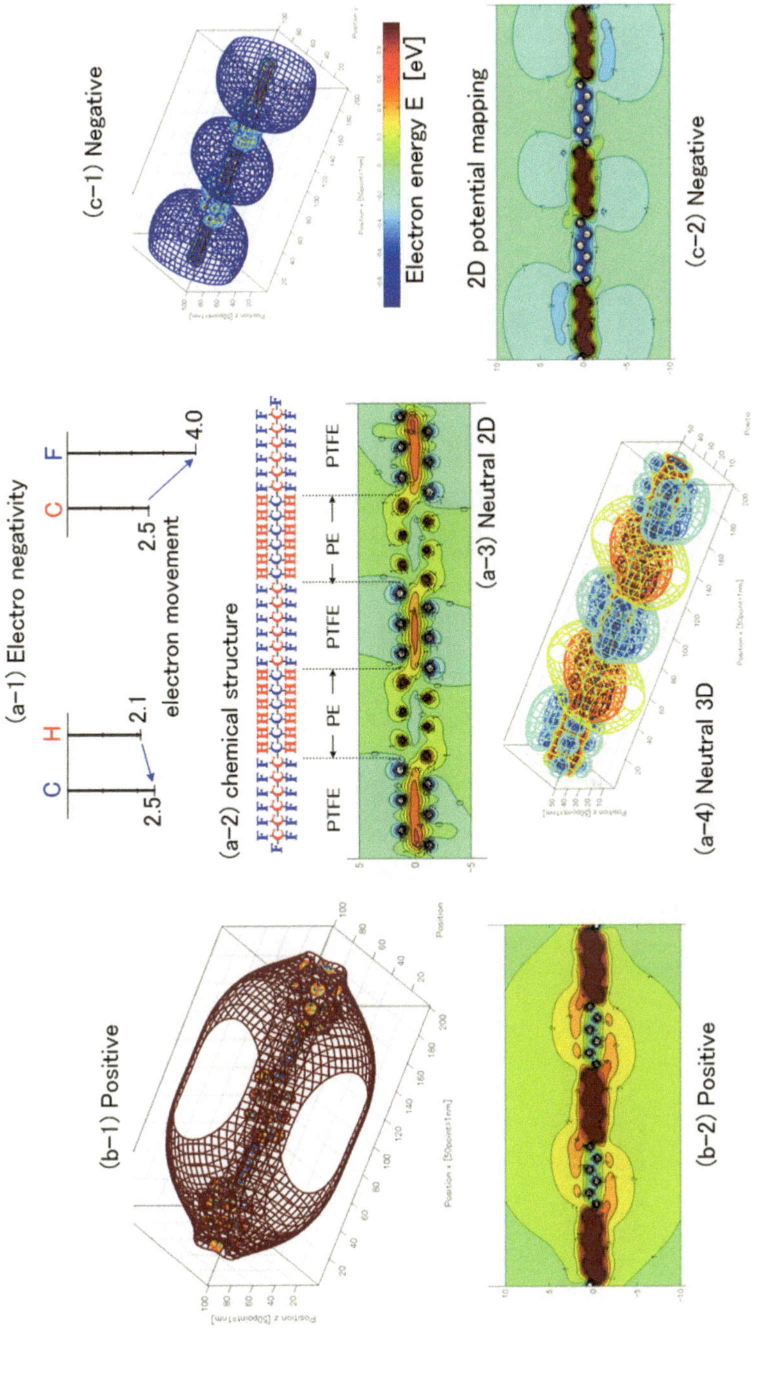

Fig. 14.30 The energy level diagram of ETFE molecule with positive or negative charge

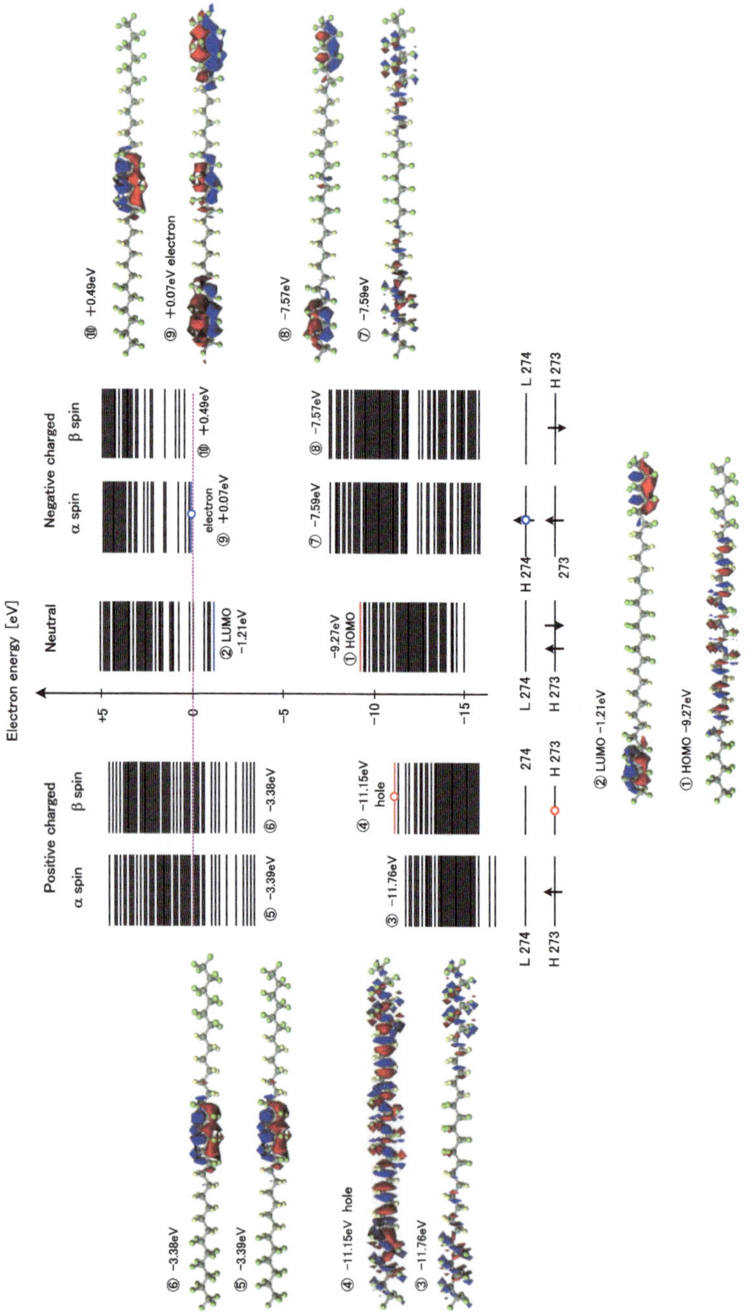

Fig. 14.31 The energy level diagram of ETFE molecule with positive or negative charge

of the uncharged one. When the molecule is charged, the electron splits into α spin and β spin, and thus the energy level diagram also separates.

Positively charged molecular orbital (cloud of electron wave): ① an electron is missed at HOMO level, ④ and thus a hole (red circle) is created. As a result, the ETFE molecule is positively charged.

Negatively charged molecular orbital (cloud of electron waves): ② Electrons are externally supplied to the LUMO level, and electron waves are generated at ⑨ L274. The ② electron waves in the neutral LUMO level exist at both ends of the molecule, while in the negatively charged case, the ⑨ electron waves also exist at both ends and the center of the molecule. As a result, the ETFE molecule is negatively charged.

14.7 Relationship Between the Needed Inception Voltage of Electrical Tree and Additives

(1) Types of Additives and Inception Voltage of Electrical Tree

It has been reported that the addition of the additive to XLPE in Fig. 14.32 increases the inception voltage of electrical tree [9]. Consider a model in which electrons are accelerated by a local electric field to break XLPE molecules and generate an electrical tree. It is thought that the electric field of the permanent dipole moment of the additive hinders the process of accelerating the electron. Therefore, the dipole moment m [Debye] of the additive obtained by DFT calculation is shown on the right vertical axis in Fig. 14.32 and compared with the inception voltage of electrical tree. Compared to XLPE ($m = 0$), the inception voltage of electrical tree increases in the presence of the additive dipole moment ($m > 0$). This dipole moment strongly depends on the chemical structure of the additive. In this section, we observe the relationship between the chemical structure, the dipole moment, and its electrostatic potential distribution. The potential distribution and electron scattering (inhibition) are then considered.

(2) Electrostatic Potential Distribution of Additive and Inception Voltage of Electrical Tree [10]

To observe the relationship between the chemical structure and the dipole moment with its electrostatic potential distribution, the (a) energy level diagram, (b) molecular orbital, (c) electrostatic potential distribution depending on the chemical structure of XLPE and additives are calculated by DFT calculation.

(a) **Energy level diagram**: Fig. 14.33 shows the energy level diagram of each sample. The blue line is the LUMO level, the red line is the LUMO level, and the difference is the energy gap ϕ_g [eV]. The difference between LUMO and HOMO levels of the PE (XLPE) material is wide. It can be seen that the additive with a narrow difference between LUMO and HOMO levels is mixed

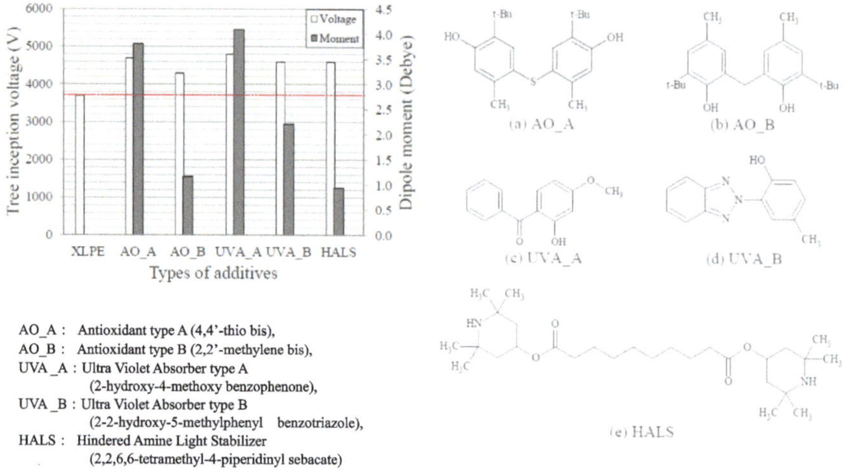

Fig. 14.32 The dipole moment *m* of the additive obtained by DFT calculation

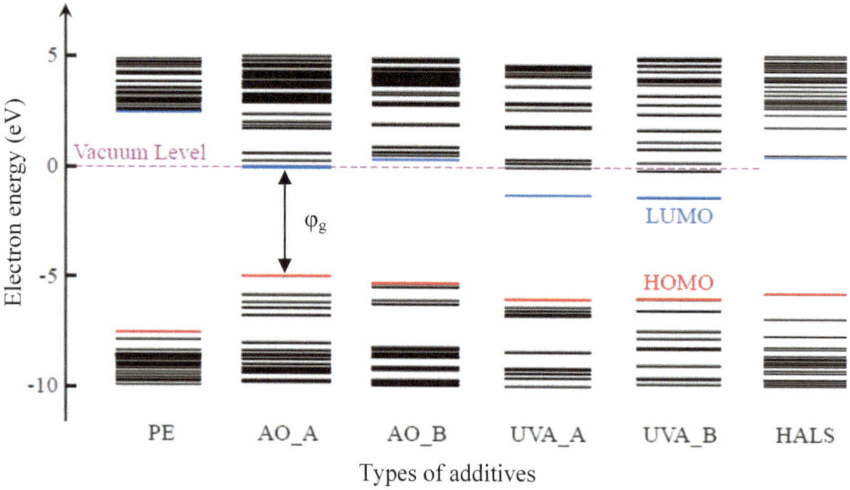

Fig. 14.33 The energy level diagram of each sample

in the PE. This indicates that there are sites for trapping electron carriers (see Sect. 13.2). The peculiarity is that the molecular orbitals of AO-B and HALS are localized and do not spread throughout the molecule. Furthermore, the LUMO and HOMO levels of HALS are close to the same level.

(b) **Molecular orbital and electrostatic potential distribution**: Fig. 14.34 shows the molecular orbital and electrostatic potential distribution of each material for

Fig. 14.34 The molecular orbital and electrostatic potential distribution of each material

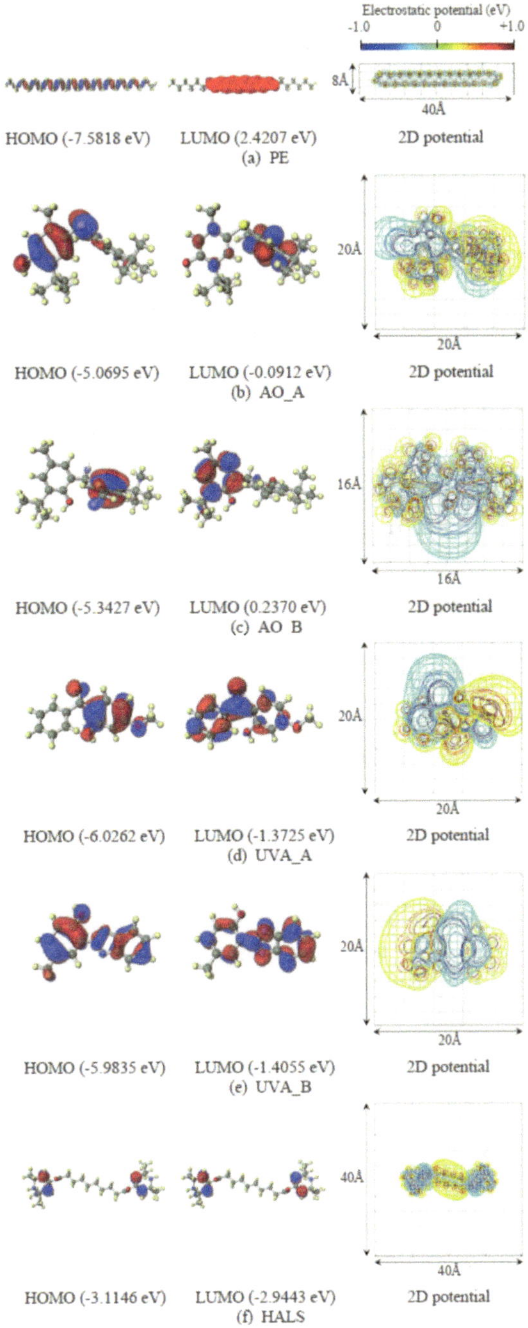

comparison. The molecular orbitals of PE are distributed throughout the molecular chain. Since the positive (hydrogen H) and negative (carbon C) potentials are periodically distributed, their dipole moments alternate in opposite directions. As a result, the dipole moment of XLPE becomes $m = 0$ Debye. In contrast, the potential distribution of AO-A ($m = 3.7$ Debye) and UVA-A ($m = 4.1$ Debye) shows that positive potential (yellow-orange) and negative potential (sky blue) are clearly separated, which thus makes dipolar case. The dipole moment m is large. In other molecules, the positive and negative potential distributions are sterically mixed, and thus the dipole moment m is small. HALS has a negative-positive-negative distribution, whose m is extremely small.

(c) **Potential distribution broadening and electron scattering**: The broadening of the electrostatic potential distribution of the additive extends over a wide range beyond the molecular size. If the electrons that have gained energy between the molecules are affected by such a wide range of potentials, the electrons will change in the acceleration direction and be scattered or trapped in the potential well with opposite polarity. As a result, it is considered that the inception voltage of electrical tree also rises because the electrons gain more energy. As described above, it can be said that the increase in the inception voltage of electrical tree is strongly influenced by the spread of the electrostatic potential distribution of the additive.

14.8 MD Simulation and DF Analysis of the Mixture of PE and AO

(1) **Molecular Structure**

Figure 14.35 shows the molecular structures of PE (n = 24, $C_{24}H_{50}$, tetracosane) and AO (Antioxidant) (4, 4'-thiobis) (details in Sect. 14.7). PE is a linear molecule, and AO is bent at the position of the sulfur S (yellow) bond. This is due to the repulsion and attraction of the electrostatic potential of the quadrupole in the benzene ring (Details in Sect. 12.5).

(2) **MO Simulation** [11]

Figure 14.36 shows a state in which 50 PE molecules (76 wt%) and 15 AO molecules (24 wt%) shown in Fig. 14.35 are arranged in a cubic pixel. MD simulation is started from this state (see Sect. 12.8). Figure 14.36a shows the initial structure of the molecular state with the initial density of 0.30 g/cm^3. Then, temperature increase and compression of cubic pixel are repeated in MD simulation. The density is increased to 0.81 g/cm^3, at which the calculation is stopped. Figure 14.36b shows the final structure of the molecule, and the spacing between the molecules is narrow.

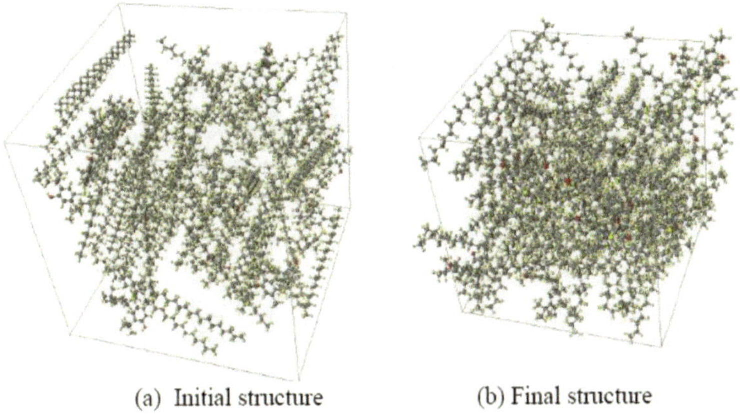

Fig. 14.35 Initial and final molecular structures by MO simulation (50 PE+15 AO)

Fig. 14.36 Respective molecular structures of PE and AO

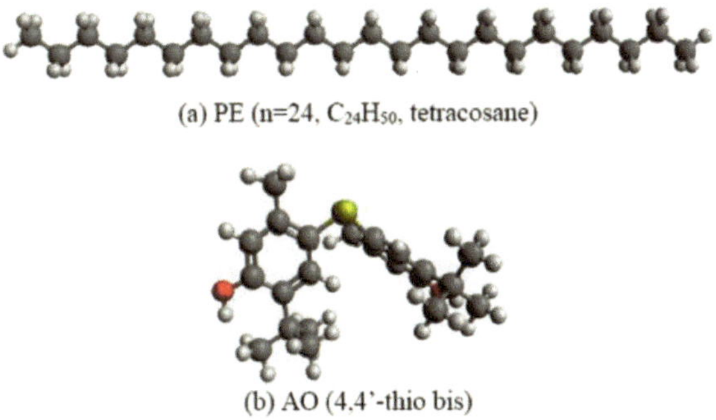

(3) **DFT Calculation**

Figure 14.37 shows the energy level and electron DOS (Density of State) of each material including PE alone, AO alone, 5PE and 5PE+3AO. PE alone has wide LUMO and HOMO levels with a large energy gap ϕ_g. AO alone has two benzene rings and they are orthogonal to each other, and thus ϕ_g is small. After the MD simulation, 5PE with 5 adjacent PE molecules are extracted, and 5PE+3A with 5 adjacent PEs and 3AOs are extracted. The results of these DFT calculations are depicted in Fig. 14.37. In each case, ϕ_g becomes smaller.

(4) **Molecular Orbital**

Figure 14.38 shows the molecular orbitals (electron clouds) of the LUMO and HOMO levels of (a) 5PE and (b) 5PE+3AO. The molecular orbitals of the HOMO level of 5PE are distributed throughout the molecule, but those of the LUMO level are present in the closest part of PE (Fig. 14.39a shows its enlarged figure). The molecular orbitals of the LUMO and HOMO levels of 5PE+3AO only exist in discrete AO molecules (Fig. 14.39b shows its enlarged figure). This is shown in the energy level of Fig. 14.37.

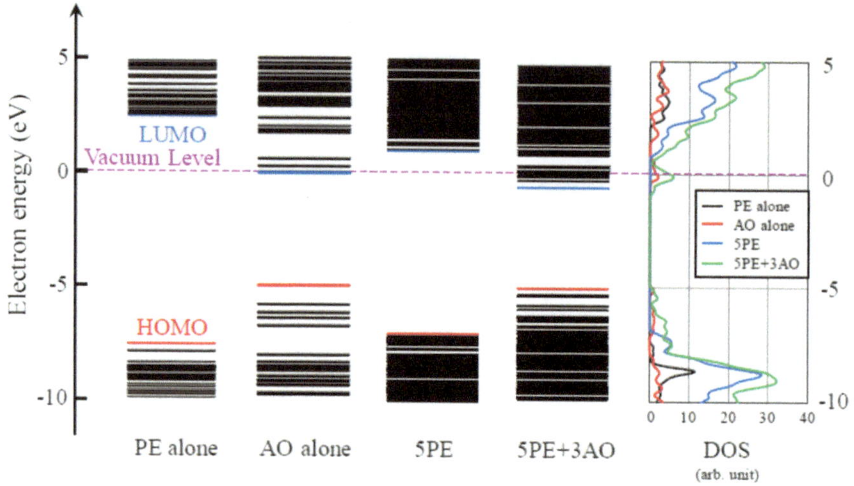

Fig. 14.37 Energy level and DOS of PE, AO, 5PE and 5PE+3AO

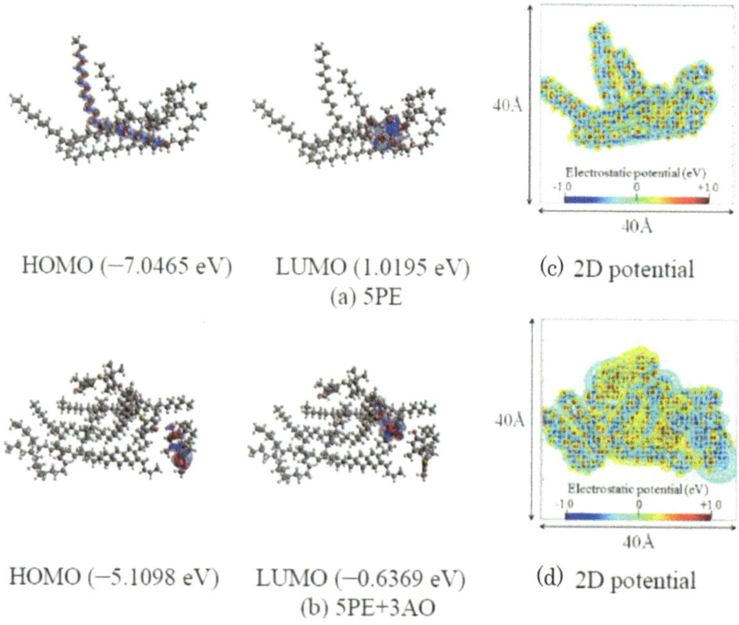

Fig. 14.38 Molecular orbital and 2D electrostatic potential distributions of 5PE and 5PE+3AO extracted after MD simulation (*x*-*y* plane)

That is, the (a) LUMO level of 5PE and (b) LUMO and HOMO levels of 5PE+3AO are localized. This localized LUMO level becomes an electron trap site, and the HOMO level becomes a hole trap site.

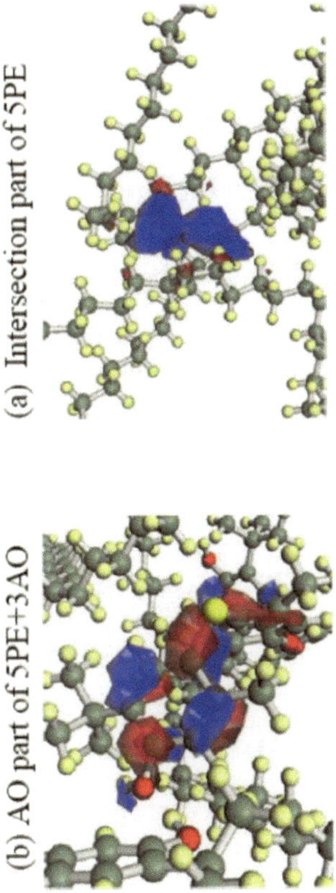

Fig. 14.39 Enlarged view of trap site (molecular orbitals of LUMO level)

(5) **Electrostatic Potential Distribution**

Figure 14.38c, d show the 2D potential distributions of 5PE and 5PE+3AO molecules. In the 5PE molecule, positive (hydrogen H) and negative (carbon C) potentials are periodically distributed, while the positive potential distribution due to the overlap of electron waves of hydrogen H in the molecular side chain is at the closest part of PE. This is the LUMO level as an electron trap site. In the 2D potential distribution of (d) 5PE+3AO molecule, the distribution of positive potential and negative potential spreads where 3AO exists (see Fig. 14.34 of AO_A in Sect. 14.7). In this case, the LUMO level of the AO molecule is an electron trap site and the HOMO level is a hole trap site.

(6) **Summary**

If AO molecule with a large dipole moment of $m = 3.2$ Debye is mixed in PE molecule with a dipole moment of $m = 0$ Debye, electrostatic potential is locally disturbed and LUMO and HOMO levels locally approach, which form electron and hole trap sites. It can be considered that the generation voltage of electrical tree rises due to this trap formation.

14.9 Mixture of PE and Surfactant

(1) Inhibitory Effect of Surfactant on Bow-Tie Tree

In the underground power CV cable using XLPE insulation materials, water trees (Bow-Tie Trees; BTTs) are discovered during the degradation process. Many researches are conducted on the mechanism and countermeasures for the occurrence of BTTs. Then, it is reported that the generation of BTTs can be significantly reduced by incorporating surfactants into the PE matrix as the insulating material (Fig. 14.40a) [12]. In the model, as shown in Fig. 14.40b, a Michelle structure model is proposed in which a large number of surfactants surround a water cluster to suppress the generation of BTTs. The XLPE is composed of PE molecules with the cross-linking agent DCP to improve heat resistance [12]. However, water (H_2O) remains as a cross-linking decomposition residue in the XLPE insulation. The measurements are taken to analyze this residual water (H_2O) surrounding with a surfactant.

(2) MD Simulation of PE+H_2O+Surfactants

The research is conducted by molecular dynamics simulation to confirm the Michelle structure in which surfactants surround water (H_2O) in PE. Each molecule of (c) PE, (d) surfactant, and (e) water in Fig. 14.40 is picked out, and molecular dynamics simulation is performed to observe the aggregated state of the molecules. For MD simulation, the three cases of Models A, B, and C shown in Table 14.1 are taken

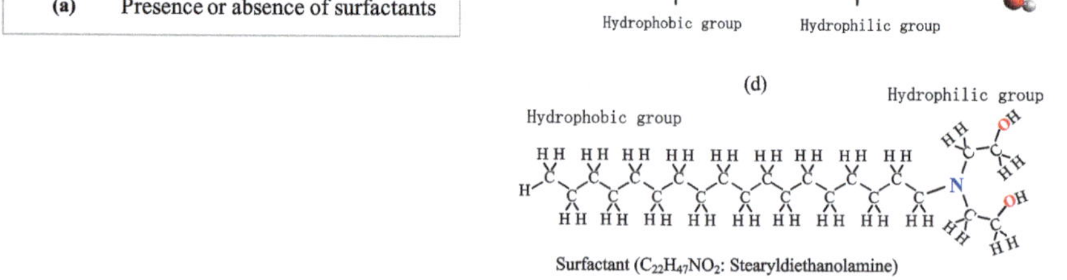

Fig. 14.40 **a** Volume density of BTTs in XLPE with and without surfactants, **b** Micelle formed in amorphous region of XLPE. Molecular structure of **c** PE, **d** surfactant, and **e** water

Table 14.1 Parameters of MD simulation models

Model	A	B	C
Number of PEs	100	100	100
Number of surfactants	0	5	10
Number of water molecules	50	50	50
wt% of PEs	97.4	92.6	88.3
wt% of surfactants	0	4.9	9.3
wt% of water molecules	2.6	2.5	2.4
Density of initial structure (kg/m^3)	200	200	200
Density of final structure (kg/m^3)	872	847	862

up. 100 molecules of PE ($C_{24}H_{50}$) and 50 molecules of water (H_2O) are used as the basic insulating material, and there are no surfactant molecules in Model A, 5 surfactant molecules in Model B, and 10 molecules in Model C. Figure 14.41 shows the results of each MD simulation. The left column of the figure shows the initial structure before the MD simulation by only mixing each molecule in the cubic pixel. Therefore, the density is 200 kg/m^3 (=0.2 g/mm^3). The middle row of the figure shows the final structure after the MD simulation. The densities at the three stages are 872 kg/m^3 (=0.872 g/m^3), 847 kg/m^3 (=0.847 g/m^3), and 862 kg/m^3 (=0.862 g/m^3), respectively. The MD simulation is ended due to the values close to those of the real material.

Model A (without surfactant molecules): PE molecules are almost parallel in the final structure and amorphous. The state is so-called amorphous LDPE. The observation of H_2O molecules in an enlarged view reveals that they form the clusters like Fig. 14.40e and the upper of Fig. 14.41. This is the shape of a water cluster in which H_2O molecules are collected.

Model B (5 surfactant molecules): The molecular structure of the surfactant is shown in Fig. 14.40d. The molecular end is divided into a hydrophilic group (Hydrophilic group) and a hydrophobic group (Hydrophobic group). The observation of H_2O molecules in an enlarged view in the middle of Fig. 14.41 shows that water clusters are gathered in the hydrophilic group of the surfactant.

Model C (10 surfactant molecules): The same effect as Model B is found. The water clusters are collected in more hydrophilic groups of the surfactant. This phenomenon can be said to be the same as the Michelle structure introduced in Fig. 14.40b. Because of the entanglement with the PE molecular net, it is quite different from the spherically symmetric Michelle structure in Fig. 14.41b.

(3) **Observation of Electrostatic Potential Distribution of H_2O and Surfactant [13]**

Both the formation of water clusters by multiple H_2O molecules and the Michelle structure by water clusters with surfactants are due to electrostatic forces. Therefore, Fig. 14.42 shows these electrostatic potential distributions.

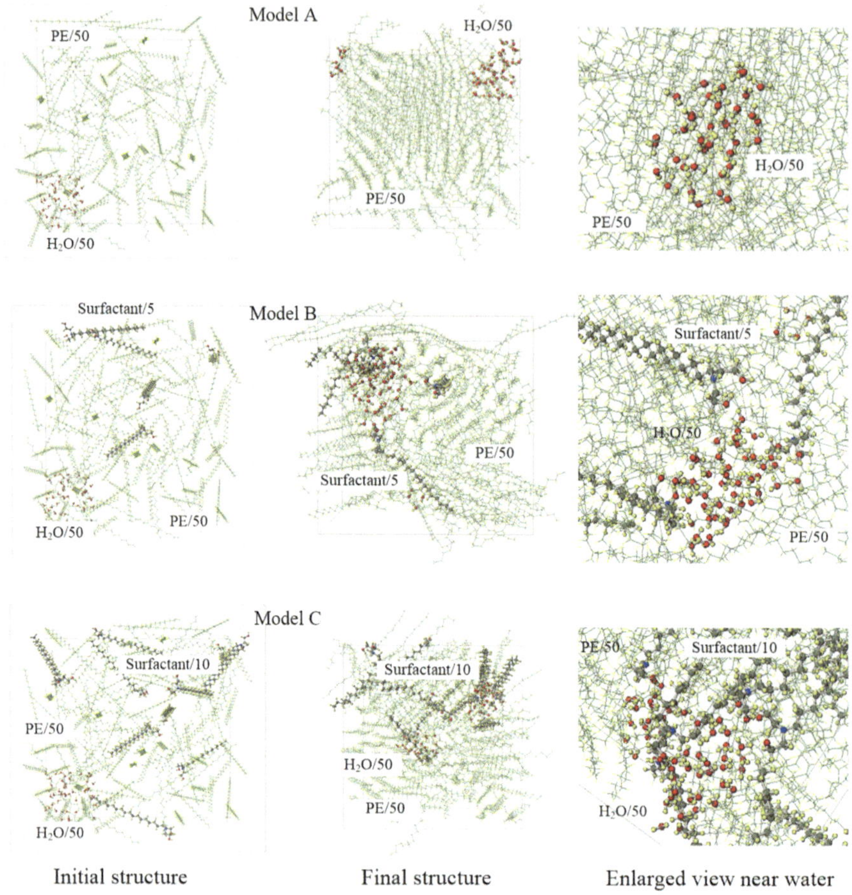

Fig. 14.41 Initial structure, final structure, and enlarged view of final structure near water cluster for each MD simulation

(a) **H_2O** (Fig. 14.42a): H_2O is a binding molecule of two hydrogen H and one oxygen O. Electronegativity causes bias in the electron orbit, which means H is positively charged and O is negatively charged. Therefore, H_2O has a strong electric dipole moment (see the part (3) of Sect. 12.2). Positive hydrogen H in a H_2O molecule and negatively charged oxygen O in adjacent H_2O molecules attract each other, and thus multiple H_2O molecules are attracted and form a cluster.

(b) **$5H_2O$** (Fig. 14.42b): The positive and negative electrostatic potential distributions of the water cluster of $5H_2O$ spread over the range of 20 Å.

(c) **Surfactant alone** (Fig. 14.42c): Positive and negative potential distributions are formed over a wide area at the Hydrophilic group of surfactant.

(d) **Surfactant+$5H_2O$** (Fig. 14.42d): It shows the electrostatic potential of the hydrophilic group, and $5H_2O$ is blocked. The range of the positive and negative

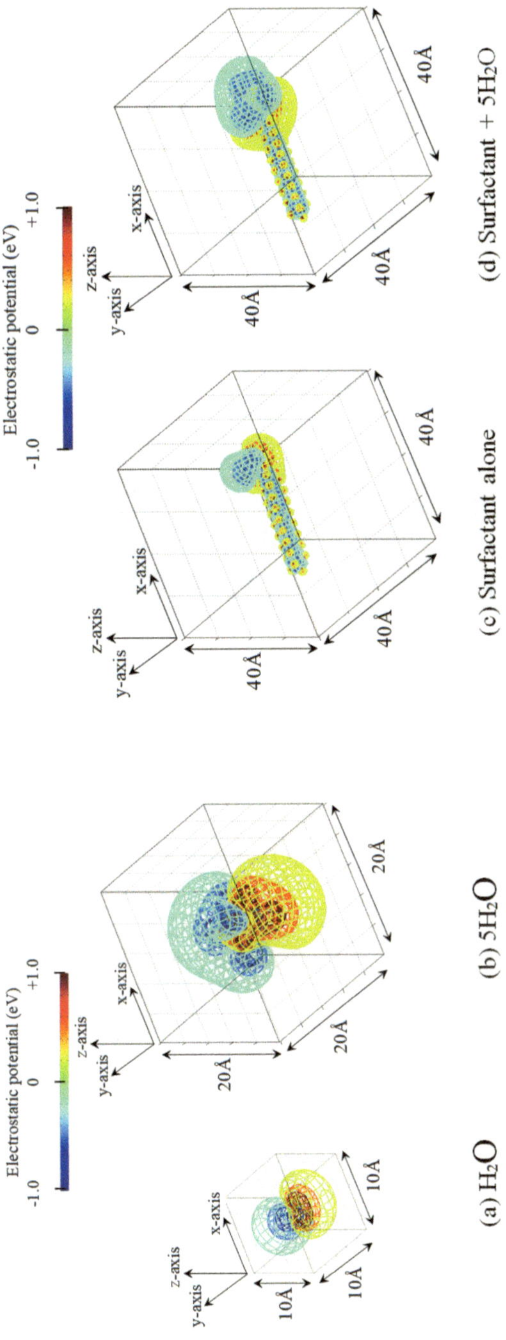

Fig. 14.42 3D electrostatic potential distributions for **a** H_2O, **b** $5H_2O$, **c** surfactant alone and **d** surfactant+$5H_2O$

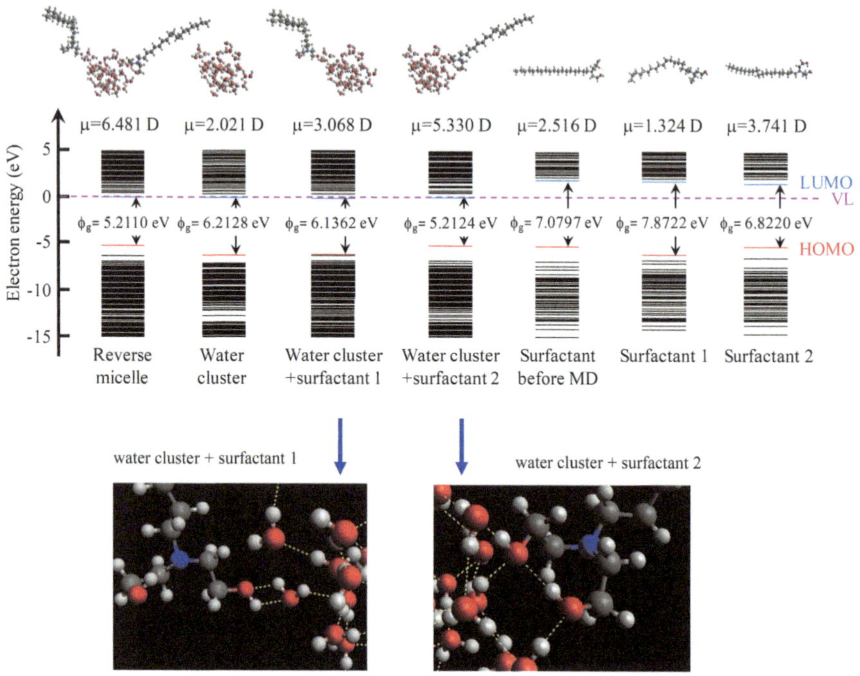

Fig. 14.43 Molecular structure and energy levels of reverse micelle, water cluster, water cluster+surfactant 1, water cluster+surfactant 2, surfactant 1, and surfactant 2

potentials is wider. It can be said that the formation of Michelle structure of water cluster and surfactant is basically caused by electrostatic force.

(4) **Electron Energy Level Observation** [14] **(Fig.** 14.43)

In the previous part (2), the Micelle structure of water cluster and surfactant is considered by using the electrostatic force. Here, we introduce the results for this issue at the atomic energy level using DFT simulation. Figure 14.43 compares the energy level diagrams for various combinations of water clusters and surfactants. There is no significant difference in energy levels due to the combination.

The energy gap is as wide as $\phi_g = 5.2$ eV or more, while charge carriers are not generated in the energy gap by this excitation. As a result of MD Simulation, surfactant appears when two hydroxyl groups (–OH) are separated from each other and when they are close to each other. The former is Surfactant 1 and the latter is Surfactant 2 (Fig. 14.43). There is a difference in the energy gap between $\phi_g = 7.88$ eV of Surfactant 1 and $\phi_g = 6.82$ eV of Surfactant 2. When a surfactant approaches a water cluster, there is a $\phi_g = 6.1$ eV and $\phi_g = 5.2$ eV in the energy gap. However, since there is no generation of charge carriers at this energy gap ϕ_g, it is not necessary to consider the deterioration of insulation due to mixing surfactant.

In summary, it can be said that mixing surfactant can suppress the generation of BBTs by adding the water clusters and the mixture does not reduce the insulating properties.

14.10 Molecular Dynamics Simulation of Oxidized EPDM Dispersion

(1) Oxidation of EPDM

Ethylene-Propylene-Diene Monomer (EPDM) is widely used as an insulating material for electrical equipment and nuclear power generation station. The EPDM is exposed to the high temperature, high electric field, light and radiation in the air. Thus, the EPDM can be oxidized, and then the insulation deterioration takes place. Molecular dynamics simulation is performed on EPDM molecular models with oxidation and cleavage, and the dispersion of oxidized EPDM is observed [15].

The molecular formula of EPDM is composed of Ethylene, Propylene and Diene (Ethylidene Norbornene), as shown in Fig. 14.44. The molecular formula after oxidative deterioration is also shown in Fig. 14.44. This is described in detail below.

EPDM/k: –CH_3 at the polyethylene end is oxidized to –CH=O. Therefore, the negative charge concentrates on oxygen O.

EPDM/l: –CH_3 at the polyethylene end becomes –CH = O, and –CH_3 at the side chain of polypropylene also becomes –CH = O, i.e. 2 positions are oxidized. Negative charge concentrates on oxygen O at these two locations.

EPDM/m: In addition to the oxidation of EPDM/*l*, –CH_2 of ENB is oxidized to –CH=O. Negative charges concentrate on the oxygen O at three locations.

EPDM/p: Chain scission occurs at the boundary between polyethylene and polypropylene, and –C*H_2 is oxidized to –CH=O. Negative charges are concentrated on this oxygen O.

EPDM/q: Similar with the previous EPDM/p, –C*H_2 at the polypropylene end is oxidized to –CH_2–OH.

Negative charges are concentrated on this oxygen O. EPDM/p and EPDM/q are generated simultaneously.

EPDM/n: –CH_2 of the previous EPDM/q ENB is further oxidized to –CH=O. Negative charges are concentrated on oxygen O at two locations.

EPDM/CL: When the above oxidized oxygen O approaches, wave electrons overlap each other and form an O–O link. This oxidative deterioration makes the insulating material lose flexibility and become hard.

(2) MD Simulation of the Dispersion State of Oxidized EPDM

In the EPDM molecular net, the molecular dispersion state due to the interaction of oxidized EPDM is observed by MD simulation. Table 14.2 shows the mixing ratio of the numbers of total EPDM and oxidized EPDM molecules in the molecular net. Figure 14.45 shows the Radial Distribution Function (RDF) $g(r)$ of the MD simulation results, and Fig. 14.46 shows the molecular coordination of the oxidized chains in EPDM system.

Definition of RDF: $g(r)$ is a distribution function that indicates the ratio of two studied atoms that are separated by an interval r [m]. Equation (14.3) is the defining equation for it. Here we observe how far the oxygen O in oxides is separated from each other. The equation represents the ratio of the other oxygen O in the volume ($4\pi r^2 dr$) of the spherical shell within the distance r [m] around the studied oxygen O. This ratio defines $g(r)$ in Eq. (14.3).

$$\text{Radial distribution function } g(r) = \frac{n(r)/4\pi\, r^2\, dr}{\rho} \tag{14.3}$$

RDF result: Fig. 14.46 shows the MD simulation result of RDF $g(r)$ of the oxidized EPDM molecule in Table 14.2. The Oxygen atoms in EPDM/k, l, m, p, q are close to each other with the distance of 3.1–5.0 Å. Oxidized EPDM has a –CH = O carbonyl group. Since the dipole moment is large, they can approach each other. The electron waves of both carbonyl groups overlap and maintain the (–CH-O* *O–CH–) approaching state. EPDM/n has a carboxyl group (HO > C = O) and a hydroxyl

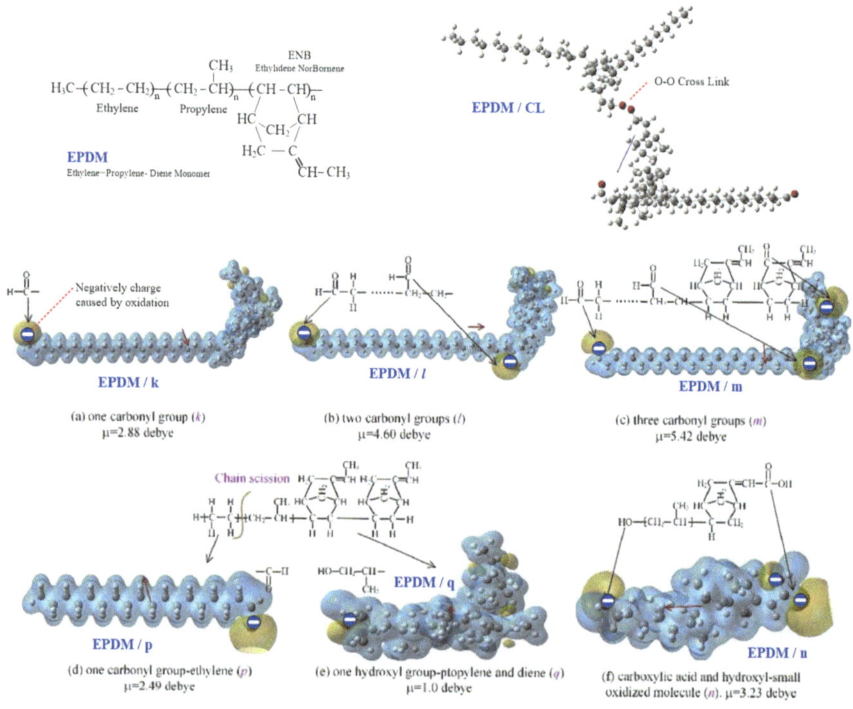

Fig. 14.44 The molecular formula of EPDM

Table 14.2 Constructions of simulation model

	Long chain molecular				Short chain molecular			
Simulation model	EPDM	EPDM/k	EPDM/*l*	EPDM/m	EPDM/p	EPDMq	EPDM/n	EPDM/CL
Number of EPDM	50	40	40	40	30	30	40	30
Number of k		10						
Number of *l*			10					
Number of m				10				
Number of p+q					15	15		
Number of n					15	15	10	
Number of cross linked chain								5

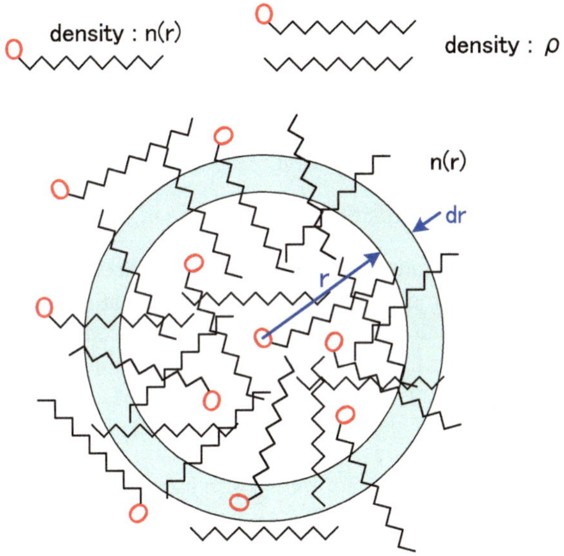

Fig. 14.45 The radial distribution function (RDF) of the MD simulation results

group (>CH–OH), and the dipole moment along the molecular axis is wide and RDF exists in the range of $r \approx 1.3$ nm. Interactions can always work even if their distances are very far.

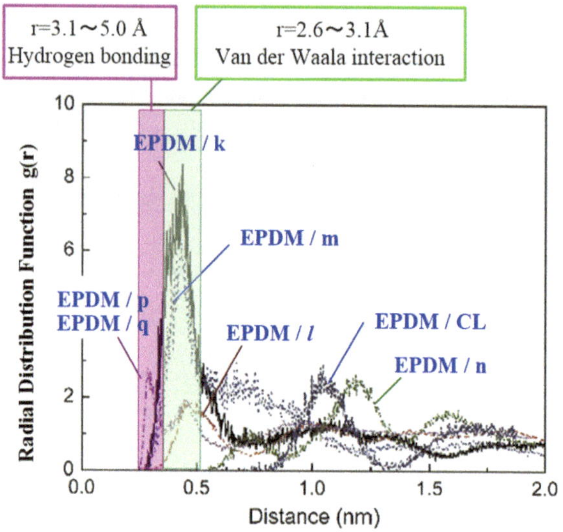

Fig. 14.46 The MD simulation result of RDF $g(r)$ of the oxidized EPDM molecule

(3) **Chain Distribution and Aggregation (Fig.** 14.47)

Observe how the oxidized EPDM molecules in Fig. 14.44 are distributed within the EPDM (non-oxidized) molecules. Long oxidized EPDM molecules (EPDM/k, EPDM/l, EPDM/m) form a "cluster" due to the dipole moment of the carbonyl group. On the contrary, short oxidized EPDM molecules (EPDM/p, EPDM/q, EPDM/n) cannot form a strong "cluster". In addition, EPDM/CL also forms a "cluster".

(4) **Oxidation and Glass Transition Temperature of EPDM**

The glass transition temperature is important for evaluating the mechanical and thermal properties of polymeric materials. In MD simulation, the glass transition temperature T_g[K] is evaluated based on the temperature dependence of the density [kg/m^3]. Figure 14.48 shows the results obtained by MD simulation about the temperature dependence of the density of various oxidized EPDMs. The value at which two of the temperature-dependent straight lines cross at the same density is evaluated as the glass transition temperature T_g [K].

The unoxidized EPDM shows $T_g = 280$ K. EPDM/p and EPDM/q, whose molecular lengths are shortened due to molecular cleavage, show a slightly low temperature of $T_g = 272$ K. EPDM/m has more carbonyl groups due to the oxidation at three positions, and thus the intermolecular interaction becomes stronger. Therefore, the temperature increases to $T_g = 308$ K. Since EPDM/CL is a long molecule and the entanglement between molecules occurs, its T_g increases to 303 K. Oxidation of EPDM has the effect of raising the glass transition temperature T_g [K].

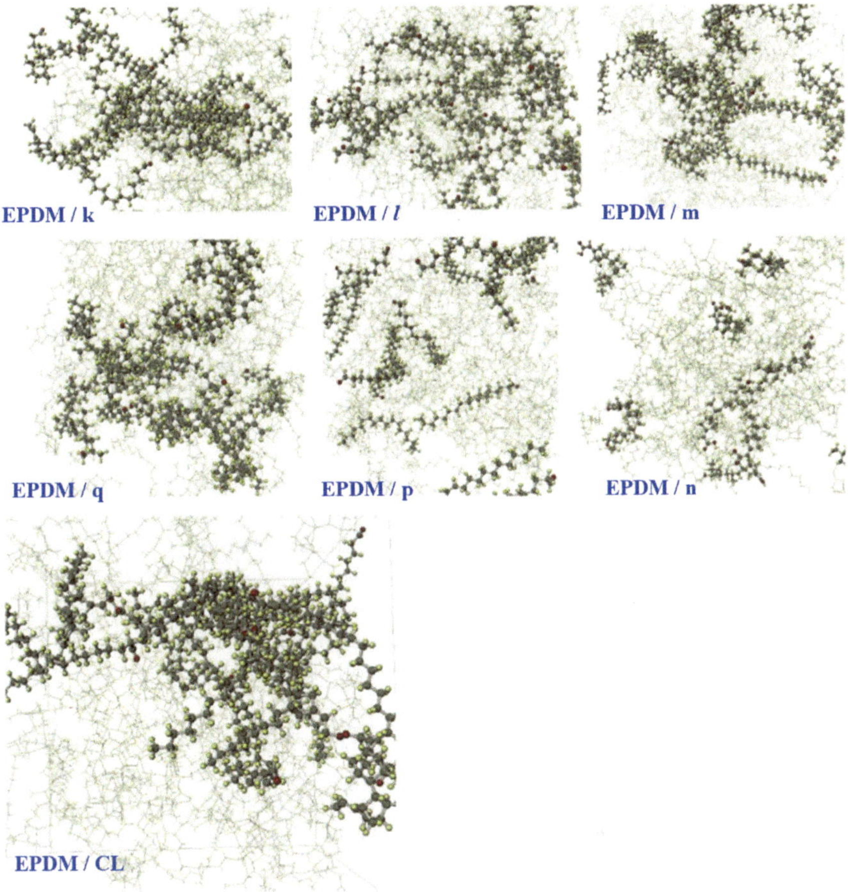

Fig. 14.47 The molecular coordination of the oxidized chains in EPDM system

Fig. 14.48 The glass transition temperature of EPDM

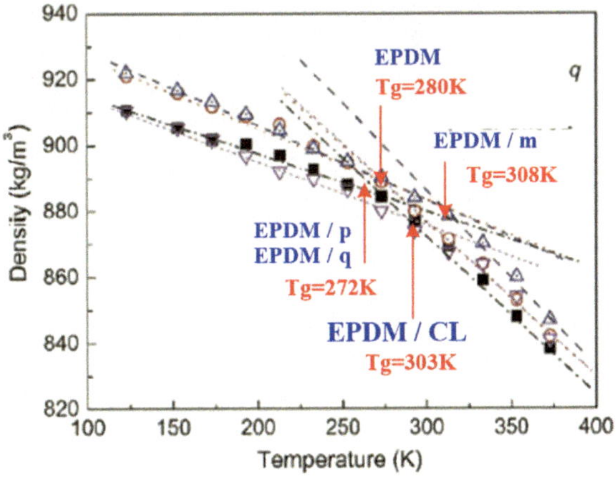

14.11 Molecular Dynamics Simulation of PE+H_2O+GMS

GMS (Glycerol Mono-Stearate) is mixed in to prevent the water tree deterioration. Its purpose is to capture and immobilize H_2O molecules by the electrostatic potential of polar groups at the ends of GMS molecules. The H_2O molecules are introduced to the following simulation to calculate diffusion coefficient of H_2O molecule by molecular dynamics [16].

(1) Molecular Dynamics Simulation of PE+H_2O+GMS

The formation of H_2O molecular clusters is observed by molecular dynamics simulation (see Sect. 12.8), which is used for the mixed system of three kinds of molecules (PE+H_2O+GMS). The molecular structural formula of Glycerol Mono-Stearate (GMS) is shown in Fig. 14.49d.

(a) **50 PE ($C_{120}H_{242}$) molecules+50 water (H_2O) molecules (1.1 wt%)** (Fig. 14.49a)
About 25 water (H_2O) molecules gather in the PE50 molecular net to form a cluster. Water is a molecule with a large dipole moment ($m = 2.40$ Debye) in which hydrogen H+ is positively charged and oxygen O- is negatively charged (see the part (3) of Sect. 12.2). Therefore, in the PE50 molecular net without dipole moment, only water molecules with dipole moment are attracted to each other due to Coulomb force and form clusters.

(b) **50 PE ($C_{120}H_{242}$) molecules+50 water (H_2O) molecules (1.0 wt%)+GMS (2.1 wt%)** (Fig. 14.49b)
Figure 14.49b shows the chemical structural formula of GMS and the electrostatic potential distribution. In this GMS, an ester bond -(C=O)-O- with a dipole

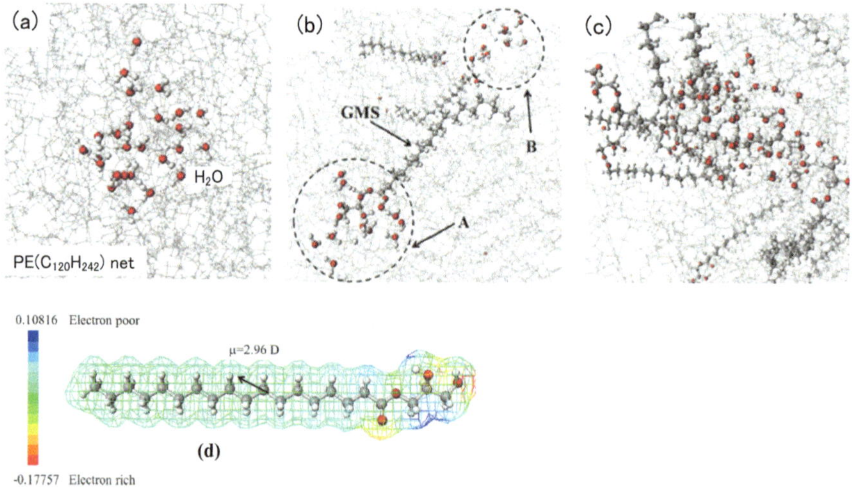

Fig. 14.49 The molecular structural formula and calculated results

moment and two hydroxyl -OH groups are present at the end of the (C17H33)-molecule. Therefore, positive and negative electrostatic potential distributions spread to the outside of the molecular ends. When GMS is mixed into the net of 50 PE molecules and 50 water molecules, the water molecules with dipole moment gather within the electrostatic potential created by the GMS edge and form clusters (part A in the figure). On the other hand, since a small amount of MS (2.1 wt%) is distant from the GMS edge, clusters are still only formed by water molecules (part B in the figure).

(c) **50 PE ($C_{120}H_{242}$) molecules+50 water (H_2O) molecules (1.0 wt%)+GMS (7.8 wt%)** (Fig. 14.49c)
When GMS is added with a larger amount (7.8 wt %), multiple GMSs gather to form a micellar structure (See Fig. 14.40b in Sect. 14.9) when incorporating water molecules. The effect of incorporating water molecules is recognized by increasing the amount of GMS.

(2) **Evaluate the Spacing Between H_2O Molecules**

The closeness of 50 H_2O water molecules in a formed cluster inside PE50 molecular net is evaluated by RDF $g(r)$ (Radial Distribution Function). The evaluation results of RDF $g(r)$ are shown in Fig. 14.50a and b. The RDF is explained in Eq. (14.3) of the part (2) of Sect. 14.10. In Fig. 14.50a and b, the vertical axis represents $g(r)$ and the horizontal axis represents the distance between oxygen of water molecules. A sharp peak appears at the distance of $r = 2.8$ Å, and a broad peak appears at $r = 4.2$ Å.

(a) The GMS concentration is changed from 0.0 to 7.8 wt% and the 50 water molecules are fixed. (Fig. 14.50a):
When the amount of GMS is 0.4 to 2.1 wt%, its effect is small. The cluster effect of water molecules without GMS (0.0 wt%) is large. However, when the amount of GMS is 4.0 to 7.8 wt%, the GMS effect becomes large. The effect of collecting water molecules within the distant region of $r = 2.8$ Å can be confirmed (see the enlarged figure).
(b) The concentration of water molecules is changed to 1.0–4.9 wt% and the 10 GMS molecules are fixed. (Fig. 14.50b):

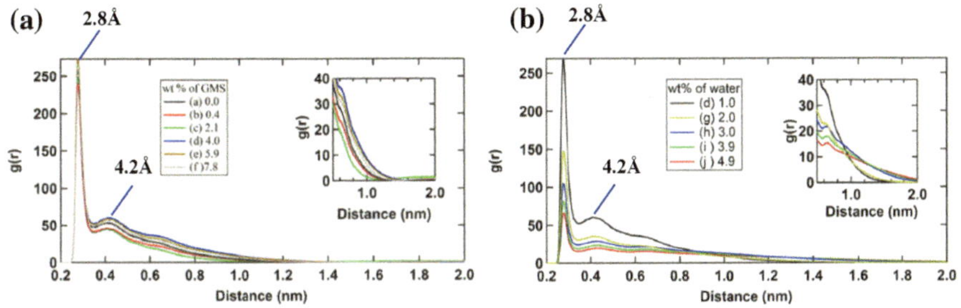

Fig. 14.50 The results RDF $g(r)$ based on changing concentration of GMS or water molecules

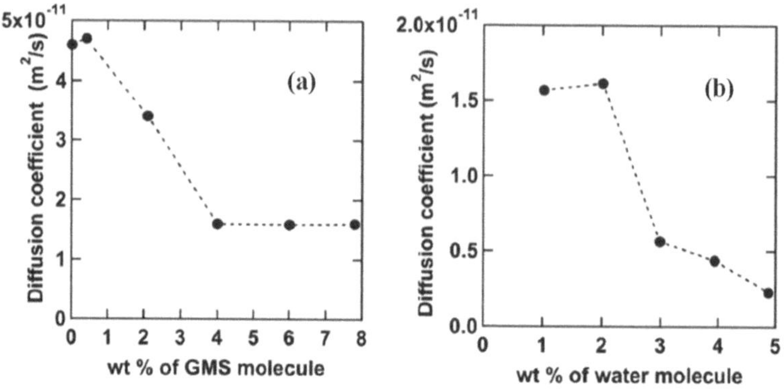

Fig. 14.51 The evaluation results of RDF $g(r)$

The size of RDF $g(r)$ is reversed at the distance of $r = 1.0$ Å (see the enlarged figure). The near water molecules ($<r = 1.0$ Å) are collected in GMS molecule. However, when the amount of water molecules is large (2.0–4.9 wt%), RDF $g(r)$ becomes small, which means it is difficult to collect the water molecules. On the contrary, the far water molecules ($>r = 10$ Å) have the opposite RDF $g(r)$ shape. When a large amount of water molecules (2.0 to 4.9 wt%) is analyzed, the water molecules in the region far than 1.0 Å are also collected in the GMS molecule, which shows a constant effect.

(3) Evaluation of Diffusion Coefficient of H_2O Molecule

The spacing of H_2O molecular cluster that finally accumulates in the PE50 molecular net in the previous part (2) is evaluated. Here, the diffusion coefficient D [m^2/s] of Eq. (14.4) in which the H_2O molecule dynamically moves is evaluated. When the I-th water molecule diffuses and moves to rI(t) from position rI(0) at the initial time ($t = 0$), the diffusion coefficient D [m^2/s] is calculated from the average of the square of the moving distance (rI(t)-rI(0)), as shown in Eq. (14.4).

(a) Diffusion coefficient of H_2O molecules depending on GMS amount (Fig. 14.51a): The 50 water molecules are fixed. The diffusion coefficient D of H_2O molecules decreases depending on the GMS amount. In other words, due to the increase of GMS up to the concentration of 4.0 wt%, H_2O molecules are less likely to diffuse. The diffusion coefficient D becomes constant at a concentration of 4.0 wt% or more.

$$D = \lim_{\Delta t \to \infty} \sum_{I}^{N} \frac{\langle (r_I(t)) - r_I(0))^2 \rangle}{6\ \Delta\ t} \tag{14.4}$$

(b) H_2O molecular weight and diffusion coefficient: The GMS with the concentration of 4.0 wt% is fixed (Fig. 14.51b). The H_2O molecules with a small amount easily diffuse, but an increased amount makes the diffusion difficult. This is

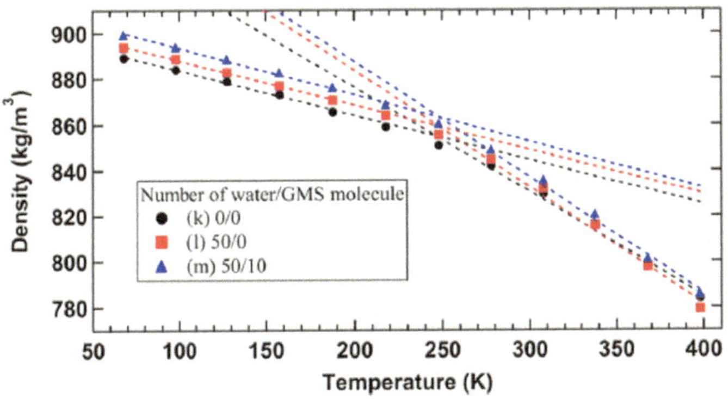

Fig. 14.52 The calculated glass transition temperature

because the diffusion of H_2O molecules is suppressed by the formed clusters of H_2O molecules due to the interaction of dipole moments.

(4) The Glass Transition Temperature Has Little Effect on H_2O+GMS

The temperature dependence of the density of the PE50 molecular system is calculated when H_2O molecules and GMS are mixed into the net to form clusters (Fig. 14.52). The density [kg/m^3] on the vertical axis is slightly higher in the H_2O/GMS mixing ratio of 50/0 and 50/10 cases than in the 0/0 case. The temperature at which the linear characteristic of the temperature dependence [K] of the density [kg/m^3] bends is the glass transition value T_g. From this result, the H_2O+GMS incorporation has little effect on the glass transition temperature.

14.12 Curing Agent Effect on Charge Accumulation of Epoxy Resin

(1) DC Insulating Materials for High Temperature And High Electric Field

The power devices such as electric vehicles and DC/AC converters have been widely used. Furthermore, there is a specification for the stable operation under severe conditions such as high temperatures and high electric fields. This is a strict specification that requires a enough heat radiation generated from electric devices and electronic elements for stable operation and insulation property retention. A guideline for developing an electrically insulating material to achieve this aim is shown in Fig. 14.53.

In Fig. 14.53, the horizontal axis represents the electric field ranging from $E =$ 0.1 to 500 kV/mm, and the vertical axis represents the temperature from 0 to 200 °C. The LDPE introduced in Sect. 14.1 remains good properties in low temperature and low electric field range, while the PEN in Sect. 14.4 is also excellent in high temperature and high electric field range. We try to propose a molecular design method by

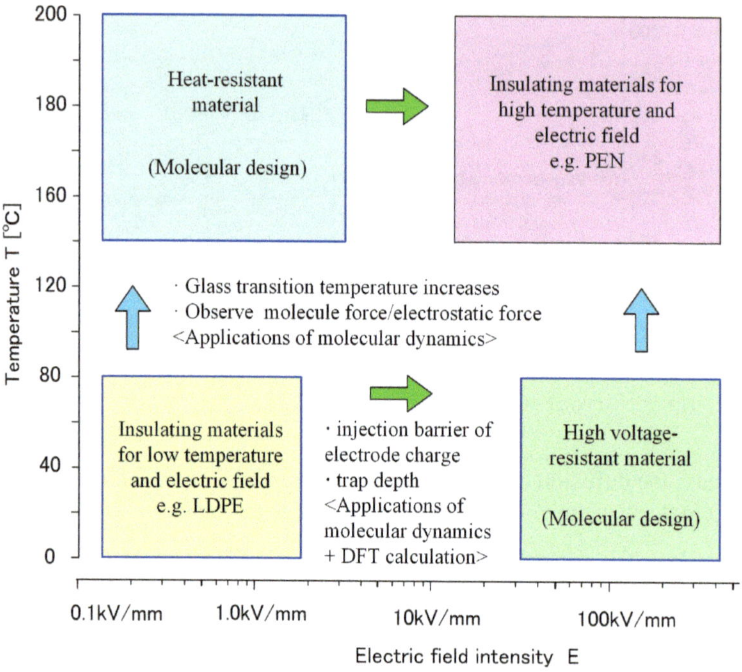

Fig. 14.53 A guideline for developing an electrically insulating material

comparing the molecular structures, electron energy levels, and electrostatic potential distribution of LDPE and PEN molecules.

(2) **Curing Agent for Epoxy Resin**

The epoxy resin with the curing agent shown in Fig. 14.54 is taken as an example. An epoxy resin is produced by a bisphenol-type liquid epoxy resin and a curing agent. As mentioned in the part (1) above, the development of epoxy resins that can be used in high temperature and high DC electric field is needed. As one of the guidelines, a report is introduced that evaluates the charge storage characteristics of various curing agents [17]. Here, the amine-type hardener and the anhydride-type hardener of the epoxy resin using the curing agents of (a) amine and (b) anhydride are analyzed, as shown in Fig. 14.54.

(3) **Hardener-Dependent Charge Accumulation in Epoxy Resin**

Figure 14.55 shows the charge accumulation characteristics of epoxy resins of EP-AM with amine-type hardener and EP-AH with anhydride-type hardener. The space charge accumulation distribution $\rho(z, t)$ [C/m^3] is measured by the PEA method under the temperature conditions of 25, 80, 140 °C with the applied electric fields from 20 to 120 kV/mm. At a temperature of 25 °C, $\rho(z, t) = 0$ and the internal electric field distribution $\rho(z)$ are constant in both samples. Thus, no charge accumulation

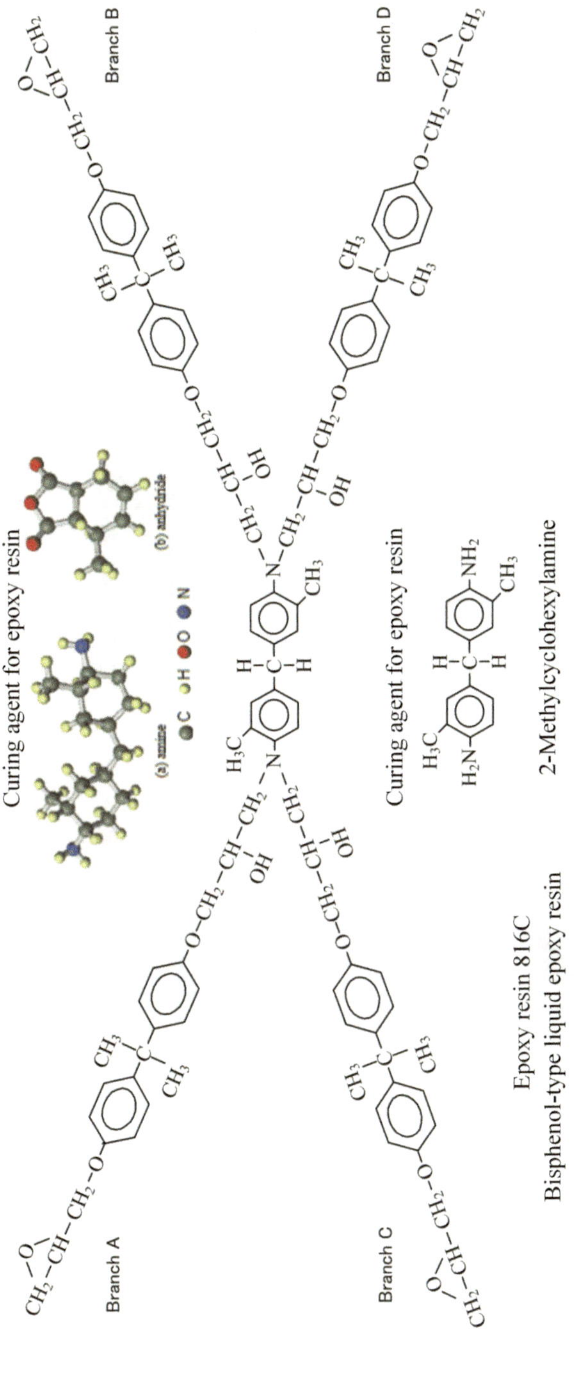

Fig. 14.54 The formula of the epoxy resin with the curing agent

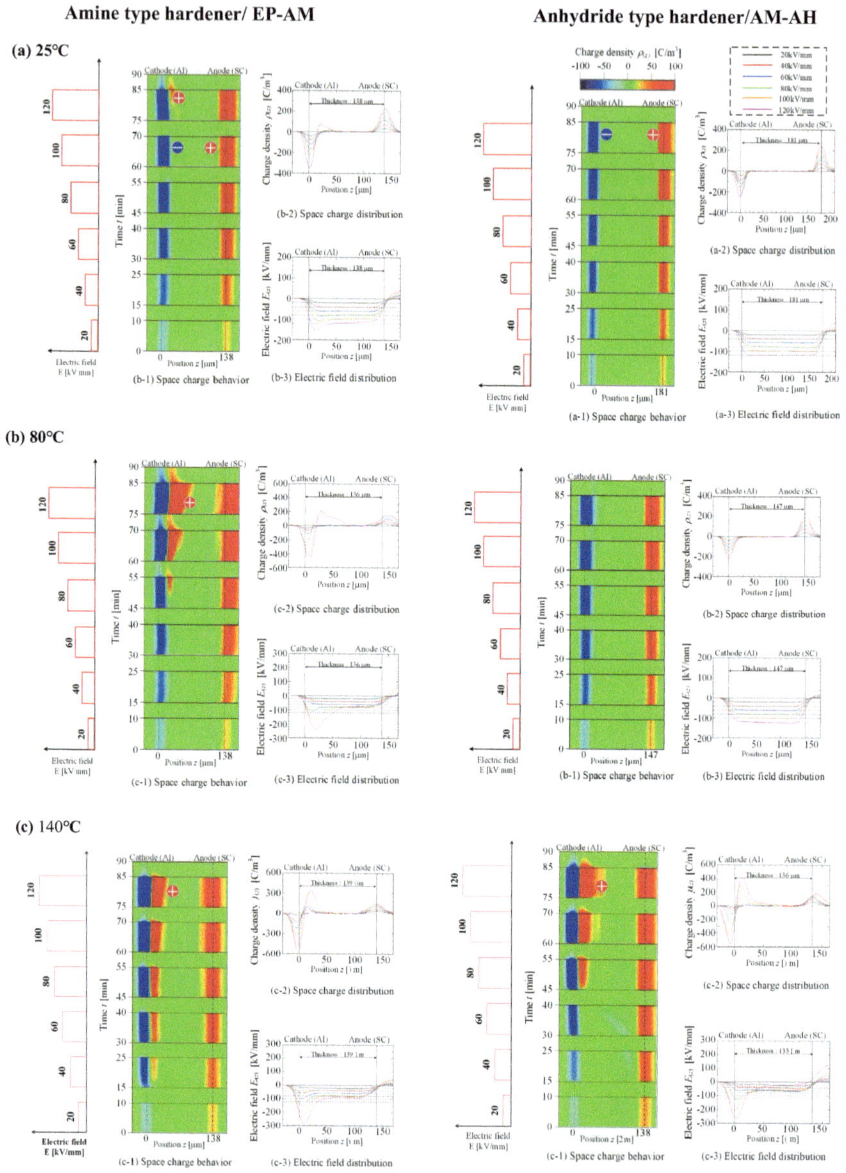

Fig. 14.55 Charge accumulation characteristics of epoxy resins of two types

is observed. At a temperature of 80 °C and $E = 120$ kV/mm, the accumulation of positive hetero-charges in the EP-AM of amine type is observed near the cathode. On the other hand, the EP-AH of anhydride type shows no charge accumulation even at $E = 120$ kV/mm. At 25 °C, both EP-AM and EP-AH samples show the accumulation of hetero-charges. In particular, charge accumulation is observed from the low electric field of $E = 40$ kV/mm in the EP-AM and from $E = 80$ kV/mm in the EP-AH. As described above, it is confirmed that the charge accumulation depends on the curing agent of the epoxy resin.

(4) Hardener-Dependent Analysis by Quantum Chemical Calculation

The anhydride-type EP-AH is superior to the amine-type EP-AM in terms of charge accumulation characteristics under different temperatures and electric fields. The molecular model is investigated by comparing the electron energy levels calculated by Quantum Chemical Calculation/DFT (Fig. 14.56). The upper part of the figure is the electron energy level of EP-AM, and the lower part is that of EP-AH. The left side of the figure shows the case before the contact between the electrode and the EP sample, and the right side corresponds to the case after the contact.

EP-AM in the upper right side of the figure shows a low injection barrier, and positive charges (holes) are present on the sample surface due to contact charging. Therefore, positive charges move and the hetero charges are accumulated. On the other hand, since EP-AH in the upper right of the figure has a high injection barrier, positive hetero-charges are accumulated at high temperature and high electric field. The difference in charge injection barrier finds that the Fermi level of EP-AM is higher than that of the electrode, while the Fermi level of EP-AH is almost the same as that of the electrode. In this way, the charge injection depends on the high or low Fermi levels. For details, see the part (4) of Sect. 14.1. In the future, I think it is necessary to study the hopping transfer of charges in these materials.

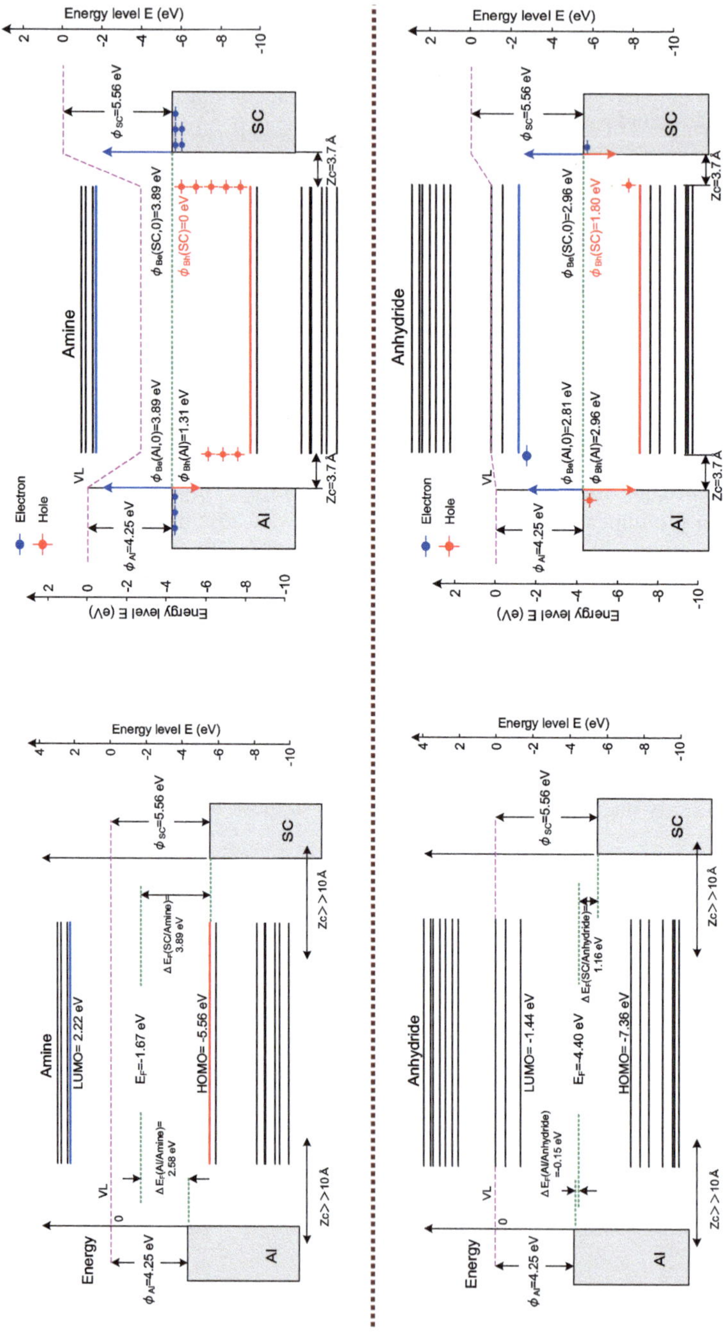

Fig. 14.56 The electron energy levels of epoxy resins of two types calculated by DFT

References

1. T. Takada, Y. Hayase, H. Miyake, Y. Tanaka, M. Yoshida, Study on electric charge trapping in cross-linking polyethylene and byproducts by using molecular orbital calculation, *IEEJ Transaction on Fundamental and Materials*, 132(2), 129–135 (2012).
2. W. Wang, T. Takada, Y. Tanaka, S. Li, Space charge mechanism of polyethylene and poly-tetrafluoroethylene by electrode/dielectrics interface study using quantum chemical method, *IEEE Transactions on Dielectrics and Electrical Insulation*, 24(4), 2599–2606 (2017).
3. T. Tadokoro, T. Motoyama, H. Harada, Y. Tanaka, T. Takada, T. Maeno, Space charge formation by irradiation of visible light in polyimide under DC electric stress, *IEEJ Transaction on Fundamental and Materials*, 129(7), 463–469 (2009).
4. T. Takada, Y. Hayase, Y. Tanaka, T. Okamoto, Space charge trapping in electrical potential well caused by permanent and induced dipoles for LDPE/MgO nanocomposite, *IEEE Transactions on Dielectrics and Electrical Insulation*, 15(1), 152–160 (2008).
5. E. Kubyshkina, M. Unge, B.L.G. Jonsson, Band bending at the interface in polyethylene-MgO nanocomposite dielectric, *The Journal of Chemical Physics*, 146, 051,101 (2017).
6. B. Dang, J. Hu, Y. Zhou, J. He, Remarkably improved electrical insulating performances of lightweight polypropylene nanocomposites with fullerene, *Journal of Physics D: Applied Physics*, 50, 455,303 (2017).
7. J. Li, H. Liang, M. Xiao, B. Du, T. Takada, Mechanism of deep trap sites in epoxy/graphene nanocomposite using Quantum Chemical Calculation, *IEEE Transactions on Dielectrics and Electrical Insulation*, 26(5), 1577–1780 (2019).
8. T. Takada, Discussion of the relationship between space charge accumulation and energy level shift by Quantum Chemical Calculation, *IEEJ Transaction on Fundamental and Materials*, 136(11), 710–716 (2016).
9. K. Hirotsu, Y. kanemitsu, N. Kamada, Y. Sekii, The influence of antioxidants on the electrical tree generation in XLPE, *IEEJ Transactions on Fundamentals and Materials*, 120(2),154–159 (2000).
10. H. Uehara, S. Iwata, Y. Sekii, T. Takada, Y. Cao, Suppression of electrical tree initiation by antioxidant and ultraviolet absorber, using a density-functional study, Paper presented at IEEE Conference on Electrical Insulation and Dielectric Phenomenon, 761–764 (2017).
11. H. Uehara, S. Iwata, Y. Sekii, T. Takada, W. Wang, Y. Cao, Molecular dynamics simulation and density-functional analysis on suppression effect of electrical tree in antioxidant-added polyethylene", Paper presented at IEEE Conference on Electrical Insulation and Dielectric Phenomena, 366–369 (2018).
12. Y. Sekii, N. Momose, K. Takatori, Y. Kanemitsu, T. Goto, Effect of surfactant on water tree generation in XLPE, Paper presented at the 7th International Conference on Properties and Applications of Dielectric Materials, 269–273 (2003).
13. H. Uehara, S. Iwata, Y. Sekii, T. Takada, Y. Cao, Molecular dynamics simulation and Quantum Chemical Calculations of surfactant having suppression effect on water trees, *IEEJ Transactions on Fundamentals and materials*, 139(2), 92–98 (2019).
14. H. Uehara, T. Okamoto, Y.Sekii, S. Iwata, T. Takada, Y. Cao, Intermolecular interaction and electric field dependence of reverse micelle on water tree initiation in polyethylene, Paper presented at IEEE Conference on Electrical Insulation and Dielectric Phenomena, 279–282 (2019).

15. W. Wang, Y. Tanaka, T. Takada, S. Iwata, H. Uehara, S. Li, Influence of oxidation on the dynamics in amorphous ethylene-propylene-diene-monomer copolymer: A molecular dynamics simulation, *Polymer Degradation and Stability*, 147, 187–196 (2018).
16. S. Iwata, Molecular dynamics simulation of effect of glycerol monostearate on amorphous polyethylene in the presence of water, *Journal of Molecular Modeling*, 23(4), 115 (2017).
17. J. Li, X. Kong, B. Du, K. Sato, S. Konishi, Y. Tanaka, H. Miyake, T. Takada, Effects of high temperature and high electric field on the space charge behaviors in epoxy resin for power module, *IEEE Transactions on Dielectrics and Electrical Insulation*, 27(3), 832–839 (2020).

Postscript

The development process of measurement technologies to evaluate the accumulated charge in polymer dielectrics is summarized at the beginning of the textbook. Looking back on this research theme that started in the 1970s, 50 years have passed.

When I was a student (since 1960), I longed for high-temperature nuclear fusion research. Actually, the research topic was a magnetic field design for high-temperature plasma confinement. The magnetic field design was based on computer software to calculate the vector potential of magnetic field, and a modified electrolysis cell was manufactured to develop an analog simulator. Then, I also developed a method for measuring the induced current in materials irradiated by atomic reactor neutrons. After that, under the guidance of Prof. Yotsuo Toshiyama and Prof. Takao Sakai, I switched to studying the charging phenomenon of electrical insulating materials.

The charging phenomenon means the accumulation of electric charge in an insulating material. When a high DC electric field is continuously applied to the polymer dielectric, electric charges are accumulated in it. The research did not proceed to the measurement of the amount and distribution of electric charges accumulated in the dielectric, which still resulted in my doctoral degree based on this project.

Challenge for Advanced Technologies

It has spent 50 years developing technology to measure the charge distribution accumulated in a dielectric and analyzing the model of the charge accumulation. The introduction of technologies from different fields was the decisive factor when working on a new research. It was difficult to read basic textbooks in different research fields. Therefore, I drastically tried to visit the book author's laboratory and asked questions. Furthermore, I also invited the author to my laboratory for a lecture and tried to understand the basics of the field. I want to introduce the famous researchers:

T. Takada et al., *Electric Charge Accumulation in Dielectrics: Measurement and Analysis*,
https://doi.org/10.1007/978-981-19-6156-4

- Reactor Neutron; Prof. Eiichi Arai (Tokyo Institute of Technology),
- High Volage Engineering; Prof. Keiichro Sugita (Tohoku University),
- Electronic circuit; Prof. Yutaka Murata (Musashi Institute of Technology),
- Technology of viscoelastic pulse sound wave; Prof. Chuubachi (Tohoku University),
  Prof. Amar Bose (MIT), Prof. Noboru Ichinose (Waseda University),
- Polymer material; Prof. Yasaku Wada (The University of Tokyo),
  Prof. Kyosoh Kanamaru (Tokyo Institute of Technology),
- Optical dielectric anisotropy measurement technology; Prof. Tomoya Ogawa (Gakushuin University),
  Prof. Markus Zahn (MIT), Prof. Kunihiko Hidaka (The University of Tokyo),
  Prof. Katsuaki Sato (Tokyo University of Agriculture and Technology),
- Signal data processing; Prof. Hiroshi Miyagawa (The University of Tokyo),
- Quantum Chemical Calculation; Prof. Masashi Yoshida (Tokyo City University),
  Prof. Fumitoshi Sato (University of Tokyo), Dr. Munetaka Takeuchi (X-Ability),
- English proofreading; Dr. Yuji Hazeyama (President of Myu Research).

International Collaboration

From 1981 to 1983, I stayed in Prof. Markus Zahn's laboratory as a Visiting Scientist (Massachusetts Institute of Technology, USA). There, we developed an electric field distribution measurement technology based on electro-optical Kerr effect for pure water. At the same time, Prof. Ziyu Liu from Xi'an Jiaotong University in China was also staying there. When we returned to China and Japan respectively, we promised to develop joint research of the US, China, and Japan laboratories in the future. Fortunately, with the support of the Japanese government's "Overseas Academic Exchange Research Program", we were able to carry out joint research and kicked off mutual visit research for faculty members and students. This overseas academic exchange research was then expanded to other universities shown below, and the number of research associates increased.

- Prof. Ziyu Liu, Prof. Demin Tu; Xi'an Jiaotong University (China)
- Prof. Markus Zahn; Massachusetts Institute of Technology (USA)
- Prof. Kwang S.Suh; Korea University (Korea)
- Prof. Robert M. Hill; King's College of London (UK)
- Prof. F.H. Kreuger, Prof. F.H. J.J. Smit; Delft University of Technology (The Netherlands)
- Prof. A.E. Davies, Prof. George Chen; University of Southampton (UK)
- Ph.D. Mats Leijon, Ph.D. Rongsheng Liu; ABB Corporate Research (Sweden)
- Prof. Zhicheng Guan; Tsinghua University (China).

The Meaning of the Tennis Ball Pyramid

The achievements of "Challenge for Advanced Technologies" and "International Collaboration" mentioned earlier can be understood by the model of the Tennis Ball Pyramid in the photograph. I was able to learn advanced technologies from the experts, discuss with many international colleagues, and open up new research fields. The development of a new science cannot be achieved by oneself. Pyramids are being built up unconsciously when they are actively seeking friends in different fields of boundaries. That was the joy of learning. Thanks for many friends all over the world.

2020/08/08; Tatsuo Takada notes

e-mail: takada@a03.itscom.net